Education 71

獠牙仙女

The Tusk Fairy

Gunter Pauli

[比]冈特·鲍利 著
[哥伦]凯瑟琳娜·巴赫 绘
李蕃 译

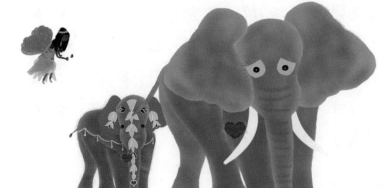

上海远东出版社

丛书编委会

主　任：贾　峰

副主任：何家振　闫世东　林　玉

委　员：李原原　祝真旭　牛玲娟　梁雅丽　任泽林
　　　　王　岢　陈　卫　郑循如　吴建民　彭　勇
　　　　王梦雨　戴　虹　翟致信　靳增江　孟　蝶

特别感谢以下热心人士对童书工作的支持：

匡志强	宋小华	解　东	厉　云	李　婧	陈　果
刘　丹	熊彩虹	罗淑怡	旷　婉	杨　荣	刘学振
何圣霖	廖清州	谭燕宁	韦小宏	李　杰	欧　亮
陈强林	王　征	张林霞	寿颖慧	罗　佳	傅　俊
胡海朋	白永喆	冯家宝			

目录

獠牙仙女	4
你知道吗？	22
想一想	26
自己动手！	27
学科知识	28
情感智慧	29
艺术	29
思维拓展	30
动手能力	30
故事灵感来自	31

Contents

The Tusk Fairy	4
Did You Know?	22
Think about It	26
Do It Yourself!	27
Academic Knowledge	28
Emotional Intelligence	29
The Arts	29
Systems: Making the Connections	30
Capacity to Implement	30
This Fable Is Inspired by	31

在肯尼亚，一只犀牛遇到了一群大象，发现象宝宝们刚刚换了乳牙，包括短短小小的獠牙。

"看到他们长大真高兴。孩子们很快就会长出自己的恒牙——獠牙。"犀牛对一头大象说。

"这些獠牙每年长15厘米，它们可有用了：挖掘、刮树皮、搜寻食物，还可以用来自卫。"大象说。

A rhino looks at an elephant herd in Kenya and realises that the elephant kids have just lost their milk teeth, including their tiny, short tusks.

"Nice to see them growing up. The young will soon have their permanent tusks," remarks the rhino to one of the elephants.

"These tusks grow 15 centimetres per year, and will come in handy for digging, scraping bark and foraging, and for defending themselves," says the elephant.

看到他们长大真高兴

Nice to see them growing up

孩子们至今还长着獠牙

They still have tusks

"那头小公象看起来像是左撇子吧，"犀牛评论道，"你难道不矫正一下小家伙们，让他们都用右边的獠牙吗？"

"讲究那些老规矩的日子早过去了，"大象回答，"现在我们正为这些獠牙忧心忡忡呢，孩子们至今还长着獠牙，这就已经令我们庆幸了。"

"It looks like that young bull is a lefty," remarks the rhino. "So you don't correct the young to make them all right-tusk users?"

"Those days are long gone," answers the elephant. "We are so worried about the tusks that we are delighted they still have them at all!"

"我听说你们的亚洲兄弟有一半都没有獠牙了。"犀牛说。

"没错。在亚洲,有獠牙的象大量遭到捕杀,所以越来越多象宝宝的父亲天生就没有獠牙。"

"I have heard that about half of your Asian brothers are tuskless," remarks the rhino.

"Indeed. Elephants with tusks were killed in massive numbers in Asia, so that more and more baby elephants were born from fathers who naturally had no tusks."

你们的亚洲兄弟有一半都没有獠牙了

Half of your Asian brothers are tuskles

非法捕猎者正在转向非洲

Poachers are turning to Africa

"天哪,在亚洲,这种为获取獠牙而杀害大象的行为已经持续很多年了。现在这些非法捕猎者正在转向非洲!"犀牛惊叫道。

"正在转向非洲?你没有开玩笑。他们正在整个非洲猎杀大象,简直要把我们都赶尽杀绝。"

"Gosh, this killing of elephants for tusks has been going on for so long in Asia. Now these poachers are turning to Africa!" exclaims the rhino.

"Turning to Africa? You're not kidding. They're poaching us to extinction all over Africa."

"为什么你不能拔掉獠牙呢？"犀牛建议。"你可以像我一样，我的角被锯掉了，但他们做得很小心，没有损坏我的颅骨。"

"呃，你的角可以被去掉是因为它只是你颅骨的一部分，但我的獠牙是一整颗牙齿，有四分之一长在我的下巴里。"

"Why don't you have your tusks removed?" suggests the rhino. "You could be like me: I have my horn sawn off, and they do it carefully, without damaging my skull."

"Well, your horn can be removed since it's part of your skull, but my tusk is a tooth, and one quarter of it sits in my jaw."

我的角被锯掉了

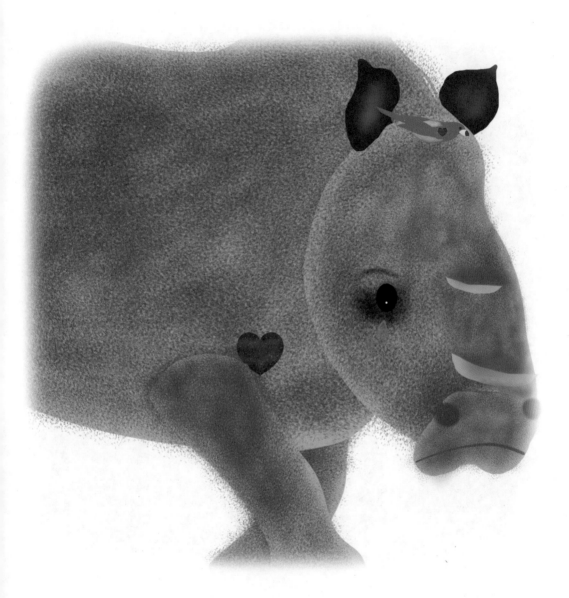

I have my horn sawn off

……会非常疼……

... very painfull ...

"那你为什么不能拔掉它呢?"

"如果只把看得到的部分拔掉,就好比去看牙医拔牙却留下了牙洞。那会非常疼,而且会疼很久。"

"So why couldn't you have it removed?"
"If you cut off what you can see, it's like going to the dentist and being left with an open cavity. It's going to be very painful for a long time."

"你的獠牙可以重新长出来吗？"犀牛问。

"是的，可以。只要牙根没有被破坏，我的獠牙就可以重新长出来。"

"Could your tusks ever regrow?" asks the rhino.

"Oh yes, my tusks could regrow as long as the root is not damaged."

你的獠牙可以重新长出来吗?

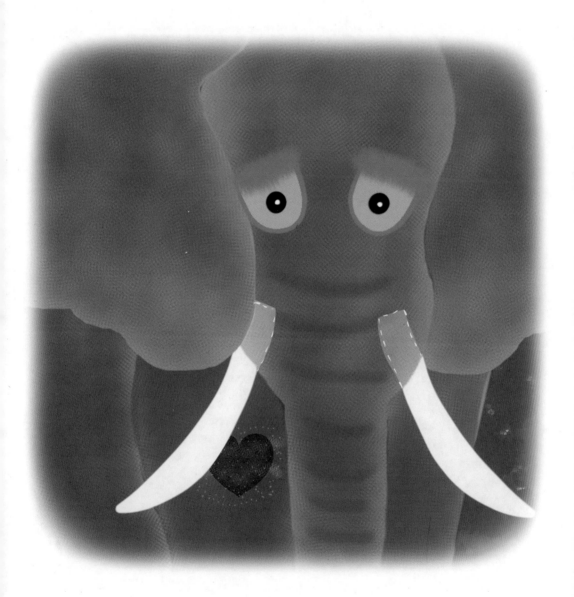

Could your tusk ever regrow?

你的獠牙只是一颗牙齿

your tusk is only a tooth

"不过,大象,你确定你的獠牙只是一颗牙齿吗?"

"绝对确定。它没有任何魔法,就像你的角一样,都只是由角蛋白组成的。"

"But, Elephant, are you sure your tusk is only a tooth?"

"Absolutely. There's no magic to it. In the same way that your horn is nothing but keratin."

"对，我的角由角蛋白组成，与构成头发和指甲的物质没什么不同。"犀牛说。"所以当人们买犀牛角粉时，他们也许只是在买碾碎的指甲呢！"

"谁说不是！但对那些相信象牙会让他们变得强壮的人有什么办法呢？他们花了一大笔钱，得到的只不过是一副用过的老牙而已。"

……这仅仅是开始！……

"That's right, my horn is made of keratin, which is the same substance that hair and nails are made of," says the rhino. "So when people buy powdered rhino horn, they might as well be buying ground-up fingernails!"

"Exactly! But what do you do with people who spend a fortune believing my tusks will make them stronger, when really they're getting nothing more than an old pair of used teeth?!"

... AND IT HAS ONLY JUST BEGUN!...

……这仅仅是开始!……

... AND IT HAS ONLY JUST BEGUN! ...

Did You Know?
你知道吗？

Elephant tusks are present at birth. Baby tusks fall out after a year, and are replaced by permanent ones.

象一出生就有牙齿。乳牙在一年以后脱落，换上永久的恒牙。

Elephant tusks grow 15 centimetres a year, and never stop growing.

象牙每年生长15厘米，而且会不停地生长。

$\mathcal{A}$ll African elephants, male and female, have tusks, whereas in Asia only the males have tusks.

所有的非洲象，无论公母，都有獠牙，然而亚洲象中只有公象有獠牙。

$\mathcal{E}$lephants have one preferred tusk, which is known as the master tusk. It is like humans being left- or right-handed.

每头大象都有自己更擅长使用的一颗獠牙，称为主牙，就像人类分成左撇子和右撇子。

While elephant tusks are made out of ivory, rhinos have horns made from compressed keratin fibres, the same material that makes up our hair and fingernails.

大象的獠牙由象牙质构成，犀牛的角则由致密的角质纤维组成，这和人类的头发、指甲的成分是一样的。

Elephants have 24 teeth: 12 upper and 12 lower. When teeth are worn down, they just fall out and are replaced by new ones that are bigger and stronger than the previous ones.

大象有24颗牙齿，上下各12颗。当牙齿磨损殆尽，它们就会自然脱落，并长出比先前牙齿的更大更结实的新牙。

A single tusk can weigh more than 60kg, so a pair is well over 100kg. It used to be common to find elephants with tusks of 100kg each, but poaching has prevented elephants from growing old enough to have such large tusks.

一颗象牙的重量超过60千克，一对象牙的重量就得有100多千克。在以往，100千克重的单颗象牙很常见，然而现在，偷猎使得大象很难活到拥有如此巨大象牙的年纪。

The Asian elephant is even more endangered than the African elephant. The threat is not poaching, but human encroachment. There are only 30,000 wild elephants left in Asia and 400,000 in Africa.

比起非洲象，亚洲象面临的灭绝危险更大。威胁不仅来自偷猎，还有人类对大象家园的侵占。在亚洲仅存30 000头野生大象，而非洲有400 000头。

Think about It

想一想

Do you think the baby elephant would enjoy a visit from the tooth fairy, or rather, a "tusk fairy"?

你觉得小象会喜欢一位牙齿仙女，或者说，獠牙仙女来拜访他吗？

一边是崭新洁白、紧致干净、自然脱落的小象獠牙，一边是又老又旧，磨损不堪而且需要杀掉大象才能获得的大象獠牙，你是否同意前者更值得收藏呢？

Would you agree that the new, white, dense and clean baby tusks that would have fallen out naturally anyway are preferable to the old, worn, damaged tusks that required killing an elephant?

What could you do to help elephants survive? Would you take the same approach in Africa as you would in Asia?

为了帮助大象生存下去你可以做些什么？你在非洲和亚洲采取的措施一样吗？

想象一下，当他们的父母长辈因为象牙被杀死，小象们会对自己的未来持什么态度？

How would you imagine young elephants would feel about their future when their parents and elders are killed for their tusks?

Do It Yourself!

Tell the story of the "tusk fairy" to a few friends. And then ask if they would rather organise a collection of baby tusks, generating jobs and income for the local communities, or if they would prefer to go poaching elephants. What do you think your friends and family will say? Is there anyone who might be against the "tusk fairy", which would offer a future for the communities and the elephants alike?

给朋友们讲讲"獠牙仙女"的故事。问问他们：是愿意一起收集小象的象牙，为当地社区带来就业和收入，还是更喜欢去偷猎大象？你觉得你的朋友和家人会说什么？会有人反对对社区和大象的未来都有好处的"獠牙仙女"吗？

TEACHER AND PARENT GUIDE

学科知识
Academic Knowledge

生物学	物种是怎样灭绝的？保证某一物种存活的生物个体最小数量是多少？
化 学	骨头、象牙和角蛋白的区别；象牙的化学组成是牙釉质、牙骨质和牙本质；时间久了，水分减少使得象牙变成白垩状。
物 理	新生小象的乳牙和旧的老獠牙在强度和性能上的区别；象牙有光滑的表面；变成化石的獠牙是象牙里最硬的。
工程学	象牙中有非常纤细的、填充了蜡质溶液的管道，它们有助于雕刻且提供了一种无法用人工材料复制的色调；对象牙进行碳测年能够确定动物的死亡时间。
经济学	稀缺性带动价格上涨；非法贸易在经济里的角色及其与非正规经济、腐败的关系；调查偷猎给旅游业带来的损失。
伦理学	饥饿和贫穷是怎样剥夺人的自尊并且刺激他们从事违法贸易的？当某些物种濒临灭绝时，我们怎么还能去追求那些只是用作赏玩和装饰的产品？
历 史	人类第一次使用象牙发生在石器时代；大象曾经出没在中国的黄河流域，但已经灭绝了；猛犸象是一种象牙化石资源；CITES（濒危野生动植物种国际贸易公约）是在1973年签订的。
地 理	找到亚洲象仅存的生存地；看看非洲大象生活的区域；找出象牙贸易的路线；找出偷猎大象现象最严重的地区。
数 学	獠牙的生长遵循正弦曲线模式：如果它持续生长，就会形成螺旋状；獠牙的横截面是椭圆形的。
生活方式	象牙在许多不同的文化里被用来做装饰品；人们购买犀牛角和象牙是因为错误地相信它们能壮阳；大象像人类一样有死亡仪式。
社会学	象牙在汉朝被用在礼服上；在清朝，象牙的使用十分普及，被做成把手、刷子、桌框、鼻烟壶、香水盒、麻将牌和印章；大多数国家已经达成一致，保护濒临灭绝的物种，尤其是大象。
心理学	当孩子们认识到成人杀害大象是为了经济利益，就可能会影响孩子的学习能力并使孩子转向暴力，孩子也许会因此失去对法律的尊重，因为他们看到了对法律和国际公约的蔑视行为没有得到惩罚。
系统论	无休止地追逐象牙将导致供应枯竭，而人类似乎无力改变自己的行为来应对这个现实；违法象牙交易价钱越高，大象就会越快灭绝。

教师与家长指南

情感智慧
Emotional Intelligence

犀牛

犀牛对大象十分友好，他一边观察象宝宝，一边和他们的父亲交流。他意识到了偏重于使用左边和右边象牙的区别，通过参与讨论展示了他对大象未来的关心。他见多识广，并且提供了适用于自己种群的方法，也就是锯掉角。犀牛仔细地聆听大象的不同观点，并且问了更多的问题，向大象表达了尊重。对话变得更具有描述性，这表明他们彼此建立起了一定程度的信任关系。犀牛获得了更多的知识，渴望为大象面临的窘境找到解决方案。

大象

成年大象花时间和犀牛分享心事，表现出了友好的态度。尽管他当时面临着自身的生存问题，但他还是耐心地与犀牛分享自己的观点，并对那些存在许久的问题（比如人们猎杀大象是为了他们的獠牙）给出自己的想法。大象使用了简洁明了的语言，让每个人都能理解。对话使他们产生了无能为力的挫折感，但他们仍试图寻找实际的解决方法。大象无法理解，为什么人类宁肯舍弃质量更好、更方便获得的东西而选择质量不好的东西。

艺术
The Arts

象牙传统上一直被用于制作多种装饰品。请你把所有能替代象牙的东西列成一个清单吧。

TEACHER AND PARENT GUIDE

思维拓展
Systems: Making the Connections

　　市场通过价格杠杆来平衡供需关系，这是一个错误的激励机制。产品越稀缺，要价就越高。象牙的价格高达每千克3 000美元，为此人们不惜铤而走险捕猎，置自身和大象的生命于不顾。因为少数人愿意付钱，这些人把个人的快乐凌驾于某些物种的生存之上，漠视生命，不顾尊严地追逐金钱。结果仅仅使极少数人得到了脱离贫穷的机会，却让一些人实现了自己的非法目的。这就造成了一种氛围：为了快速挣钱可以不择手段。个人的自由意志受到了胁迫；强取豪夺地赚钱成为生活态度，而这样就破坏了社会尤其是乡村的固有结构。

　　不过，小象乳牙的自然脱落给出了一个解决方案，这些乳牙不为人注意，更不用说被收集起来产生收入。"乳牙再发现"会成为一个有价值的活动：按照今天的象牙价格，一对小象乳牙可以赚500美元。全世界每年有22 000头小象出生，那么乳牙的总值就高达1 000万美元了。如果再加上配套拨款、集资和公众捐款，该项目就可以在保证农村可持续发展的同时，也为象群的生存留出空间。

动手能力
Capacity to Implement

　　和你的父母及老师讨论，并在学校或社区发起一个"獠牙仙女"活动。孩子们喜欢为乳牙掉落而收到礼物或钱，让我们也为小象这样做吧。分享这个信息，并联系非洲各地的非政府组织。你可以让这个创意变得非常诱人：每找到一颗小象乳牙，就可以发起一个资金众筹项目。

教师与家长指南

故事灵感来自
This Fable Is Inspired by

吉尔达·莫拉蒂
Gilda Moratti

　　吉尔达·莫拉蒂出生在意大利，在纽约的苏富比拍卖行工作了十几年。但她真正的热情并不在艺术，而是关注动物。她受到格利高里·考伯特（一位关注人与自然和谐的摄影师）的作品的启发，决心投身动物保护，从此踏上了新的人生旅途。最初，莫拉蒂致力于拯救流浪狗，后来又扩展到救助猫。为了帮助那些年老无依、被人弃养的宠物，她在意大利成立了"四条腿组织"。在毕业于牛津大学的动物学家、"拯救大象"活动发起者伊恩·道格拉斯－汉密尔顿博士的探索性工作鼓舞下，吉尔达加入了拯救大象的行列。她和专业机构以及非专业组织展开合作，"胡萝卜与大棒"兼施，确保大象成为我们共同未来的一部分。

图书在版编目(CIP)数据

冈特生态童书.第二辑修订版:全36册:汉英对照/
(比)冈特·鲍利著;(哥伦)凯瑟琳娜·巴赫绘;
何家振等译.—上海:上海远东出版社,2021
书名原文:Gunter's Fables
ISBN 978-7-5476-1759-5

Ⅰ.①冈… Ⅱ.①冈…②凯…③何… Ⅲ.①生态
环境-环境保护-儿童读物—汉、英 Ⅳ.①X171.1-49

中国版本图书馆CIP数据核字(2021)第213075号

著作权合同登记号图字09-2021-0823

策　　划　　张　蓉
责任编辑　　祁东城
封面设计　　魏　来　李　廉

冈特生态童书
獠牙仙女
[比]冈特·鲍利　　著
[哥伦]凯瑟琳娜·巴赫　　绘
李　蕃　译

记得要和身边的小朋友分享环保知识哦!
八喜冰淇淋祝你成为环保小使者!

Education 72

用空气射击
Shooting with Air

Gunter Pauli

[比]冈特·鲍利 著
[哥伦]凯瑟琳娜·巴赫 绘
李蕃 译

上海远东出版社

丛书编委会

主　任：贾　峰

副主任：何家振　闫世东　林　玉

委　员：李原原　祝真旭　牛玲娟　梁雅丽　任泽林
　　　　王　岢　陈　卫　郑循如　吴建民　彭　勇
　　　　王梦雨　戴　虹　翟致信　靳增江　孟　蝶

特别感谢以下热心人士对童书工作的支持：

匡志强　宋小华　解　东　厉　云　李　婧　陈　果
刘　丹　熊彩虹　罗淑怡　旷　婉　杨　荣　刘学振
何圣霖　廖清州　谭燕宁　韦小宏　李　杰　欧　亮
陈强林　王　征　张林霞　寿颖慧　罗　佳　傅　俊
胡海朋　白永喆　冯家宝

目录

用空气射击	4
你知道吗？	22
想一想	26
自己动手！	27
学科知识	28
情感智慧	29
艺术	29
思维拓展	30
动手能力	30
故事灵感来自	31

Contents

Shooting with Air	4
Did You Know?	22
Think about It	26
Do It Yourself!	27
Academic Knowledge	28
Emotional Intelligence	29
The Arts	29
Systems: Making the Connections	30
Capacity to Implement	30
This Fable Is Inspired by	31

两只工蚁在到处跑来跑去寻找食物。他们听到了一声爆炸声。

"那是在向我们开枪吗？"一只蚂蚁感到诧异。

Two worker ants are running around looking for food. They hear a popping sound.

"Is that a shot gun firing at us?" wonders one ant.

那是在向我们开枪吗?

Is that a shot gun firing at us?

"这周围没有枪……"

"There are no guns around here ..."

"这周围没有枪，"另一只蚂蚁说，"就算有，人类也绝不会用枪来射击我们蚂蚁——我们太小了，子弹打不着。"

"听，爆炸声又响了。呃，我希望周边没有人在打猎。你永远不会了解这些人。"

"There are no guns around here," says the other. "And even if there were, humans would never use guns to shoot us ants – we're way too small to be hit by a bullet!"

"Listen, there's that popping sound again. Well I hope there's no hunting going on. You never know with these people."

"我觉得这是只甲虫。"
"带着猎枪的甲虫?"
"不,一只用空气射击的甲虫!"

"I think it was a beetle."
"A beetle with a shotgun?"
"No, a beetle shooting with air!"

"我觉得这是只甲虫。"

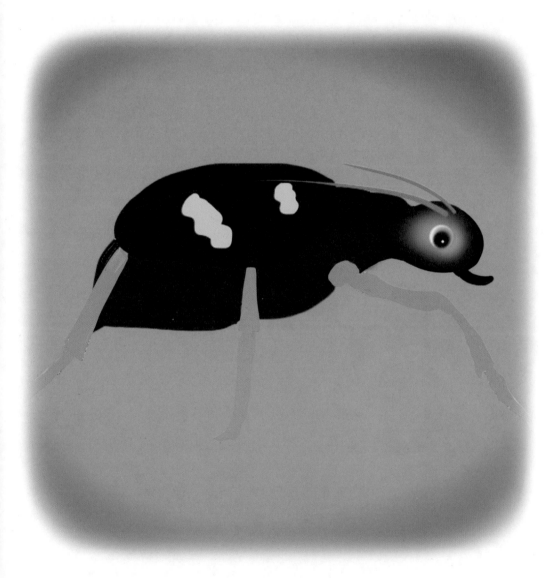

"I think it was a beetle."

化学喷雾像是滚烫的空气

Chemical spray like boiling-hot air

"我知道子弹是很危险的,但是我没有明白用空气射击是什么意思。"

"就是热空气——呃,是一种化学喷雾,像是滚烫的空气。"

"I know that bullets are dangerous, but I don't get the point of shooting with air."

"It's hot air – well, it's a chemical spray like boiling-hot air."

"这些甲虫喷射这样滚烫的东西时，不会烫到自己的手指或者嘴巴吗？"

"唔，这些东西是从这些射炮步甲的臀部喷射出来的。"

"射炮——你的意思是救火队？但是听上去它们是点火的人呀。"

"Don't these beetles burn their fingers or their mouths handling such hot stuff?"

"Well, it comes out of these bombardier beetles' rear ends."

"Bombardiers – you mean fire brigade? But it sounds like they're the ones making the fire."

听上去它们是点火的人呀

It sounds like they're the ones making the fire

精确地瞄准捕食者

Precisely target predators

"你知道吗，非洲的射炮步甲在臀部有一个喷口，可以精确地瞄准270度范围内的捕食者。"

"哇，了不起。"

"这种甲虫不是只能喷射出一股热空气。它可以在一秒钟内喷射500次。"

"You know, the African bombardier beetle has a nozzle on its rear that can precisely target predators within a 270° range."

"Wow, impressive."

"This beetle doesn't just pop out one flush of hot air. It can repeat it 500 times in one second."

"那可是在一瞬间呀——真是无法测量。听上去这种甲虫好像可以运行火箭。"

"事实上,这些甲虫随身带着一个完整的工厂,并且储存着它们发射炽热空气所需的所有东西。"

"它们能飞吗?"

"That's a split second – impossible to measure. It sounds like this beetle could operate a rocket."

"Actually, these beetles carry a whole factory with them, and store everything they need to shoot with hot air."

"Can they fly?"

它们能飞吗?

Can they fly?

蜻蜓可以即刻起飞

The dragonfly can instantly take off

"当然,但与蝴蝶或者蜻蜓可以即刻起飞不同,这种甲虫需要时间来展开它的翅膀。那就到了我们蚂蚁,还有青蛙,开始咬它们的腿的时候了。"

"因此它们可以像火箭一样运行,但不能像火箭一样飞!"

"是的,蚂蚁老弟。就像你不应该试图玩枪一样,你也不应该梦想着发射火箭。"

"Sure, but while a butterfly or dragonfly can instantly take off, this beetle needs time to unfold its wing. That's when we ants – and also frogs – start biting its legs."

"So they operate like rockets, but they can't fly like rockets!"

"True, brother Ant. And just as you shouldn't think about playing with guns, you shouldn't dream about launching rockets either."

"这种甲虫只使用大自然提供的资源来变魔术吗?"

"除了利用我们现有的资源,我们别无选择!只有人类才会做出非自然的行为,比如用子弹来杀戮。"

"但是这些甲虫被设计得比我们蚂蚁能想象到的还要好得多。"

"山外有山,人外有人!"

……这仅仅是开始!……

"Does this beetle only use what nature provided to do its magic?"

"We have no choice but to use what we have! Only humans do unnatural things like shooting with bullets to kill."

"But these beetles are designed better than we ants could ever imagine."

"There's always someone who is better than us!"

... AND IT HAS ONLY JUST BEGUN!...

……这仅仅是开始！……

... AND IT HAS ONLY JUST BEGUN! ...

Did You Know?

你知道吗？

Air guns powered by compressed air can shoot metal projectiles. Firearms use chemicals that produce pressurised gas to propel a bullet.

气枪依靠压缩空气提供动力，可以射击金属发射物。热兵器依靠化学物质产生高压气体来推进子弹。

There are hundreds of types of bombardier beetles with various defense mechanisms, from non-spraying beetles to those with foam excretions. The African bombardier beetle is the only one that can aim explosive spray in any direction.

射炮步甲有好几百种，各自有着不同的防御机制，从非喷雾甲虫到那些产生泡沫分泌物的甲虫。非洲的射炮步甲是唯一有着可以瞄准任何方向的爆炸喷雾的甲虫。

The Chinese first invented black powder in the 9th century, and went on to design rockets in the 13th century in their quest to produce fireworks. A monk filled sections of bamboo with black powder, making firecrackers to drive away evil spirits.

在9世纪时，中国人首先发明了黑火药。到了13世纪，为了制作烟花，中国人又设计了火箭来燃放烟花。一名和尚在竹子里填入了黑火药，制作出爆竹来驱赶邪神。

The four great inventions of China that have impacted the development of civilisation throughout the world are: the compass, gunpowder, paper and printing.

中国的四大发明分别是：指南针、火药、纸张和印刷术，这些发明影响了世界文明的发展。

𝒞hina is the world's biggest producer of fireworks, responsible for 90% of world output. The City of Liuyang in Hunan Province produces 60% of all fireworks used around the world.

中国是世界上最大的烟花制造国，烟花产量占全球生产量的90%。湖南省浏阳市生产了全球烟花用量的60%。

𝑀arco Polo brought fireworks to Europe in 1292, and the Italians developed fireworks into art.

马可·波罗在1292年将烟花带入欧洲，之后意大利人将烟花发展成了一种艺术。

Fireworks are poisonous and cannot be recycled. Birds are so startled by fireworks that they get disoriented, fly into objects and get killed.

烟花有毒并且不能被回收。鸟类会因受到烟花惊吓而迷失方向，撞上物体而死。

Because synthetic clothing can create sparks that may detonate fireworks, those who produce fireworks must wear cotton clothing, including socks and underwear.

因为合成面料会产生引爆烟花的火花，那些制造烟花的人必须穿棉制服装，包括袜子和内衣。

Think about It
想一想

Do you enjoy a noisy, explosive fireworks display, knowing that it burns chemicals?

知道烟花表演是在燃烧化学品的情况下，你还会喜欢喧闹的、爆炸的烟花表演吗？

如果射炮步甲可以制造对大气无害的推进气体，人类为什么不能做同样的事呢？

If the bombardier beetle can make a propulsion gas that is harmless to the atmosphere, why can't humans do the same?

If you think you're the best at something, are you sure there will never be anyone who can do better?

如果你觉得自己擅长某事，你确定没有人能做得更好？

你可以做到某些自然界做不到的事情吗？

Are you able to do things that nature cannot do?

Do It Yourself! 自己动手!

When we discuss propulsion, guns, explosives, black powder and fireworks we must first think about our safety and the safety of the people around us. Accidents happen fast, and it's better to be safe than be sorry. Ask your parents, teachers or even the fire brigade what to look out if you are in touch with any of these exciting toys, and before you do anything else, make certain that you understand how they work.

当我们讨论喷气式推进、枪支、爆炸、黑火药和烟花时，我们必须首先考虑我们的安全和周围人的安全。事故会瞬间发生，保障安全比后悔更为有用。问问你的父母、老师或者消防队，如果你接触任何这些令人兴奋的玩具，应该注意些什么。在你进行任何操作之前，一定要确保知道它们该如何操作。

TEACHER AND PARENT GUIDE

学科知识
Academic Knowledge

生物学	射炮步甲是一种步行虫；臀板的囊袋；射炮步甲的捕食者包括蚂蚁、青蛙和蜘蛛；昆虫通过血压而非肌肉来完成折叠和展开的动作。
化学	过氧化氢是昆虫新陈代谢的副产物，醌被用来使昆虫的角质层变硬，对苯二酚的气味是捕食者不喜欢的，并且使得一些臭虫有臭味；过氧化氢分解为氧气和水；对苯二酚氧化分解成刺激性化学物质；火箭推进剂发生放热化学反应；黑火药由硝酸钾、碳和硫磺组成；烟花火药由木炭、硫磺、硝酸钾组成，颜色则由特殊的化学物质产生：碳酸铜是蓝色的，铁屑是金闪闪的，硝酸钡是绿色的，钾是紫色的，硝酸锶是红色的，冰晶石是黄色的，镁是白色的。
物理	射炮步甲采用了脉冲推进系统；甲虫展开翅膀所需的压力与翅膀长度、体重成正比。
工程学	极小尺寸的工厂；脉冲燃烧和喷射机制；如何在高空实现飞机引擎的重新点火；有振荡阀门的间歇式脉冲喷射器；火箭燃料经过化学反应产生推力；离子推进是指用电场力推动带电粒子加速。
经济学	间歇燃烧与连续燃烧的对比。
伦理学	人类用枪杀害无辜，做某些别的动物不会做的事情。
历史	金属火箭最早发明于18世纪初的印度迈索尔；火箭被用于滑铁卢战役。
地理	中国人发明了黑火药和火箭，蒙古人将这项技术输出到俄国、欧洲东部和中部。
数学	作用在物体上的力等于物体的质量乘以加速度（牛顿第二定律）；黑火药是一种低速爆炸物；炸药爆炸速度大于1 000米/秒，是一种高速爆炸物。
生活方式	中国人发明了烟花，意大利人把它变成了一门艺术。
社会学	英语中"火箭"这个词来自意大利语rocchetta（小雷管），是一种小爆竹的名字。
心理学	承认天外有天、人外有人是很重要的。
系统论	推进气体对大气产生了极大破坏且不受国际公约限制；射炮步甲仅利用当地自然资源，只向空气释放氧气，为人类提供了一种新的选择。

教师与家长指南

情感智慧
Emotional Intelligence

蚂蚁

蚂蚁们好奇心很强。他们了解自己的身体大小，并且知道他们不太可能是枪击的目标。蚂蚁们见多识广，并且乐于提问。他们乐于彼此分享信息（包括专业细节），表现出了良好的智力。他们的洞察能力表明他们受过良好的教育。作为射炮步甲的捕食者，蚂蚁们分析了如何抓住甲虫（它们需要时间展开翅膀），并且知道自身的优点和甲虫的缺点。蚂蚁尊重甲虫，推断甲虫仅仅借助当地可利用的自然资源构建了自己的防御机制。即使蚂蚁们因团队合作的能力而受到称赞，并且为自己收获的褒奖感到骄傲，他们仍然承认这次甲虫更胜一筹。

艺术
The Arts

是时候拿出你的铅笔和蜡笔了。不要看图片，凭你的记忆力画出你看到过的10种不同形状和颜色的焰火。焰火是一门艺术，你可以制作一个印象集。现在，在你向任何人展示之前，看一看真实的焰火图片，然后拿着铅笔，在你画的画上作出一些调整。

TEACHER AND PARENT GUIDE

思维拓展
Systems: Making the Connections

　　面对健康、能源和废物威胁，非洲射炮步甲提供了一种全新的解决之道。射炮步甲会产生废物和过剩物质，分别是过氧化氢和醌，前者是其新陈代谢的副产物，因此是一种容易得到的化学物；后者是甲虫自然产生的一种物质，能使它的壳变硬。关键是如何使这两种主要的可用成分相互协调发挥出作用。这种甲虫因此在阀门系统的基础上建立了一个有效的防御机制：甲虫能以每秒500次的频率执行这个反应，并能够按需求开始或者停止这个反应。工业化生产需要持续的批量生产。然而，只有在拥有推进系统及充足的可再生的当地原材料时，才能更有效地实现批量生产，因为能随时快速地开始或停止生产。这个灵感来自长久以来被测试和证实的甲虫，它能通过释放热能和氧气生产100摄氏度的热水，这与人类生产过程会产生如氯氟烃或氯氟化碳等有害气体不同。人类生产了好几种哮喘喷雾器，其中的推进气体并不完全相同，但是自然界特别是射炮步甲为医疗行业提供了另一种选择。我们应该学习甲虫利用废弃物和剩余材料来生产最好的推进气体，这个卓越的进展激发人们重新设计在工业和医疗领域，特别是哮喘方面的许多标准程序。

动手能力
Capacity to Implement

　　没有什么比自己动手制作爆竹更让人激动的了，但这是个危险的过程，因此你必须认真负责。在美国，有5万多人在自制爆竹。你必须获得家长和你所在地区法规的准许，并且一定要有一位成年人监督你——你不能自己独自制作。步骤很简单，但是要采取正确的预防措施，享受你自己的爆竹吧！

教师与家长指南

故事灵感来自
This Fable Is Inspired by

拉斯–尤诺·拉森
Lars-Uno Larsson

拉斯–尤诺·拉森在药物界卓有成就。起初他是一名员工，后来发展成一位企业家。1988年，他建立了奥丰国际公司，专门从事罕见病症的医学治疗，并生产相应的医疗产品。当时环境立法禁止生产破坏环境的推进器喷雾，拉斯–尤诺·拉森就专心于发展新的哮喘喷雾器，其灵感来自射炮步甲。他决定支持由安迪·麦金托什（Andy McIntosh）教授带领的利兹大学研究团队来研发解决方案。他创建了Biomimetics 3000来投资发展这项事业以及其他传统工业还未涉及的新方法。

图书在版编目（CIP）数据

冈特生态童书.第二辑修订版:全36册:汉英对照／
(比)冈特·鲍利著;(哥伦)凯瑟琳娜·巴赫绘;
何家振等译.—上海:上海远东出版社,2021
书名原文:Gunter's Fables
ISBN 978-7-5476-1759-5

Ⅰ.①冈… Ⅱ.①冈… ②凯… ③何… Ⅲ.①生态
环境-环境保护-儿童读物—汉、英 Ⅳ.①X171.1-49

中国版本图书馆CIP数据核字(2021)第213075号

著作权合同登记号图字09-2021-0823

策　　划	张　蓉
责任编辑	祁东城
封面设计	魏　来　李　廉

冈特生态童书
用空气射击
[比]冈特·鲍利　著
[哥伦]凯瑟琳娜·巴赫　绘
李　蕃　译

记得要和身边的小朋友分享环保知识哦！
八喜冰淇淋祝你成为环保小使者！

Education
68

植物会唱歌吗？

Can Plants Sing?

Gunter Pauli

[比] 冈特·鲍利 著
[哥伦] 凯瑟琳娜·巴赫 绘
李蕃 译

上海远东出版社

丛书编委会

主　　任：贾　峰
副主任：何家振　闫世东　林　玉
委　　员：李原原　祝真旭　牛玲娟　梁雅丽　任泽林
　　　　　王　岢　陈　卫　郑循如　吴建民　彭　勇
　　　　　王梦雨　戴　虹　翟致信　靳增江　孟　蝶

特别感谢以下热心人士对童书工作的支持：

匡志强	宋小华	解　东	厉　云	李　婧	陈　果
刘　丹	熊彩虹	罗淑怡	旷　婉	杨　荣	刘学振
何圣霖	廖清州	谭燕宁	韦小宏	李　杰	欧　亮
陈强林	王　征	张林霞	寿颖慧	罗　佳	傅　俊
胡海朋	白永喆	冯家宝

目录

植物会唱歌吗	4
你知道吗？	22
想一想	26
自己动手！	27
学科知识	28
情感智慧	29
艺术	29
思维拓展	30
动手能力	30
故事灵感来自	31

Contents

Can Plants Sing?	4
Did You Know?	22
Think about It	26
Do It Yourself!	27
Academic Knowledge	28
Emotional Intelligence	29
The Arts	29
Systems: Making the Connections	30
Capacity to Implement	30
This Fable Is Inspired by	31

一只母鸡和一只公鸡正在灌木丛里寻找昆虫和新长出来的嫩草。这时,他们发现有个人在和一株植物讲话。

A hen and a rooster are foraging in the bush, looking for insects and freshly sprouted grass. They notice someone talking to a plant.

寻找昆虫

Looking for insects

植物还能说话甚至唱歌呢

Plants can talk and even sing

"难道这位女士真的认为植物能听懂她的话？"公鸡问道。

"植物不但能听懂，它们还能说话甚至唱歌呢。"母鸡回答。

"你一定是在开玩笑！植物没有嘴，没有舌头，没有声带，它连大脑都没有！"

"Does that lady really believe that plants listen to her?" asks the rooster.

"Plants don't only listen, they can talk and even sing," responds the hen.

"You must be joking! A plant does not have a mouth or a tongue or vocal cords, it doesn't even have a brain!"

"没错,亲爱的。你每天早上喔喔喔地叫我们起床,可植物并不见得就得跟你一样做呀!"母鸡反驳公鸡。

"好吧,那么你呢?你下蛋的时候,你也发出很大的声音,好像要让全世界知道你的感受。可植物怎么能说话或者唱歌,来让我们知道它们的想法和感受呢?"

"True, dear. But just because you crow your heart out every morning waking all of us, it doesn't mean that plants have to do it the way you do!" The hen shoots the rooster back.

"Well, what about you? When you are laying eggs you make a lot of noise too, letting the whole world know how you feel. How can plants possibly talk or sing, or let us know what they think and feel?"

反驳公鸡

shoots the rooster back

植物不会下蛋

Plants do not lay eggs

"不管怎样,植物不会下蛋,那你告诉我它们应该怎么让我们知道它们的想法和感受呢?"

"Anyway, plants do not lay eggs, so tell me how are they supposed to let us know what they think and feel."

"让我这么跟你解释吧：你是不是有时候会觉得脚冷？"母鸡问道。

"嗯，冬天的时候每天早上都会。"

"那么在夏天，你会不会又觉得太热了呢？"

"是啊，每个夏天都会感到很热，瞧我这身羽毛！"

"Let me put it to you this way: Do you sometimes have cold feet?" asks the hen.

"Yes, every morning in winter."

"And in summer, do you ever feel too hot?"

"Yes, every summer – with all these feathers!"

冬天觉得脚冷

Cold feet in winter

小鸡窝在你身边

Chicks huddle around you

"那你喜欢咱们的小鸡窝在你身边取暖么？"

"当然，那是我一天中感觉最棒的时刻。"

"那为什么植物不会有同样的感受呢？"

"因为它们是植物，不是动物——我们属于大自然的不同界。"公鸡回答说。

"And do you like it when our chicks huddle around you to keep warm?"

"Of course, that is the best moment of my day."

"Now why would plants not feel the same?"

"Because they are plants, not animals – we belong to different kingdoms of nature," replies the rooster.

"你说得对，但是我们都依靠同样的自然法则生活和成长。当植物的根是冷的，而叶子温暖的时候，它就会长得更快。这就像你感到脚冷和嘴热的时候一样。"

"哼，让我告诉你吧：我宁愿有一双暖暖的脚和一张冷冷的嘴！起码这样更舒服。"

"Yes, but we all live and thrive by the same natural laws. A plant grows faster when its roots are cold and its leaves warmer. It is like you having cold feet and a warm beak."

"Well, let me tell you: I prefer to have warm feet and a cold beak! It is simply so much more comfortable."

冷冷的根和暖暖的叶子

Cold roots and warm leaves

植物会长得更好更快

Plants will grow better and faster

"那样你是觉得舒服了，但对植物来说并不如此。冷冷的根和暖暖的叶子会让植物长得更好更快，而且长大以后味道更甜、颜色更绿。如果我们提供了合适的条件，它们会用更多的营养和更好的口感来回报我们的。"

"For you, but not for plants. With cold roots and warm leaves plants will grow better and faster and will be sweeter and greener. If we provide the right conditions, they will reward us with more nutrients and a better taste."

"我猜那是它们说谢谢的方式吧。"公鸡不情愿地说。"但是植物会说话?我才不信呢。"

"爱信不信!我知道植物是会唱歌和发光的,只是我们不要期待植物用我们的方式表达自己!"

……这仅仅是开始!……

"I suppose that is their way of saying 'thank you' to us," says the rooster. "But that plants can really talk, I'm not buying that."

"You don't need to! I know plants sing and shine – as long as we don't expect them to do it the way we do it!"

...AND IT HAS ONLY JUST BEGUN!...

······ 这仅仅是开始！······

... AND IT HAS ONLY JUST BEGUN! ...

Did You Know?

你知道吗?

Plants grow better when there is a difference in temperature between the roots and the leaves.

当植物的根和叶子有温差时，植物会生长得更好。

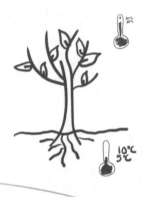

Plants need soil to grow. We should not consider soil dirt, but a life-giving mixture of nutrients that supports self-renewing nourishing plants.

植物需要土壤来生长。我们不应当认为土壤是脏的，而要把它看成支持生命的各种养分的混合物，有助于植物自我更新和获得营养。

Willow and poplar trees can warn each other against insect attacks.

柳树和杨树会互相警告小心昆虫袭击。

When insects chew leaves, plants respond by releasing chemicals into the air. When other plants detect these airborne signals, they increase their defences against invaders.

当虫子啃食叶子时，植物会向空气中释放化学物质。当其他植物察觉到空气里传播的信号时，他们就会增强对入侵者的防卫。

Plants are able to transmit information via electrical pulses and a system of voltage-based signals.

植物可以通过电脉冲和一系列基于电压的信号来传播信息。

Maize attacked by beet worms releases chemicals that attract wasps. The wasps then kill the caterpillar before laying their eggs in the caterpillar. In this way plants also 'talk' to insects.

玉米被甜菜夜蛾幼虫袭击时，会释放化学物质来吸引胡蜂。胡蜂会杀死这些幼虫，并把它们的卵产在幼虫里。植物就是通过这种方式来跟昆虫"说话"的。

In order to understand plants we have to think the way they do. If you can imagine what a plant may be thinking, you will be able to understand more about it.

为了了解植物，我们要以它们的方式来思考。如果你可以想象植物在想什么，你就更能理解它们了。

When you stand very close to a plant and talk, the carbon dioxide you release is food for the plant. So you can feed a plant and help it grow by talking to it.

当你跟一株植物站得非常近并且跟它说话，你释放的二氧化碳就是植物的食物。所以你可以养一株植物，并且通过跟它说话来帮助它生长。

Think about It

想一想

If vegetarians do not eat meat because they do not want to have sentient beings killed for food, will they eat plants when they realise that plants are also sentient beings?

如果素食者不吃肉，是因为他们不想杀掉有感觉的生物作为食物，那么当他们意识到植物也是有感觉的生物时，他们还会吃植物吗？

Could we learn about defence from plants, and should we use this knowledge to control pests instead of relying on harmful chemicals to do so?

我们可以从植物那里学习防卫吗？我们应该运用这个知识取代有害的化学品来控制害虫吗？

How much scientific proof does one need before accepting new information and adding it to one's knowledge base?

一个人需要掌握多少科学证据，才能接受新信息并且将这个信息加入自己的知识库里？

如果科学已经证明了植物可以与昆虫交流，那我们为什么不能跟植物和昆虫交流？

If science has proven that plants are able to communicate with insects, why are we not able to communicate with plants and insects?

Do It Yourself!
自己动手！

Plant two beans in two separate pots, using the same soil. Wrap each pot in foil so that the plant is protected from the wind. Leave a wide opening at the top and water the plants equally. Place them together in the same spot so they get the same amount of light. As soon as the beans start sprouting, form a team and take turns to talk to only one of the bean plants for at least one hour per day. The other bean plant is left to grow on its own. What do you notice after a month?

分别在两个花盆里用同样的土壤种两颗豆子。用箔纸分别将两个花盆包起来，这样植物就不会被风吹跑了。在上方留出一个大大的口子并且给植物们浇等量的水。将它们放在一起，让它们能得到同样的光。当豆子开始发芽，组织一个团队，轮流跟其中一株豆芽说话，每天至少说上一个小时。另外一株豆芽就让它自己生长吧。一个月之后，你注意到了什么？

TEACHER AND PARENT GUIDE

学科知识
Academic Knowledge

生物学	植物跟植物交流的证据；植物与昆虫的交流；真菌在土壤交流里的角色；食草动物与食肉动物的区别；植物没有神经系统或大脑；植物的防御机制：植物受到袭击时会让叶子变得不那么可口，这样昆虫和吃嫩叶的动物将会去别的地方找食物；植物激素；由于过度使用化学品，害虫的抗药性进化了；用声带说话的方法。
化学	挥发性有机化合物（VOC）；生物化学信息；蛋白酶抑制剂（扰乱昆虫的消化能力）；酶；植物与大气的氧气-二氧化碳循环。
物理	基于电压的信号；基于电脉冲的神经系统；薄膜和带电离子；根部寒冷而叶子温暖能促进植物产生更强的渗透作用；鸟类尤其是鸡的羽毛的作用；非语言表达的交流。
工程学	农作物借以抵抗害虫的农业系统是以植物所用的交流手段为基础的。
经济学	杀虫剂：农场主的开支和给社区带来的健康风险；有机水果和蔬菜的定价与使用了化学品催长的水果和蔬菜的定价的对比。
伦理学	我们能够继续只是通过人类的视角（人类中心主义）去看这个世界吗？或者，我们有力量和洞察力通过其他物种的视角去审视现实吗？
历史	新的科学领悟是如何连续不断地被当权者否决的？从曾经的地球是平的观点和地球是宇宙中心的观点开始回顾。
地理	法国把公鸡作为国家的象征。
数学	在公式里用Δ表示差值，比如用Δt（温度差）来计算植物潜在的生长；温度差越大，从根部来的营养就会越多；大量实验证明结果中存在统计学意义上的差异，也有大量实验证明相同的结果可以从统计学意义上充分多元化的样本中获得。
生活方式	人类中心主义：我们采取的生活方式是为人类的需求服务的，因此人类很难意识到对环境的破坏。
社会学	行为与交流的定义；相互竞争的有机体都认为彼此适合分享知识；不断发展的科学与教条之间的差异；人们抗拒变化。
心理学	我们会不会认为植物像人一样（拟人化）？还是说我们会像植物一样思考和表现（植物化）吗？当你对一个学科没有一个清晰的理解时，你如何形成看法？
系统论	我们生活在这样一个生态系统里：生物的五界创造了生命之网。我们正在不断了解各个界为了相互的利益和人类的利益，是怎么互相影响的。

教师与家长指南

情感智慧
Emotional Intelligence

公 鸡

公鸡很顽固，并且有先入为主的想法。他将他的判断建立在他所知道的以及他一直这么认为的事情上，没有留下一点容纳新见解的空间。他从自己的视角评判所有事物，更糟的是，他不能容忍任何不同的视角，即使这种新视角是出现在新的情景里，而且有着科学数据表明新科学正在诞生。公鸡在交流中始终坚持自己的观点。即使母鸡激起他的情感（与小鸡们窝在一起），他还是坚持自己的观点。母鸡提出的论点没有创造出什么机会，反而好像再次证明了他的固执己见。这使他看不见新的事物。

母 鸡

母鸡对新出现的事物采取了赞同的观点。她耐心倾听公鸡的话，虽然公鸡狭隘、顽固，有时甚至会出言不逊地说出轻率的评论。母鸡展开带有感情的辩论，企图使公鸡受到影响，指出小鸡们在冬天使得公鸡保持温暖，但是这并没有打动公鸡。母鸡仍然坚持努力，没有显示出任何焦虑，并且把公鸡的论点记下来，将新的逻辑引入她的推理。她对公鸡保持尊敬，并且留下空间让公鸡自行其是，而没有任何评判。她分享了她的智慧，而且并不指望她的思想能改变公鸡的心。

艺术
The Arts

像"挥发性有机化合物"（VOC）和"空气传播交流"这样的词，不在我们日常用词之内。是不是每一个人都理解这些词呢？弄懂像这样一些专业术语的一个办法，是将它们置于合适的语境中。一个好办法是用铅笔或者颜料画一些画。试着通过艺术来解释这些词是什么意思。

TEACHER AND PARENT GUIDE

思维拓展
Systems: Making the Connections

大自然是生态系统的组合。每一个生态系统都服从同样的物理规律，靠可以获得的资源生长，并不断进化。自然系统已经从几个活细胞进化出各种各样的生命。我们可以说，我们都是这些第一代单细胞物种通过突变和共生发展出的后代，这种进化导致了生物学家所称的大自然的五界（原核生物、原生生物、植物、真菌和动物）。不久以前生物学家仍然相信只有三个界：矿物、植物和动物。但随着科学家们对物种间的相似性和差异有了更好的理解，并且形成了关于生态系统如何运转的更丰富的知识库，普通的共识已经从三个界进化成五个界了。很可能在将来我们会将地球上的生命分为七界，或者抛弃"界"这个名称而用"域"这个概念。随着科学证据的累积，我们不断完善自己的知识。现实不断发生改变证实了知识的丰富，而我们还没有发现物种和生态系统是如何真正运作的。这还证明了科学的力量，而科学永远在寻求新的真理。科学经常聚焦于单一的、往往定义得很细的研究课题，但我们实际上需要的是一个对万物如何关联的更好的理解——万物是如何适应、运作、相互作用和动态演化的。我们知道，自然里唯一不变的是变化，我们还知道人类活动使变化加快了——不管是变得更好还是更糟。人类只是众多共享这个星球的物种之一。如果我们尝试通过真菌、植物甚至细菌的视角来审视现实，然后再审视当代社会并观察我们的行为，我们就可以采取非常不同的视角，来思考如何解决我们现今面对的问题。

动手能力
Capacity to Implement

上网搜索"来自植物的音乐"。你会感到吃惊，你甚至能够找到全部由植物组成的整个"管弦乐队"，能用不同乐器创作出长达一小时的音乐。将这个新发现分享给朋友和邻居。他们的第一反应是什么？然后再问，你是否可以向他们展示一个视频，或者他们是否愿意聆听你从网上下载下来的录音？他们的第二个反应是什么？我们的行动能力取决于我们展示新现实的能力，即使不是每个人都被说服，至少确保你自己会从中受益。

教师与家长指南

故事灵感来自
This Fable Is Inspired by

凯瑟琳娜·巴赫
Katherina Bach

凯瑟琳娜·巴赫的职业生涯始于获得一次青少年网球比赛的冠军，之后她决定当一名律师。她获得了商业法的高级文凭，并继续专攻气候变化条约，代表她的国家哥伦比亚参与了全球性公约的多方谈判。在时任国际自然保护联盟主席阿谢克·科斯拉（Ashok Khosla）博士向她介绍了一个把植物的电脉冲转化为音调的机器之后，凯瑟琳娜着手向孩子们展示人类与植物的新关系。因为确信人类已经改变了他们对动物的认识，她觉得是时候也改变人类与植物的关系了。她发现，一旦孩子因为某个草莓丛可以发出最美的曲调而选择了它（从而改变了他们与植物的关系），他们在吃草莓的时候就会带着尊敬和感激。凯瑟琳娜正在准备一台大型"音乐机器"，让世界各地学校里的孩子可以欣赏这些通过他们自己与植物、环境的互动而生成的声音。这个过程也能帮助孩子们了解植物在为我们提供生存所需的氧气时的角色。

图书在版编目(CIP)数据

冈特生态童书.第二辑修订版:全36册:汉英对照 /
(比)冈特·鲍利著;(哥伦)凯瑟琳娜·巴赫绘;
何家振等译. —上海:上海远东出版社,2021
书名原文:Gunter's Fables
ISBN 978-7-5476-1759-5

Ⅰ.①冈… Ⅱ.①冈…②凯…③何… Ⅲ.①生态
环境–环境保护–儿童读物—汉、英 Ⅳ.①X171.1-49

中国版本图书馆CIP数据核字(2021)第213075号
著作权合同登记号图字09-2021-0823

策　　划　　张　蓉
责任编辑　　祁东城
封面设计　　魏　来　李　廉

冈特生态童书
植物会唱歌吗？
[比]冈特·鲍利　著
[哥伦]凯瑟琳娜·巴赫　绘
李　蕃　译

记得要和身边的小朋友分享环保知识哦！
八喜冰淇淋祝你成为环保小使者！

Education
69

数学启蒙

Maths for Beginners

Gunter Pauli

[比]冈特·鲍利 著
[哥伦]凯瑟琳娜·巴赫 绘
李蕃 译

上海远东出版社

丛书编委会

主　任：贾　峰
副主任：何家振　闫世东　林　玉
委　员：李原原　祝真旭　牛玲娟　梁雅丽　任泽林
　　　　王　岢　陈　卫　郑循如　吴建民　彭　勇
　　　　王梦雨　戴　虹　翟致信　靳增江　孟　蝶

特别感谢以下热心人士对童书工作的支持：

匡志强　宋小华　解　东　厉　云　李　婧　陈　果
刘　丹　熊彩虹　罗淑怡　旷　婉　杨　荣　刘学振
何圣霖　廖清州　谭燕宁　韦小宏　李　杰　欧　亮
陈强林　王　征　张林霞　寿颖慧　罗　佳　傅　俊
胡海朋　白永喆　冯家宝

目录

数学启蒙	4
你知道吗？	22
想一想	26
自己动手！	27
学科知识	28
情感智慧	29
艺术	29
思维拓展	30
动手能力	30
故事灵感来自	31

Contents

Maths for Beginners	4
Did You Know?	22
Think about It	26
Do It Yourself!	27
Academic Knowledge	28
Emotional Intelligence	29
The Arts	29
Systems: Making the Connections	30
Capacity to Implement	30
This Fable Is Inspired by	31

老鼠走过一棵大树，抬头望见猫头鹰正在树上沉思，回味着它以往所有的知识。老鼠很明白，如果他胆敢惹恼猫头鹰，他就会成为猫头鹰的晚餐。

"晚上好，教授。您解决了苹果怎么长到树上去的问题了吗？"

A mouse passes by a tree, looks up and sees an owl lost in thought, contemplating all the wisdom of the past. The mouse is well aware that he could end up on the owl's dinner if he dares annoy him.

"Good evening, Professor. Have you figured out yet how the apple gets up on the tree?"

望见猫头鹰正在树上沉思

Sees an owl lost in thought

为什么我们需要学习那么多的公式

Why we have to study so many formulae

"啊，是你呀！"猫头鹰说。"你准备好学习更多的数学知识了吗？"

"我只是很纳闷，为什么我们需要学习那么多的公式，还要用心记住它们。"

"每个人都要懂得基础的数学及几何知识。学会如何去计算圆的面积、立方体的体积、管道的长度……"

"Oh, it's you!" says the owl. "Are you ready to learn some more mathematics?"

"I was just wondering why we have to study so many formulae and learn them off by heart."

"One needs to know the basics of mathematics and geometry. How to calculate the size of a circle, the volume of a cube, the length of a pipe ..."

"但是为什么要学呢?这些不都可以在维基百科上找到吗?"

"维基百科是个很好的工具。"猫头鹰说。"但如果没有电或者上不了网,你怎么办呢?"

"要是有一天有人能够提出一个可以计算任何东西的公式:面积、圆周长,甚至星星、硅藻和无线电波,那该多么美妙。"

"But why? Isn't it all on Wikipedia?"

"Wiki is a great tool," says the owl, "but what if there is no electricity and you cannot get on to the net?"

"It would be great if one day someone could come up with one formula that allows me to calculate everything: squares, circles, even stars, diatoms and radio waves."

一个可以计算任何东西的公式

A formula to calculate everything

类似于海浪

Like the waves of the ocean

"噢,那将是一个超级公式。但在它变成现实之前,你只能先去学习更多的基础知识。"

"那您能告诉我电波是怎么传播的吗?"

"正如它的名字所说,这是一种波,类似于海浪。"

"Well, that would be a superformula. But before that becomes a reality, you will simply have to study a great deal more."

"Can you tell me how a radio wave moves?"

"As the word says, it is a wave, like the waves of the ocean."

"那电波在传播过程中如果遇到大石头或者高山会怎么样呢?"

"电波将从那些障碍物上面或者四周绕过去。这或许会形成一些湍流。"

"And if there is a big stone or a high hill in its way?"

"Then it will have to flow up and over, or around it and perhaps generate some turbulence."

从上面或者四周绕过去

Flow up and over, or around it

用冲浪板来捕捉电波

Catch some waves with a surfboard

"那为什么电话公司不用冲浪板来捕捉电波呢？"

"冲浪板？"

"是啊，现在都是利用天线去收集并发射这些电波！假使无线电波真的像冲浪时的波涛一样，那利用固定的板去传递这些信号并不够好，对吗？但如果这些板可以动的话……"

"So why don't telephone companies catch some waves with a surfboard?"

"A surfboard?"

"Yes, look at all these antennae used to pick up radio waves and transmit them! If a radio wave is like a wave in the surf then a fixed board isn't much good, is it? But if it could move …"

"嗯……"猫头鹰疑惑地说:"固定天线随处可见。这是标准,因此它必定是正确的。"

"如果天线这么好用,那为什么手机会经常没有信号或者无法通话呢?"

"Hmmm ..." wonders the owl. "One sees fixed antennae everywhere. It is the standard, so it must be good."

"But if they are so good, why do mobile phones lose signal and drop calls all the time?"

为什么手机会没有信号

Why do mobile phones lose signal

使通话声音更洪亮

Make them scream louder

"为了处理更多的通话，通信公司应该架设更多的天线，这会使通话声音更洪亮。"

"更洪亮？您的意思是放大声音并且消耗更多的能量吗？"

"To handle more calls, the mobile companies should put up more antennae and make them scream louder."

"Scream louder? You mean turn up the sound and burn more energy?"

"通信公司过去很多年一直都这么做。"

"噢,我对现在怎么做不感兴趣。"老鼠回答道。"我只想知道,如果有超级公式,未来又会怎么做!"

……这仅仅是开始!……

"That is the way it has worked for years."

"Well, I am not that interested in how it works now," answers the mouse. "I want to know how it will work in future – with the superformula!"

...AND IT HAS ONLY JUST BEGUN!...

……这仅仅是开始！……

... AND IT HAS ONLY JUST BEGUN! ...

Did You Know?

你知道吗？

Mathematics is used in all the sciences, such as biology, chemistry, physics, psychology, sociology and engineering (civil, mechanical and industrial).

数学被运用在所有的科学门类中，诸如生物学、化学、物理学、心理学、社会学和工程学（土木工程、机械工程和工业工程）。

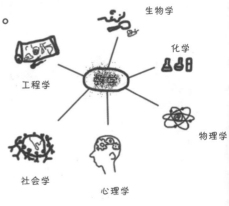

Anything that uses a computer uses mathematics.

所有用到电脑的东西都用到了数学。

Scientists consider mathematics the 'second microscope' of biology as it offers us new insights into biology.

科学家认为数学是生物学的"第二个显微镜",因为它能为生物学展现新的视野。

The Voyager's space journey to distant planets could not have been calculated without the mathematics of differential equations.

"旅行者"号探测器探索外太空星球时的路线计算离不开数学中的微分方程。

Geometry (how to deal with spatial relations) and the study of numbers were the first aspects of mathematics.

几何（如何处理空间关系）以及对数字的研究是数学最重要的方向。

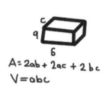 

The mathematicians of Babylon used 60 (and not 10) as the basis of numbers. From this we derived 60 seconds in a minute, and minutes in an hour, and 360 degrees in a circle.

古巴比伦数学家用 60 进制（而不是 10 进制），由此衍生出了一分钟 60 秒、一小时 60 分钟以及一个圆周 360 度。

The number zero was invented independently 4,000 to 5,000 years ago by Babylonians, Mayans and Indians.

4 000 至 5 000 年前，古巴比伦人、玛雅人以及印度人分别独立发明了数字 0。

Mobile phones depend greatly of mathematics to send a digitised version of speech with data compression. However the only way to ensure that calls are not dropped is if the antennae transmit amplified sound, using more energy.

手机的正常使用非常依赖数学，需要把语音通过数据压缩转换成数字信号才能发送。而唯一确保手机通话畅通的方式是通过天线传递经过放大的声音，这会消耗更多的能量。

Think about It

想一想

Why do we need to learn some things off by heart when it is readily available on the internet?

为什么即便在网络上可以搜索到知识，我们还是需要用心学习？

If all information is digital, what happens the day that there is no electricity?

如果所有的信息都是数字化的，那么断电的时候会发生什么？

Why are all the antennae the same, even though the landscape and the structure of buildings is different?

为什么所有的天线都是相同的，即使景观和建筑的结构不同？

If there is a problem, what is the best solution: using more strength and power doing more of the same as has always been done, or to invent a new way of reaching the same objectives?

面对问题，什么是更好的解决方案？是利用之前的解决方式但花费更多的力量和精力，还是创造一个新的解决方式来达到相同的目的？

Do It Yourself!
自己动手!

Take a portable radio and play music. Walk around and go stand behind a wall, then go inside a building. Now take the elevator and go to the basement, or the garage, and close the doors. What do you notice? The music does not change at all, and the sound is stable too. Now take a mobile phone and tune into your favourite radio station on it. Now follow the same circuit you did with the radio. What happens? If you have good reception throughout, you are lucky. As soon as you are in the elevator or get down to the basement, the connection is likely to be poor or interrupted. What is the difference between the radio and the mobile phone?

带着你的收音机播放音乐并随处走走。你可以走到墙后，再走进一栋建筑。现在坐电梯到地下室或车库，关上大门，你发现了什么？你听到的音乐并没有什么变化，声音还是很稳定。现在，打开你的手机并调到你喜欢的电台。遵循之前的步骤，发生了什么？如果自始至终信号稳定，那说明你很幸运。一旦你走入电梯，进入地下室，信号很可能会变差甚至中断。想想收音机和手机的区别是什么？

TEACHER AND PARENT GUIDE

学科知识
Academic Knowledge

生物学	生物数学：将蛋白质的相互作用转换成精确的数学模型；植物（尤其是竹子）的生长方式启发人们设计出了新的数学模型。
化 学	化学现象和化学反应的数学建模；代数学能用于分析实验的反应速率，向量帮助人们理解晶体结构，对数有助于理解pH值；概率论和统计学使我们可以从实验中得出一般结论。
物 理	数理物理学应用数学来解决物理问题，也运用数学来构建物理理论；波在空气及水中的传播；天线的形状；电和磁组合在一起就成了电磁学；存在一些无线电波，其频率低于可见光；将电磁辐射用能量、波长（米）或频率（赫兹）来表达。
工程学	工程数学由微分方程、向量和张量分析、线性代数和应用概率论组成。
经济学	监督、设计和控制生产过程；商业模式的变化使得天线从全球统一标准变成根据空间或时间来设计，并在当地用3D打印生产出来；将一个参数的结果最大化与对整个系统的结果进行优化的对比。
伦理学	新的数学使得我们可以将通信所需的能量减少至十分之一，但这需要将商业模式从全球生产转变为本地生产；随着无线电波数量和强度的增长，我们必须发现它们会造成哪些意想不到的后果，以免带来其他伤害。
历 史	1901年，马可尼让无线电信号跨越了大西洋，他利用了一根垂直到地面的电线，而在大洋彼岸则是一根由风筝牵引到200米高的电线。
地 理	短波无线电可以环绕整个地球，之所以能做到这一点是因为存在电离层，电离层反射某些频率的无线电波，这些电波回到地面并绕着地球运动；无线电波在晚上比在白天传播得更远。
数 学	效率是用方差或均方差来定义的。
生活方式	我们已经习惯了运用指尖上的通信工具，它取代了书信、固定电话、使用宽频的短频电波以及数字卫星无线电。
社会学	新的社交模式消除了个人隐私，在我们不知情的情况下，关于我们喜欢什么、买了什么、阅读什么的数据常常未经同意就被出售给第三方，以便用有针对性的方式向我们推销货物。
心理学	我们用即时沟通的能力创造归属感，与此同时，人与人之间真正的沟通被流于表面的远距离数字交互所替代。
系统论	天线"声音"越大，来自使用类似波长的系统的干扰越大，从而需要安装更多的天线，增加更多的能量来对抗来自各种障碍物的背景噪音。

教师与家长指南

情感智慧
Emotional Intelligence

猫头鹰

　　猫头鹰被视为年迈而智慧的象征,而他的反应正印证了这点。尽管老鼠在之前对他有所冒犯,猫头鹰也没有表现出任何怨恨。猫头鹰也许认为老鼠的开场白有些无礼,但他仍一如既往地坚持数学的重要性和懂得基础知识的必要性。他没有阻止老鼠进行不拘一格的思考,反而认同如果新的公式被提出,那将会是一个超级公式。猫头鹰表明了他的观点,他遵从通行标准(如果它是标准,那它必定是正确的),并相信保持现状对所有人都有好处。

老　鼠

　　老鼠喜欢冒险。即使他之前冒犯过猫头鹰,并知道他有可能无法逃脱后者的攻击,却还是用一种近乎傲慢的自信和猫头鹰对话。老鼠坚持认为学习数学没有什么用处,因为一切都能在互联网上搜索到。当猫头鹰用停电的情况来回答他的问题后,他很快转变了话题,并提出了新的问题。猫头鹰很愉快地分享了他的知识。老鼠促使猫头鹰思考了更多的东西,但仍然持续用柔和的方式来提出问题,直到老鼠表示,他对事物现在如何运作没有任何兴趣,而是想知道这些问题在将来如何解决。他信奉探索者的逻辑:决不安于现状。

艺术
The Arts

　　在互联网上搜索曼德勃罗(Benoît Mandelbrot),看看那些利用分形创造出的令人惊叹的图形。分形是一组点集,其边界是一些极易识别的二维图形,它将每个数的实部和虚部都看成图像坐标,并且是彩色像素。曼德勃罗因为他创造的绚丽的分形图案而在数学界之外也赫赫有名,他用这些分形图案作为例子,说明应用简单的数学规则也会出现复杂的结构。现在,运用数学来创造属于你自己的艺术品吧。

TEACHER AND PARENT GUIDE

思维拓展
Systems: Making the Connections

　　我们的生活中到处都渗透着数学，但是如果拒绝变革，数学就变得保守了。用3D打印方式制造为家庭、办公室或学校量身定做的小型天线，不仅使信息传输更加高效，费用也更低。而且这种小型天线生产设备还有可能用来创建本地企业。设计这些独特的、能以低于1美元的价钱生产的天线，所需要的数学知识是现成的。但是目前，为了解决通话中断和网络信号不佳的问题，我们还是依靠安装更多同类的大型天线。

　　解决问题的关键，不在于提供更多过去使用（耗费了更多材料和能源）的解决方案，而是要找到方法来同时解决多个问题，并且兼顾不同方面的利益。目前手机通话中断和网络信号不佳的改善是通过增加能耗实现的。因此，要同时从用户、供应商（能耗）和社会的角度来解决问题。这种数学模型称为"系统动力学"，是20世纪60年代由麻省理工学院的杰·弗雷斯特博士提出的。他的模型可以通过研究交叉影响来寻找最优解决方案，而不是仅仅考虑单一参数的最优化。

动手能力
Capacity to Implement

　　让我们来制造自己的天线。这个新型的天线叫罐形天线，由一个金属罐和一根天线组成。它是怎么工作的呢？我们吃的食物常常是用金属盒保存的，有时候也用内部涂铝膜的纸盒容器。这两者都可以用。只要你插上一根电缆，这些容器都可以转换成天线。这个简单的装置会增加你的Wi-Fi网络的范围，扩大手机信号的覆盖面，不用增加用电就能改善信号接收并减少噪音。你可以在互联网上找到一些指导，但是一定要自己先尽力去寻找最佳方法。你可以为你的手机或者Wi-Fi做天线，也可以用几个金属衣架做一根强力的电视天线。完全行得通！你准备好开始销售你的高能效天线了吗？

教师与家长指南

故事灵感来自

This Fable Is Inspired by

约翰·吉利斯
Johan Gielis

约翰·吉利斯的第一个学位是园艺学方面的。出于对竹子用于再造森林和碳汇的兴趣,他开发了适用于各类竹子的创新性组织培养技术。他对竹子的兴趣促使他研究世界各地有关竹子的科学和文化。尽管没有在数学方面受过学术训练,他仍然着手建立植物的一般数学模型,特别是关于竹子的生长模式。2003年他第一次在《美国植物学杂志》上发表了他的发现,对抽象的天然形态和各种相关特征(形状、尺寸、进化和变异)都给出了全新的视角。他的数学公式被誉为"超级公式",现在更多地被称为吉利斯公式。这个公式已经让人们得以开发出新的软件系统,其重要特点是可以依照顾客需求去设计天线,减少视频数据流所需的带宽。

图书在版编目(CIP)数据

冈特生态童书.第二辑修订版:全36册:汉英对照/
(比)冈特·鲍利著;(哥伦)凯瑟琳娜·巴赫绘;
何家振等译.—上海:上海远东出版社,2021
书名原文:Gunter's Fables
ISBN 978-7-5476-1759-5

Ⅰ.①冈… Ⅱ.①冈…②凯…③何… Ⅲ.①生态
环境－环境保护－儿童读物—汉、英 Ⅳ.①X171.1-49

中国版本图书馆CIP数据核字(2021)第213075号

著作权合同登记号图字09-2021-0823

策　　划　　张　蓉
责任编辑　　祁东城
封面设计　　魏　来　李　廉

冈特生态童书
数学启蒙
[比]冈特·鲍利　著
[哥伦]凯瑟琳娜·巴赫　绘
李　蕃　译

记得要和身边的小朋友分享环保知识哦!
八喜冰淇淋祝你成为环保小使者!

Education 70

和蚕丝一样结实

Strong as Silk

Gunter Pauli

[比] 冈特·鲍利 著
[哥伦] 凯瑟琳娜·巴赫 绘
李蕃 译

上海远东出版社

丛书编委会

主　任：贾　峰

副主任：何家振　闫世东　林　玉

委　员：李原原　祝真旭　牛玲娟　梁雅丽　任泽林
　　　　王　岢　陈　卫　郑循如　吴建民　彭　勇
　　　　王梦雨　戴　虹　翟致信　靳增江　孟　蝶

特别感谢以下热心人士对童书工作的支持：

匡志强　宋小华　解　东　厉　云　李　婧　陈　果
刘　丹　熊彩虹　罗淑怡　旷　婉　杨　荣　刘学振
何圣霖　廖清州　谭燕宁　韦小宏　李　杰　欧　亮
陈强林　王　征　张林霞　寿颖慧　罗　佳　傅　俊
胡海朋　白永喆　冯家宝

目录

和蚕丝一样结实	4
你知道吗？	22
想一想	26
自己动手！	27
学科知识	28
情感智慧	29
艺术	29
思维拓展	30
动手能力	30
故事灵感来自	31

Contents

Strong as Silk	4
Did You Know?	22
Think about It	26
Do It Yourself!	27
Academic Knowledge	28
Emotional Intelligence	29
The Arts	29
Systems: Making the Connections	30
Capacity to Implement	30
This Fable Is Inspired by	31

一位年轻女孩正在为她的婚礼做准备。她的新郎给她买了一条漂亮的丝绸裙子。村里的每一个人都很好奇地期待着她从父母家走出来的情景。

所有人都是这样,只有桑树上的一群蚕除外。他们全都处于震惊之中。

A young lady is getting ready for her wedding. Her groom has brought her a beautiful silk dress. Everyone in the village is curious to see the bride emerging from her parents' home.

That is everyone, except a school of silkworms in the mulberry tree. They are in shock.

为她的婚礼做准备

Getting ready for her wedding

……至少100万被煮死……

... more than a million were boiled ...

"至少100万个兄弟姐妹被煮死了,用来做那条裙子。"一只蚕大声哭喊道。"人类对我们太残忍了! 他们这么做只是为了看起来更漂亮!"

"呃,"桑树说,"那是因为蚕丝只有被扔进热水里才会变软,容易抽出来。人们不想等你慢慢爬出蚕茧,因为当你从丝房爬出来时会钻出个洞,那就弄坏了丝的纤维。得花费更多的时间和精力去把它卷起来。"

"More than a million of our brothers and sisters were boiled to death to make that dress," one of them sobs loudly. "Humans are so cruel to us! And they do it just to look beautiful!"

"Well," says the mulberry tree, "that is because your silk only gets soft and ready to reel if thrown into hot water. People don't want to wait for you to leave your cocoons, because when you make a hole to crawl out of your silky homes, you break the silky fibres. It takes so much more time and effort to roll it up."

"唔,这些告诉了你,人类是多么无知。我们制造了最好的纤维和最牢固的建筑材料,所以我们用不着让它变软。"

"建筑材料!没有人会用你的蚕丝建造一座房子,那太贵了。"

"Well, that shows you how little they know. We make the best fibres. We make the strongest building material, so there is no need for us to make it soft."

"Building material! No one is going to make a house out of your silk. That would be too expensive."

我们制造了最好的纤维!

We make the best fibres!

我们爱吃你的叶子

We love to eat your leaves

"但那就是我们正在做的!我们爱吃你的叶子,并且用我们排泄出来的粪便给土地施肥。我们还可以制作骨头和软骨,甚至可以再造神经以帮助人们在意外事故之后康复。"

"我非常感谢你们的粪便;确实要谢谢你们喜欢吃我的叶子,你们让我周围的土地可以种出卷心菜,这在十年前是行不通的。"

"But that's what we do! We love to eat your leaves, and we drop all our poop to fertilise the soil. We can also make bones and cartilage, even rebuild nerves to help people get better after an accident!"

"I am very grateful for your poop; indeed thanks to your great appetite for my leaves, you make it possible to grow cabbages on land around me that was of no use just ten years ago."

"有了蚕丝的帮助,那些原本无法承受整个身体重量的膝盖,现在却能站直了!"蚕说。

"听上去真不可思议。人们绝对不应该将你们煮死,而应该像我这样赞美你们!"

"With the help of our silk, knees that could not hold the weight of a body, now stand tall!" says the silkworm.

"That sounds magical. People should definitely not boil you to death, but celebrate you like I do!"

像我这样赞美你们!

Celebrate you like I do!

所有的表层土壤正在流失

I am losing all that top soil

"你知道,我们过去每年生产100万吨蚕丝,但是现在人们更喜欢把塑料变成某种看起来像是丝绸的东西,而不是以一种人道的方式和我一起工作。"

"最糟糕的是由于用石油代替你,所有的表层土壤,那养育了众多生物的富含碳的表层,正在流失。最重要的是,我想念你!"

"You know that we used to produce one million tons of silk every year, and now people prefer to turn plastics into something that looks silky instead of working with me in a humane way."

"The worst is that by using petroleum instead of you, I am losing all that top soil, the rich carbon cover that feeds everyone. The worst is that I am missing you!"

"噢，这些人类知道什么！我实际上是个毛虫，他们却叫我蠕虫。"

"你知道，生物学不是他们的强项。他们相信靠胡乱摆弄我们的基因，或者在他们的身体里放进金属，就能拯救世界。"

"Oh, what do these people know! They even call me a worm, when I am really a caterpillar."

"You know that biology is not one of their strong points. They believe they will save the world by fumbling around with our genes, or putting metals in their bodies."

我实际上是个毛虫

I am really a caterpillar

比钢还结实

Stronger than steel

"唔，我很高兴有一些科学家非常了解我的基因。只有我只需通过调整压力和湿度水平就可以生产几千吨丝，而且蚕丝经过处理后可以与世界上最好的蜘蛛丝相媲美。"

"蜘蛛丝，那可比钢还结实啊！"桑树惊叫。

"Well, I am glad that a few scientists understand my genes very well. I am the only one that can produce thousands of tons of silk so by tweaking the levels of pressure and moisture a bit, my silk can be transformed to equal the best spider silk in the world."

"Spider silk, but that is stronger than steel!" exclaims the mulberry tree.

"噢，我知道。"毛虫叹息着，露出了一个嘲讽的微笑。"所有人类的食物都有保质期，但是他们的学位证书却没有！现在是时候给他们接受的教育也定个保质期了：如果过了保质期，那就扔掉再换新的。"

……这仅仅是开始！……

"Oh I know," sighs the caterpillar with a wry smile. "All their food has an expiry date, but their diplomas don't! It is about time they apply what they do with their food to their education: If it is beyond the sell-by date, throw it out and get something new."

... AND IT HAS ONLY JUST BEGUN! ...

……这仅仅是开始！……

... AND IT HAS ONLY JUST BEGUN! ...

Did You Know?
你知道吗？

Silkworms are not worms, but caterpillars.

蚕不是蠕虫，而是毛虫。

3,000 silkworms eat 100 kilograms of mulberry leaves to produce 1 kilogram of silk as well as 80 kilograms of excrement, which fertilises top soil.

3 000条蚕要吃掉100千克的桑叶才能生产1千克的蚕丝，同时会产生80千克的粪便，粪便可以肥沃表层土壤。

Silk is a protein fibre that is shaped in a triangular prism structure, producing different colours with incoming light.

蚕丝是一种三棱柱结构的蛋白质纤维，会使入射光线折射出不同的颜色。

Crickets, spiders, beetles, bees, wasps, ants and even fleas and flies produce silk.

蟋蟀、蜘蛛、甲虫、蜜蜂、黄蜂、蚂蚁，甚至跳蚤和苍蝇都产丝。

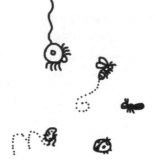

Trade in silk between Europe and Asia was so prominent that the route followed became known as the Silk Road.

欧洲和亚洲之间的丝绸贸易非常引人注目，以至人们将其路线称为丝绸之路。

亚洲

India is the largest consumer of silk in the world due to the tradition of wearing silk saris for ceremonies.

印度是世界上最大的丝绸消费国，因为印度人有在仪式上穿丝绸纱丽的传统。

When silkworms start pupating their cocoons are boiled (with the caterpillar inside) and the long fibres are then reeled in one by one.

当蚕开始化蛹，人们会用水煮蚕茧（毛虫还在里面），然后长长的纤维才会被一根一根缠起来。

A single silk thread from one cocoon can be 300 to 900 metres long.

一个茧里的一根丝可以长达300—900米。

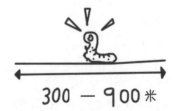

Think about It

想一想

One million caterpillars are boiled to death to produce the fibre required for one elaborate silk dress. How does this make you feel?

制作一条精致的丝绸裙子需要煮死100万只毛虫来生产纤维。你对此有什么感觉?

It is possible to take the silk from silkworms and convert it to a stronger silk like that produced by spiders by using pressure and moisture control. Or would you prefer relying on petroleum?

通过控制压力和湿度,有可能从茧里抽出蚕丝并将它改造成像蜘蛛丝一样结实的丝。那你还会不会愿意依赖石油?

If you use petroleum to make fibres, then you replace not only silk, you also eliminate the excrement that fertilises the soil and sequesters greenhouse gas emissions. Is that smart?

如果用石油来制造纤维,那么代替的不仅是蚕丝,你还损失了能给表层土壤施肥并且隔绝了排放出的温室气体的粪便。你觉得这么做聪明吗?

The world is changing fast, but our learning system is still the same. Do you think what you learned today will be relevant thirty years from now?

世界正在快速变化,但是我们的学习机制还是没变。你认为你现在学到的东西,30年后还有用吗?

Do It Yourself!
自己动手!

Have a look at the labels of the clothing in your wardrobe. Make a list of the variety of fibres used. How many are 100% natural? Then ask yourself: What amount of top soil has been generated by my choice of clothing? Or, perhaps you should ask yourself: How much top soil could not be generated because of my choice of clothing materials? Have a look at the clothing your mom, dad, brothers and sisters wear. List the fibres used and estimate what the impact is that one family has. Also estimate the impact of all the families associated with the school where you study. Write to me and tell me how many tons of top soil did we lose?

看看你衣柜里衣服上的标签。将所使用纤维的种类列成清单。有多少是100%天然的？然后问问自己：因为我选择的衣服，能产生多少表层土壤？也许更应该问问自己：因为我选择的衣服面料，有多少表层土壤因此而不能产生？看看你妈妈、爸爸、兄弟姐妹穿的衣服。列出使用的纤维，并估计一个家庭可以造成多大影响。再估计一下与你学校有关的所有家庭会造成怎样的影响。给我写信并告诉我：我们损失了多少吨表层土壤？

TEACHER AND PARENT GUIDE

学科知识
Academic Knowledge

生物学	组成蚕丝的蛋白质是由植物蛋白转变而来的动物蛋白；毛虫经变态发育过程，成为蛾或者蝴蝶；蛾和蝴蝶之间的区别；如何重新接合被切断的神经；如何利用人工软骨使神经再生；关节中软骨的作用；蠕虫和毛虫之间的区别。
化 学	生蚕丝没有有害的细菌和真菌；蛋白质转化成高分子材料；粪便转化为表层土壤；表层土壤在固定碳元素和稳定气候方面的重要性；从可再生能源中扩展蛋白质原料，以生产各种功能性高分子材料。
物 理	在周围的压力和温度下生成高分子；树木的阴影降低地表温度，使在树阴中种植变为可能；表面温度的改变使雨水比土壤更温暖，所以水能更好地渗透进土壤；蚕丝的拉力强度与其他纤维的对比；非晶体和晶体结构的蚕丝的区别。
工程学	尝试使用许多短的（被破坏的）纤维而不是长的细纤维；在医疗领域中以蚕丝作为金属的替代品；如何将纤维纺织成布料；织布机的发展；技术进步提高了纺纱和纺织的速度；村民或者企业用什么能源煮水；人工膝盖与髋关节的安装；保鲜技术和保质期；钛在植入物与假肢中的运用；替换身体上已死部分的植入物和再生这些部分的植入物的区别。
经济学	贸易作为经济发展工具的重要性；为了运作生产和消费系统而免费提供生态系统服务的重要性；现代经济中公地的作用及其私有化；丝绸贸易中的知识产权。
伦理学	我们需要为人类的快乐牺牲动物的福利吗？我们能证明用会产生碳排放的石油化工产品代替表土再生等生态系统服务的合理性吗？
历 史	马可·波罗寻找去东方的新贸易路线；对公地和公共物品的承认。
地 理	丝绸之路上的城市；世界上传统的蚕丝产地。
数 学	供应链管理：用100万只蚕制作一条裙子。
生活方式	世界各地的婚礼礼仪和在特殊活动中要求穿丝绸服装的原因；世界各地不同的文化中与新娘、新郎有关的角色、规则和传统；继续教育的需求。
社会学	村庄在婚姻证明中的角色；故意用错误的名字称呼某物（用蠕虫称呼毛虫），并且不更正错误的称呼。
心理学	为得到受人欣赏的美丽面容怂恿残暴行为；有必要质疑自己拥有的知识，并接受基于数据、信息和知识的新见解，不断提高自己的智慧。
系统论	干旱气候里桑树的角色。

教师与家长指南

情感智慧
Emotional Intelligence

桑 树

桑树评估了形势，并思考了这一铁的事实：如果你不煮死蚕，你就得不到高质量的蚕丝。成千上万只毛虫在造丝的过程中死了，这个赤裸裸的事实起初被桑树无视了，但是，因为蚕提供了表土，桑树和蚕之间有一个感情上的纽带，他甚至对蚕表示感谢。随着对话逐渐展开，桑树并不相信这些事实，但是产生了更多共鸣，并且认识到了蚕的焦虑。桑树对蚕表示坚定支持，并且对当代文明以丧失诸多生态系统服务为代价，用合成物代替天然产品的做法，表达了批评意见。

蚕 虫

蚕非常焦虑：这是关乎一整代蚕生死的事情。恐惧驱使着蚕寻求结盟。第一步战略就是激发与他最近的盟友——桑树的感情。蚕仔细倾听桑树的逻辑，并且意识到这么做没什么用。蚕的第二个战略是解释他对社会独一无二的贡献，用逻辑回应逻辑。蚕使桑树意识到他们是彼此依赖的，于是桑树对蚕的困境产生了共鸣。即使蚕没有完全说服桑树，但他成功地重新定义了契约的界限。最后，蚕又重新使用游说策略，获得了桑树的全力支持。

艺术
The Arts

找一块木板。随机敲100个小钉子在木板上。买三个不同颜色的纱线球，线要尽可能细。最便宜和最牢固的线或许是可以用来做刺绣的。开始时，将纱线从一个钉子绕过另外一个,不停地绕圈,并从一种颜色换到另外一种。开始编织细线和颜色，编织后的图案就会开始显现。在纱线用光之前，你就会找出规律来。同样的图案不可能出现两次，让它成为一件独一无二的艺术品吧。

TEACHER AND PARENT GUIDE

思维拓展
Systems: Making the Connections

　　一个世纪以前，全世界的丝绸产量约为100万吨。现在，产量只有十分之一。当由石油制成的合成纤维的诞生使得纺织品更便宜时，我们并没有意识到，丝绸业的死亡使依赖养蚕业的地区的土地不再因蚕的粪便而变得肥沃。过去，桑树被种在半干旱的地区，这样几十年后就会出现肥沃的表土层，使土地的肥沃程度上升。用桑树养蚕，是与在阴地种植的水果和蔬菜相配合的。现在，因为缺少丝绸需求，我们已经砍掉了大量桑树，大量碳被释放，导致大气层中的碳逐渐增多。当我们一边庆祝将蚕从痛苦的死亡中拯救出来时，碳排放为零甚至大于零的企业将碳排放为负的企业替代了。这不是一个刻意的决定——人类在追求效率的过程中会忽略由此产生的影响。现在，一个新的机会出现了：替代假肢和植入物中的钛。这创造了对丝的潜在的新需求。现在的问题是：如果拥有可再生的蛋白质来源，丝是否能通过转基因获得，并在高科技园区集中生产？还是说这会是一个重塑现代养蚕业的机会呢？我们能否确保这个新应用会强化生态系统服务（如表土的生成和碳的固定）？有没有可能利用基于可再生原材料和取之不尽的太阳能的可用资源和天然过程，来创造一种经济增长模式？

动手能力
Capacity to Implement

　　研究一下钛在医学和畸齿矫正中的运用。什么是开采原料的碳足迹？收集信息来制作一张流程图，从采矿、熔炼一直到金属的使用和处理。然后制作一个复杂的流程图，它包含反馈回路，允许使用蚕丝等有多重效益的可再生、可持续资源来代替金属。用绿色圈出对环境的好处，用蓝色圈出对商业的好处，用红色圈出对社会的好处。某些地方可能会有多种颜色的圈。把所有信息收集起来，用一分钟的视频来进行阐述。

教师与家长指南

故事灵感来自
This Fable Is Inspired by

弗里茨·沃尔拉夫
Fritz Vollrath

弗里茨·沃尔拉夫教授在弗赖堡大学获得了动物学博士学位。他曾在位于巴拿马的史密森热带研究所工作,在那里他发现了蜘蛛的诸多奇特本领,尤其是蜘蛛丝的妙用。他曾在英国、瑞士和丹麦工作,现在,他在牛津大学动物学部门领导丝绸小组的工作。沃尔拉夫教授在过去十年一直专注于研究丝作为天然合成材料的应用。他认为丝在许多产业中都可以得到可持续应用,对发展中国家的包容性增长机遇有重大的社会经济影响。他的研究成果已经催生了多家企业。

图书在版编目（CIP）数据

冈特生态童书.第二辑修订版:全36册:汉英对照／
（比）冈特·鲍利著；（哥伦）凯瑟琳娜·巴赫绘；
何家振等译. —上海：上海远东出版社，2021
书名原文：Gunter's Fables
ISBN 978-7-5476-1759-5

Ⅰ.①冈… Ⅱ.①冈… ②凯… ③何… Ⅲ.①生态
环境－环境保护－儿童读物—汉、英 Ⅳ.①X171.1-49

中国版本图书馆CIP数据核字（2021）第213075号
著作权合同登记号图字09-2021-0823

策　　划	张　蓉
责任编辑	祁东城
封面设计	魏　来　李　廉

冈特生态童书

和蚕丝一样结实

[比]冈特·鲍利　著
[哥伦]凯瑟琳娜·巴赫　绘
李　蕃　译

记得要和身边的小朋友分享环保知识哦！
八喜冰淇淋祝你成为环保小使者！

水在说话

Water 37

Talking Water

Gunter Pauli

［比］冈特·鲍利 著
［哥伦］凯瑟琳娜·巴赫 绘
郭光普 译

上海远东出版社

丛书编委会

主　任：贾　峰

副主任：何家振　闫世东　林　玉

委　员：李原原　祝真旭　牛玲娟　梁雅丽　任泽林
　　　　王　岢　陈　卫　郑循如　吴建民　彭　勇
　　　　王梦雨　戴　虹　翟致信　靳增江　孟　蝶

特别感谢以下热心人士对童书工作的支持：

匡志强　宋小华　解　东　厉　云　李　婧　陈　果
刘　丹　熊彩虹　罗淑怡　旷　婉　杨　荣　刘学振
何圣霖　廖清州　谭燕宁　韦小宏　李　杰　欧　亮
陈强林　王　征　张林霞　寿颖慧　罗　佳　傅　俊
胡海朋　白永喆　冯家宝

目录

水在说话	4
你知道吗?	22
想一想	26
自己动手!	27
学科知识	28
情感智慧	29
艺术	29
思维拓展	30
动手能力	30
故事灵感来自	31

Contents

Talking Water	4
Did You Know?	22
Think about It	26
Do It Yourself!	27
Academic Knowledge	28
Emotional Intelligence	29
The Arts	29
Systems: Making the Connections	30
Capacity to Implement	30
This Fable Is Inspired by	31

尼尔森抿了一口杯子里的凉开水。他生活在比勒陀利亚，这会儿正在自己的花园里欣赏着美妙的非洲音乐。这时他的孙子来看望他并问道："爷爷，你知道吗？水有话对你说哦！"

Nelson sips from a glass of cool water. He is listening to wonderful African music in his garden in Pretoria when his grandson visits and asks:

"Granddad, did you know that water has a message for you?"

水有话对你说哦

Water has a message for you

水真的在听你的话,它还知道你的感受呢!

Water listens to what you say and feel!

"什么?水有话对我说?"沉思了一会儿,聪明的老人答道:"啊啊,是的,水告诉我,喝干净的水,我才会健康。"

"不是的,爷爷!水真的在听你的话,他还知道你的感受呢!"

"Water has a message for me?" replies the wise old man after pondering the question for a moment. "Well yes, drinking clean water tells me that I will be healthy."

"No, no, Granddad. Water does listen to what you say and feel!"

"噢！我是个老头，我的身体里差不多有60%是水……而当你还是个小婴儿的时候，身体里有80%是水呢！因此我最迫切的希望就是每个人，特别是孩子们，都能喝上干净的水！"

"Well, I am an old man and my body consists of about sixty percent water... and when you were just a little baby, your body was made up of about eighty percent water. That is why I feel so strongly about everyone - but especially children - having clean water."

我的身体里差不多有60%是水

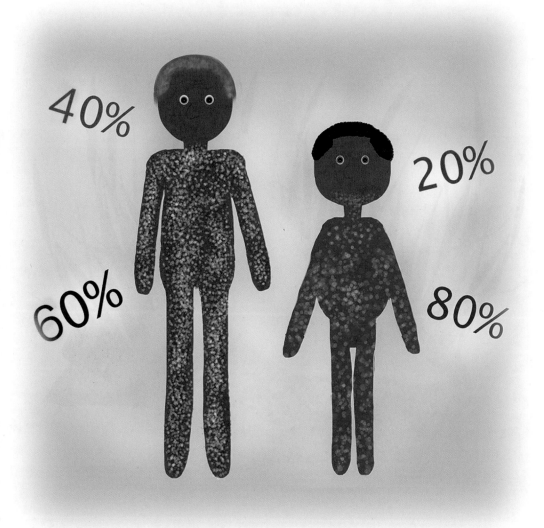

My body is about 60% water

水有话对你说

Water has a message for you

"但是爷爷,请听我说,水真的有话对你说!"

"怎么回事?难道有一滴水给你寄了明信片?还是给你打了电话?"

"当然没有,爷爷!你知道的,水又不会说话或写字,但是它可以感觉和变形。"

"But Granddad, please listen to me, water does have a message for you."

"What do you mean? Did a drop of water send you a postcard? Or did it talk to you over the phone?"

"Of course not, Granddad! You know water cannot speak or write, but it can feel and take on form."

"你这是啥意思呢?'感觉'和'变形'?"

"哦,爷爷,那你知道水有哪些形态吗?"

"噢!我知道这个杯子里的水是液体,在叶子上的就是水滴,而在空气中就是水蒸气了!"

"那么,爷爷,固态的水看起来会是怎样的呢?"

"Now what do you mean, 'feel' and 'take on form'?"

"Well, Granddad, in what form do you know water?"

"Well, I know it as a liquid in this glass, or as a droplet on a leaf, or even as vapour in the air."

"Come on, Granddad, what does water look like when it's solid?"

那你知道水有哪些形态吗?

In what form do you know water?

冰是由晶体构成的

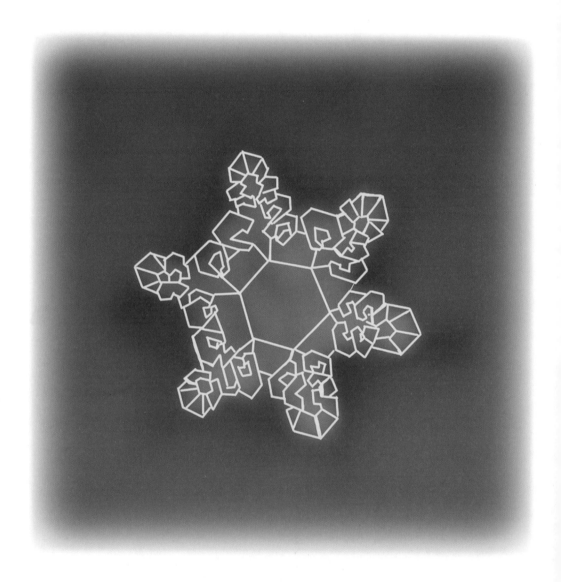

Ice is made of crystals

"好了，我不开玩笑了！固态的水就是冰，而冰是由晶体构成的！"

"那么爷爷，你是不是知道每一朵雪花都有独特而美丽的图案呢？"

"是的，我听说过。所以，你认为水会变成晶体来说话？"

"I'll stop teasing now! Solid water takes the form of ice and ice is made of crystals."

"Well, Granddad, did you know that every snowflake has its own unique, beautiful pattern?"

"Yes, I have heard that. So are you suggesting that water speaks in crystals?"

"那会不会当我们产生不同的情绪时,水就会形成不同的晶体形状呢?"

"噢!那就是说,如果你淘气而我因此生气时,我们周围的水的晶体就会发生变化喽。当然了,它们会看上去很生气,因为你是个小淘气包!"

"Could it be that water will form different crystal shapes when we feel differently?"

"Ah, that will mean when I am angry with you because you are naughty, the water crystals around us will look different. Of course, they will look cross, as you are such a naughty boy!"

水就会形成不同的晶体形状

Water will form different crystal shapes

……和你在一起总是很开心!

...always such fun to be with you!

"噢，爷爷！你知道在你身边时我几乎从不淘气！和你在一起总是很开心！"

"Oh, Granddad, you know I am nearly never naughty when you are around! It is such fun to be with you."

"是的,我们在一起时的确很开心,不是吗?但是,奶奶觉得你在旁边的时候我却变成老淘气包了。来吧,让我们把我们的非洲音乐的音量调大,并尽情地跳舞。我和你打赌,水也会和我们一起跳舞!"

……这仅仅是开始!……

"Yes, we do have fun, don't we? But Grandma thinks I get naughty when you're around. Tell you what, let's turn up our African music really loud and dance. I bet you that the water will start dancing with us!"

... AND IT HAS ONLY JUST BEGUN!...

······ 这仅仅是开始！······

...AND IT HAS ONLY JUST BEGUN!...

Did You Know? 你知道吗？

Water is the only substance on Earth that expands when it gets colder than 4 degrees Celsius. All other materials will contract when exposed to colder temperatures.

水是地球上唯一可以在4摄氏度以下随温度降低而膨胀的物质，而其他物质的体积都随温度降低而缩小。

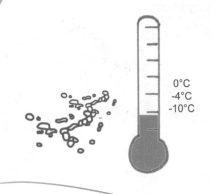

Water boils at 100 degrees Celsius, but only when you are at sea level. If you are higher up, for instance in the mountains, then water will boil at a much lower temperature.

当处于海平面高度时，水在100摄氏度沸腾。如果海拔升高，比如在山上，水的沸点就会降低。

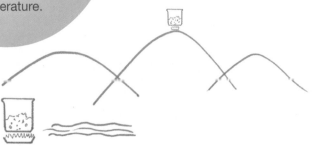

水是地球上最珍贵的物质。即使每桶原油价值100美元，一瓶水也比一瓶原油要珍贵。

*W*ater is the most precious substance on Earth. A bottle of water is more expensive than a bottle of crude oil, even at the price of US$ 100 per barrel of oil.

人体的卵子和精子结合后产生的细胞，其含水量达98%。这就是说，生命其实是从水里诞生的。

A cell made from an egg and a sperm consists of 98% water. That means that life in fact emerges in water.

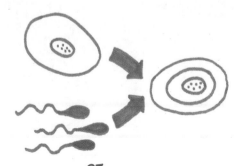

*A*ll life is generated in a water environment. This has been a major challenge for reproductive systems on land, which had to, over millions of years, invent a system simulating a water world.

所有的生命都是在有水的环境中产生的。因而水资源问题就成了陆地繁殖系统面临的主要挑战，经过很多年的时间，陆地繁殖系统形成了一套模拟水环境的系统。

*W*ater crystals are 100 to 200 times smaller than anything the naked eye can see.

水的晶体分子的体积相当于我们眼睛能看到的最小的东西的两百分之一到一百分之一。

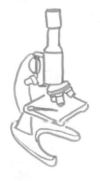

Water in the air is the main absorber of sunlight. The 13,000 billion tons of water in the atmosphere removes about 70% of the radiation from the sun.

空气中的水分是太阳光的主要吸收者。大气层中有13万亿吨的水，吸收了70%的太阳辐射。

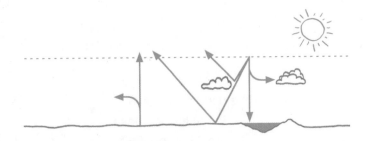

Water is the only material on Earth that is commonly found as a liquid, a solid and a gas.

水是地球上唯一一种液态、固态和气态都很常见的物质。

Think about It

想一想

How did the grandson feel when his granddad at first pretended not to understand the question?

当爷爷一开始假装不理解那个问题时，小孙子是什么感受？

What technique is used by the grandson to get his granddad to think about his concerns?

小孙子采取了什么方法让爷爷思考他关心的问题？

当爷爷说他为小孙子的淘气而生气时，小孙子是什么感受？

How does the grandson feel when his granddad says that he gets angry with him when he is naughty?

How does the grandson react when his granddad wants to be naughty by turn up the music?

当爷爷想开玩笑而调高音乐声的时候，小孙子是什么反应？

Do It Yourself!

自己动手！

Take a small amount of water and freeze it. Now get a powerful microscope that can magnify at least 100 times, but ideally 200 times. Chop off a small piece of ice and quickly place it under the microscope. This may not be easy, but see if you can be fast enough to have a look at its beauty before it melts.

取一点水并冻成冰，再准备一台能放大100倍，最好是放大200倍的高倍显微镜。敲下一小块冰，迅速放到显微镜下观察。这可能不容易，但你可以试试看能不能在冰融化前迅速看到它那美丽的结构。

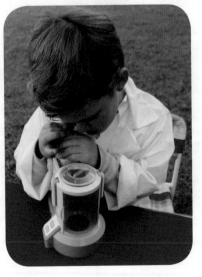

TEACHER AND PARENT GUIDE

学科知识
Academic Knowledge

生物学	所有生命都离不开水；水是矿物质和细菌的携带者；胃肠道疾病；通过沸水来灭菌；细菌能在冻水中繁殖。
化 学	用化学方法净化水。
物 理	水的冷凝和露珠；用物理方法净化水；声音和光在水中传播得比在空气中快；冰晶体的几何学；水有固态、液态和气态；水的第四相；空气中的水蒸气能阻挡紫外线；每朵雪花的独特性；声音如何影响水的运动；说话对冰晶形状的影响；显微镜和显微摄影。
工程学	如何供应城市中数百万人的饮用水？如何大规模净化水质？如何处理污水？
经济学	瓶装水生意。
伦理学	水是公共资源，但是却越来越商业化，并且不是对每个人都免费。
历 史	水何时成为商品？非洲音乐在世界的影响。
地 理	纳尔逊·曼德拉的出生地和家乡；主要供水来源；旱灾发生地区。
数 学	晶体的几何学；晶体的生长顺序。
生活方式	所有人的饮用水储备；停止利用水运输废物。
社会学	音乐和舞蹈在社会中的角色；创造一个幽默的学习环境；科学家何时才会承认一项创新得到了充分的证明？
心理学	爷爷的重要性；"淘气"的定义及其对性格形成的重要性；鼓励人们超越已有的、被广泛接受的认知，探索新的问题；苏格拉底式的追问式学习法。
系统论	水和健康的关系。

教师与家长指南

情感智慧
Emotional Intelligence

爷爷

爷爷乐于和自己的孙子共度开心的时刻。他用问题启发小孙子,并且用更多的问题回应他。他很欣慰的是小孙子愿意对问题进行更深刻的理解,并且他非常注意让小孙子在学习中吸取经验教训。他感受到了小孙子的努力和振奋,并不断引导小孙子对讨论的问题进行更深的理解,同时也创造了一个紧密的家庭纽带。爷爷创造了一个适合学习、欢快轻松的气氛,同时了解家庭成员可接受的边界。

小孙子

小孙子对自己的新发现感到非常兴奋,并且非常乐于和别人分享。虽然他还不能完全理解科学,但他还是准备在爷爷那里验证一下他对那些迷人现象的有限了解,并提高自己的认识。即使爷爷似乎没有完全理解他的意思,但爷爷还是很有耐心地和他探讨,一点儿也不嫌烦。小孙子想要探索现象背后的问题,但同时表示了对爷爷的尊重。同时,他非常享受这种积极向上的情绪,以及家庭各个成员间的亲密关系所形成的欢乐气氛。

艺术
The Arts

咖啡不仅仅是一种饮品,还是纺织品的原材料以及一种重要的着色剂。因此,收集一些咖啡渣(咖啡冲泡之后留下的废弃物)吧。用手挤干其中残留的水或者油脂。现在,拿起画笔在T恤上画出你喜欢的图案,用咖啡作为你的油墨。确保你使用的是旧T恤,这样你妈妈就不会因为T恤被染上去不掉的颜色而介意了。

思维拓展
Systems: Making the Connections

水是生命之源。水决定着我们的健康，如果没有水就没有农业。水无处不在，但水现在却成了稀缺商品，尤其是在城市里，还得花大价钱才能获得。水成为了一种产品，但也和疾病产生了关联，因为很多水被污染了。几乎所有物质在低于4摄氏度的时候体积都随温度降低而缩小，只有水是个例外——它随温度降低而膨胀。然而，因为水是地球上含量最丰富的物质，这又不是一个例外，也许水的变化才是规则，而世界上其他东西才是例外。我们仍然在继续认识水的特性。水形成晶体的能力不仅仅取决于水分子的数量多少，还取决于一些我们尚不了解的因素。

动手能力
Capacity to Implement

我们可以创造一种新的产业模式，就是在可循环使用的水瓶上画上美好的愿望，为人们和他们的家庭带来积极的氛围。我们可以尝试用水的晶体创作一幅画，并把它变成数字明信片。

教师与家长指南

故事灵感来自
This Fable Is Inspired by

江本胜
Masaru Emoto

江本胜运用来自世界各地的水制造晶体，他发现不同来源的水会形成不同的晶体。当他用取自拉斯加维奥塔斯（那个地区重新长出了森林）的水样制作晶体时，发现了一种简单而美丽的结构。他用精巧的几何形状描绘出了山泉水的晶体结构。之后在2007年，江本胜开创性地将一瓶水放在了非洲音乐（索韦托弦乐四重奏）的环境中，并在显微摄影中发现晶体一直在随着音乐变化，就像跟着鼓点的节奏摇摆一样。那些"听非洲音乐的水"的照片作为礼物送给了南非总统纳尔逊·曼德拉。虽然现代科学认为江本胜的方法并不科学，但他的确拍到了令人惊叹的晶体照片，这说明除了温度，还有很多因素会影响水的晶体结构。

图书在版编目(CIP)数据

冈特生态童书.第二辑修订版:全36册:汉英对照/
(比)冈特·鲍利著;(哥伦)凯瑟琳娜·巴赫绘;
何家振等译.—上海:上海远东出版社,2021
书名原文:Gunter's Fables
ISBN 978-7-5476-1759-5

Ⅰ.①冈… Ⅱ.①冈… ②凯… ③何… Ⅲ.①生态
环境-环境保护-儿童读物—汉、英 Ⅳ.①X171.1-49

中国版本图书馆CIP数据核字(2021)第213075号

著作权合同登记号图字09-2021-0823

策　　划	张　蓉
责任编辑	祁东城
封面设计	魏　来　李　廉

冈特生态童书
水在说话
[比]冈特·鲍利　著
[哥伦]凯瑟琳娜·巴赫　绘
郭光普　译

记得要和身边的小朋友分享环保知识哦!
八喜冰淇淋祝你成为环保小使者!

Water 38

水藻衣裳

Dressed up in Algae

Gunter Pauli

[比] 冈特·鲍利 著
[哥伦] 凯瑟琳娜·巴赫 绘
郭光普 译

上海远东出版社

丛书编委会

主　任：贾　峰
副主任：何家振　闫世东　林　玉
委　员：李原原　祝真旭　牛玲娟　梁雅丽　任泽林
　　　　王　岢　陈　卫　郑循如　吴建民　彭　勇
　　　　王梦雨　戴　虹　翟致信　靳增江　孟　蝶

特别感谢以下热心人士对童书工作的支持：

匡志强　宋小华　解　东　厉　云　李　婧　陈　果
刘　丹　熊彩虹　罗淑怡　旷　婉　杨　荣　刘学振
何圣霖　廖清州　谭燕宁　韦小宏　李　杰　欧　亮
陈强林　王　征　张林霞　寿颖慧　罗　佳　傅　俊
胡海朋　白永喆　冯家宝

目录

水藻衣裳	4
你知道吗？	22
想一想	26
自己动手！	27
学科知识	28
情感智慧	29
艺术	29
思维拓展	30
动手能力	30
故事灵感来自	31

Contents

Dressed up in Algae	4
Did You Know?	22
Think about It	26
Do It Yourself!	27
Academic Knowledge	28
Emotional Intelligence	29
The Arts	29
Systems: Making the Connections	30
Capacity to Implement	30
This Fable Is Inspired by	31

在中国的一片棉花田里，一颗小棉籽坐在地上，盯着地平线。

$\mathcal{A}$ tiny cottonseed sits in a cotton field in China, staring at the horizon.

中国的一片棉花田

A cotton field in China

你过去一直都是那么洁白明亮!

You were always so white and bright!

一片小水藻来到这个看起来有些伤感的小棉籽前,问道:"你怎么了?你过去一直都是那么洁白明亮!"

$\mathcal{A}$ minute piece of algae approaches the sad looking seed and asks:
"What is happening to you? You were always so white and bright!"

"那个农民伯伯过去一直种植棉花，人们可以用棉花来做衣服。可是我刚听他说现在他打算用有限的水来种粮食了。"小棉籽呜咽着说。

"噢，这是有道理的！你喝的水太多了，可还有许多人要养活，保证没人挨饿才是明智的。"小水藻争辩道。

"I have just heard that the farmers, who used to plant cotton for making clothing, now plan to use the little water available for growing food," sobs the cottonseed.

"Well, that makes sense. You have been guzzling too much water. Now that there are millions of extra mouths to feed, it's only smart to make sure that no one starves," argues the algae.

喝的水太多了

Guzzling too much water

二氧化碳和化学品

Carbon dioxide and chemicals

"你也需要水，可没人阻止你呀！"小棉籽反驳道。

"的确，我们是需要水和很多二氧化碳，而就是二氧化碳在大气中积累太多才导致气候变化的。但我听说人类用于农业的水中，有四分之一被你用掉了。不仅如此，你还用了太多的化学品。"小水藻回答道。

"You are also in need of water, and no one is stopping you!" the cottonseed snaps back at him.

"True, we do need water and a lot of carbon dioxide. That is the gas that has been causing climate change, as too much of it ends up in the atmosphere. But I'm told you use one quarter of all the water needed for farming, and that is not all: You use a lot of chemicals too," replies the algae.

"我得承认我总是觉得渴。但你也要知道,我从来没有要求用那些化学品。在白人想要所有棉花越来越白之前,我们本来有很多天然色彩。"小棉籽叹息道。

"I must admit that I am always very thirsty. But you should know I never asked for those chemicals. We had many natural colours until white people wanted us all to be whiter than white," laments the cottonseed.

我们本来有很多天然色彩

We had many natural colours

人们有的吃没的穿

People with food but nothing to wear

"来，让我们一起努力创造一个更美好的世界！"小水藻说。

"难道这意味着人们将要有的吃没的穿？"小棉籽迷惑地问道。

"Look, we should all work together to make this a better world," says the algae.

"Does that mean that people will now have food but then have nothing to wear?" wonders the cottonseed.

"人们还可以用很多其他很好的纤维，比如荨麻、亚麻。但我很可能是最好的选择之一。"小水藻大笑着说。

"你？你不适合用来做衣服！"小棉籽喊道。

"There are also many other great fibres they can use, like nettles, flax. But I may very well be one of the best solutions!" says the algae with a wide smile.

"You? You are no good for making cloth!" cries the cottonseed.

很好的纤维，比如荨麻、亚麻

Great fibres like nettles, flax

水藻做的聚酯纤维

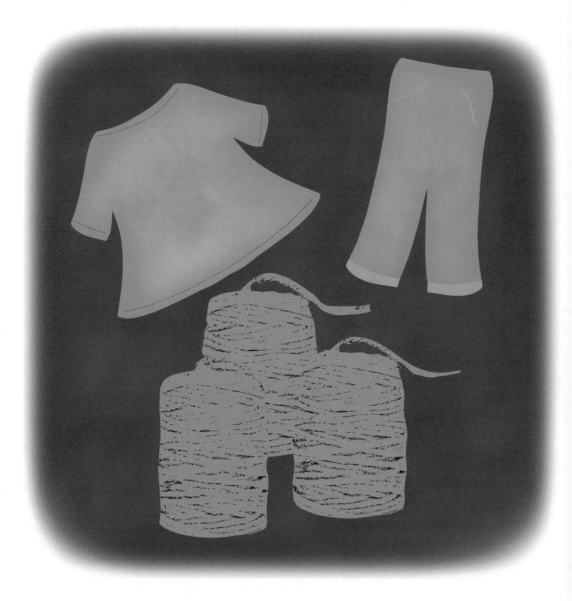

Polyesters out of algae

"你不知道吗?
人们可以用我制造聚酯纤维。"
"聚酯纤维！所有这种聚合物都是从石油中提取制造出来的，而且是人工合成的，它们不能取代我！"小棉籽大喊道。

"Didn't you know that you can make polyester from me?"

"Polyester! All this poly stuff is made from petroleum, it's synthetic, and no substitute for me!" shouts the cottonseed.

"不！你真的可以用我制造出聚酯纤维，而且我还是天然的。我生活在海里的姐妹——海草，也含有丰富的油脂，被称为海藻酸钠。人们可以从海洋植物中提取比陆地上的植物更多的纤维来做衣服。"

"噢！如果这样，那你的姐妹为什么会叫'草'呢？"

……这仅仅是开始！……

"No, no, you can make it from me; I am natural. My sisters from the sea, known as seaweed, are also rich in oils called alginates. They can make more fabric for clothing in salt water than you could ever imagine making with your fresh water on land!"

"Well, if that is so, why are your sisters called a weed?"

... AND IT HAS ONLY JUST BEGUN!...

……这仅仅是开始!……

...AND IT HAS ONLY JUST BEGUN!...

Did You Know?

你知道吗？

Cotton is composed of a string of cells of nearly pure cellulose, the same building material used by trees.

棉花由一串丝状的细胞组成，这些细胞几乎只含有纤维素，树木也是由这种物质构成的。

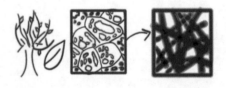

The cotton we wear today is a mix of a plant from Africa and a plant from America. The mix offers longer, stronger and softer fibres than any other single plant.

我们现在穿的棉质衣服所用的棉花是非洲的一种植物和美洲的一种植物杂交而产生的。这种杂交种长出的纤维比单独一种植物的纤维更长、更结实、更柔软。

It has been estimated that 25% of all water used for farming is used to quench the thirst of the cotton plant. One T-shirt produced from conventional cotton consumes 2,700 litres of water.

据估计，农业用水的 25% 都用来为棉花解渴了。一件用传统的棉花做的 T 恤要用掉 2 700 升水。

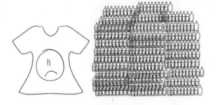

Bugs love to eat cotton. Therefore growers douse their plants with chemicals that are dangerous to people and harmful to the planet. Altogether 10% of all agricultural chemicals in the world and 25% of all pesticides are used on cotton plants. One T-shirt requires 100 grams of chemicals.

有些虫子喜欢吃棉花，所以棉农就往棉花株上喷洒化学药剂，这些药剂威胁人们的健康，也危及地球。世界上 10% 的农用化学品和 25% 的杀虫剂都用于棉花。生产一件 T 恤要用 100 克化学品。

Cotton has natural colours including red, green and brown. Natural colours do not fade when exposed to the sun.

自然界的棉花有红色、绿色和棕色。天然颜色在太阳下曝晒是不会褪色的。

Cotton farming is believed to have started some 5,000 years ago in the Andes mountains, in what is now called Peru.

人们认为棉花种植开始于大约5 000年前的安第斯山脉，那里现在叫秘鲁。

One ton of brown seaweed can be turned into 200 kilograms of fibre for cloth. And one hundred grams of algae fibre can produce one square meter of cloth.

一吨褐藻可以生产200千克做衣服用的纤维；100克水藻纤维可以生产1平方米的布匹。

Algae fibre does not burn and has numerous metal ions.

水藻纤维不会燃烧，并且含有大量金属离子。

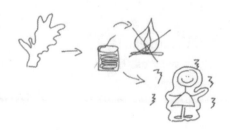

Think about It

想一想

How would you feel if you needed a large amount of water and chemicals to grow?

如果你需要大量的水和化学品才能成长,你会是什么感觉呢?

What would your reaction be if someone claimed they did not need any water, while everyone has been using so much water for so many years?

这么多年来每个人已经用了这么多的水,如果有人说他们不再需要水,你会是什么反应呢?

如果你知道你可以生产出很多天然的颜色,但人们依然只想要白色,你会是什么感受呢?

How would you feel if you know you can make many colours naturally, but people still want white only?

Would you like to wear clothing that is made from nettles or flax?

你想穿荨麻或亚麻做的衣服吗?

Do It Yourself!

自己动手!

Take some clothes made from cotton. Look at the label inside the garment. Does the label give you any information about how much water or chemicals were needed to produce each item? Probably not. Now weigh different garments, like a shirt or jacket, and calculate how much water was needed to produce each one, as each garment required water to produce the cottonseeds, to grow and harvest the cotton plants, to process it into yarn, dye it in different colours and prepare it for sale. Now create your own label and share it with your friends!

拿几件棉花做的衣服,看看衣服内侧的标签,上面有没有生产一件衣服需要多少水或多少化学品的信息呢?大概没有吧!现在来称量一下不同的衣服(比如一件衬衫或夹克),计算一下生产一件衣服需要多少水:每件衣服需要多少水来生产棉花种子、种植和收获棉花;还需要多少水把棉花做成线、染成不同颜色并上市出售。现在做一个自己的标签并和你的朋友们分享一下。

TEACHER AND PARENT GUIDE

学科知识
Academic Knowledge

生物学	灌溉需要集约式种植；棉花的天然颜色；从亚麻和苎麻提取的天然色素；淡水水藻和海藻的区别；小型藻类和大型藻类的区别；杂草在生态系统中的作用；从植物提取的天然纤维；红藻、绿藻和褐藻的区别。
化　学	硫酸在人造纤维生产中的作用；植物和水藻从空气中捕获二氧化碳的循环；如何在纤维中把颜色固定下来；化学品在棉花生产中的作用；多种多样的石油化学纤维及其功能用途。
物　理	如何检测纤维的强度？如何在太阳紫外线照射下使色彩不消褪？
工程学	如何利用海藻酸盐生产纤维？如何使石油中的分子分裂形成纤维？
经济学	社会为缺水付出的代价；如何在用水来种植粮食还是种植棉花之间进行选择？为什么天然纤维慢慢被人工合成纤维替代？棉花生产从美洲转移到了亚洲；和棉花生产中的劳动力相比，水和化学品的重要性；人口增长对稀缺资源的影响；创新的代价；传统产业通过锁定制度和技术来对抗变革；由于产品设计中没有考虑循环（或回收）利用而使产品使用寿命短，从而导致了资源利用效率较低。
伦理学	如果当地人民需要水来灌溉农田、生产食物和保持个人健康，我们还能支持为了生产衣服以供出口而使用水源吗？
历　史	谁把棉花带到了中国？棉花出现以前人们用什么纤维？
地　理	棉花生长在温暖、光照充足的气候中，苎麻和亚麻生长在温带大陆性气候中。
数　学	计算生产成本。
生活方式	以前购买的衣服要穿一些年头；现在服装的购买被不断变化的潮流所左右，衣服也被设计得不耐穿。
社会学	世界范围内社会对棉花的需求；把白色作为标准的需要；马斯洛的需求金字塔。
心理学	如何处理依赖性（棉花需要水和化学品）？
系统论	对棉花需求的增长如何导致了对水的需求的指数增长？化学品如何污染土壤，并使种植粮食的土地面临多年的贫瘠状态？

教师与家长指南

情感智慧
Emotional Intelligence

小棉籽

小棉籽很在意别的生物关心的事情,那些生物也担心他们的未来。小棉籽也明白一个事实,就是棉花的种植和生产模式要和粮食生产而不是衣服生产竞争水和资源。小棉籽很了解其他天然纤维和合成纤维的区别,但是他觉得自己的优点(比如天然的色彩)被忽视了。小棉籽起初还没有强大的心智能量思考出新的解决方案,而且还有很大的抵触情绪。然而,他从观察中慢慢明白了对用词的选择(把水藻叫作水草)并不能反映水藻作为节水型制衣纤维这一替代资源的真实潜力。因此,在故事结尾小棉籽也表现出了与水藻的共鸣。

水藻

水藻感觉到了小棉籽的焦虑并很关心他的幸福。即使小棉籽开始时不够友好,水藻也表示理解,并花时间解释不仅仅是棉花,而是整个人类所面临的真正挑战。通过对这些挑战的详细说明,水藻提出了包括水藻,但又不限于水藻的替代方案。水藻显得很有自我意识;他明白自己的潜力,并对自己的未来充满信心,而不受"水草"之类的蹩脚名字的影响。

艺术
The Arts

旧的棉质衣服能做什么呢?搜集多种多样的衣服碎片并把它们缝在被子上。这是一种传统的缝纫方式,用边角料创造一件色彩绚丽、设计精美的艺术品。如果你花精力搜集许多不同的布片,甚至可以做一个大艺术品,把整个床都罩起来。

TEACHER AND PARENT GUIDE

思维拓展
Systems: Making the Connections

棉花消耗了大量的水,这是第一个挑战。然而,第二个挑战是棉花产品的设计使用寿命越来越短。这成了一个单向产品,一旦穿上并洗几次之后就被丢弃了。现代消费者在流行趋势改变时,就觉得正在穿的衬衣或夹克一文不值,便扔在一边,也不回收利用。这种情况,再加上人口爆炸,给用来种植棉花的土地和水增加了很多压力。缺水和高昂的补充用水的代价已经使棉花种植带从美国转移到了亚洲,目前中国和印度成了世界上最大的棉花生产国。然而,这两个国家也正在忍受长期缺水的困境。因此,应该掀起一场关于水的保护和食物生产(也需要大量的水)之间的讨论。还应该掀起一场可用资源和大自然提供的替代资源之间的讨论。许多其他选择(如荨麻和亚麻)已经存在上千年了,另外一些则是最近出现的(水藻),这些都提供了节约土地和减少用水量的机会。不幸的是,棉花已经被时尚产业所支配,那个行业中大多数人对大众的粮食需求不敏感。因此,现在是时候来学会利用那些在种植过程和使用过程中都耗水较少的纤维了。

动手能力
Capacity to Implement

去做一份市场调查吧!问问你学校里的朋友:"谁愿意穿由水藻纤维做成的衣服?"然后问问他们是否能想象穿上一件用水藻纤维做成的T恤,皮肤会是什么感觉。这种衣服的质地感觉像什么?容易洗吗?根据他们的回答,写出你的营销手册来销售你想象中的一款独一无二的水藻纤维T恤。看看你是否能说服你的父母、家人和老师买一件。

教师与家长指南

故事灵感来自
This Fable Is Inspired by

苏珊娜·李
Suzanne Lee

苏珊娜·李一心想推进生物材料设计的服装和时尚产品的生产,在2003年创立了生物服装公司(BioCouture)。她曾是伦敦艺术大学和中央圣马丁艺术设计学院时装与纺织学院的高级研究人员。她的人生目标是研究新的资源来替代碳氢化合物纤维和高水耗的棉花,并影响、启发和激励时装业和设计师也这样做。她已经建立了一个由致力于实施这些突破性概念的创新者们组成的网络。

图书在版编目(CIP)数据

冈特生态童书.第二辑修订版:全36册:汉英对照 /
(比)冈特·鲍利著;(哥伦)凯瑟琳娜·巴赫绘;
何家振等译. —上海:上海远东出版社,2021
书名原文:Gunter's Fables
ISBN 978-7-5476-1759-5

Ⅰ.①冈… Ⅱ.①冈…②凯…③何… Ⅲ.①生态
环境−环境保护−儿童读物—汉、英 Ⅳ.①X171.1-49

中国版本图书馆CIP数据核字(2021)第213075号

著作权合同登记号图字09-2021-0823

策　　划	张　蓉
责任编辑	祁东城
封面设计	魏　来　李　廉

冈特生态童书
水藻衣裳
[比]冈特·鲍利　著
[哥伦]凯瑟琳娜·巴赫　绘
郭光普　译

记得要和身边的小朋友分享环保知识哦!
八喜冰淇淋祝你成为环保小使者!

Water
39

人人都爱的宝座

A Throne for All

Gunter Pauli

[比]冈特·鲍利 著
[哥伦]凯瑟琳娜·巴赫 绘
郭光普 译

上海远东出版社

丛书编委会

主　任：贾　峰

副主任：何家振　闫世东　林　玉

委　员：李原原　祝真旭　牛玲娟　梁雅丽　任泽林
　　　　王　岢　陈　卫　郑循如　吴建民　彭　勇
　　　　王梦雨　戴　虹　翟致信　靳增江　孟　蝶

特别感谢以下热心人士对童书工作的支持：

匡志强	宋小华	解　东	厉　云	李　婧	陈　果
刘　丹	熊彩虹	罗淑怡	旷　婉	杨　荣	刘学振
何圣霖	廖清州	谭燕宁	韦小宏	李　杰	欧　亮
陈强林	王　征	张林霞	寿颖慧	罗　佳	傅　俊
胡海朋	白永喆	冯家宝			

目录

人人都爱的宝座	4
你知道吗？	22
想一想	26
自己动手！	27
学科知识	28
情感智慧	29
艺术	29
思维拓展	30
动手能力	30
故事灵感来自	31

Contents

A Throne for All	4
Did You Know?	22
Think about It	26
Do It Yourself!	27
Academic Knowledge	28
Emotional Intelligence	29
The Arts	29
Systems: Making the Connections	30
Capacity to Implement	30
This Fable Is Inspired by	31

当安娜遛狗的时候,她会捡起狗的便便。她的弟弟彼得喜欢猫,他会打扫猫舍,清理猫砂。一天晚上,爸爸告诉他们,城里缺水了。

When Anna takes her dog for a walk, she picks up the poop. Her brother, Peter, loves cats and he cleans out the litter box. One evening, their dad talks to them about the lack of water in their town.

告诉他们，城里缺水了

Talks to them about the lack of water

我在下雨前不洗车了

I won't wash the car until it rains.

"我们应该节约用水。我们这里3个月都没有下雨了。"爸爸解释说。"所以我在下雨前不洗车了。孩子们,你们能做些什么呢?"

"我要和狗狗一起上厕所。"安娜提议道。

家里每个人都转头看着她。

"We have to save water. We have had no rain for three months now," explains Dad. "So I won't wash the car until it rains. Children, what will you do to help?"

"I will go to the toilet with my dog," proposes Anna.

Everyone in the family turns their heads to look at her.

"你这是什么意思？"爸爸纳闷地问。

"噢，每天早晨我醒来后就去上厕所并用水冲马桶。我想我最好在遛狗时到花园里找个地方和狗狗一起上厕所。"

"What do you mean?" wonders Dad.

"Well, every morning after I wake up, I go to the toilet and then I use water to flush it. I think I'd better take my dog for a walk and find a place in the garden for both of us."

到花园里找个地方

Find a place in the garden

"不行！"

"No way!"

"不行!"爸爸提高了嗓门。"你不能把花园变成公共厕所!"

"但是,爸爸,就像我清理狗的便便一样,我也可以清理自己的便便呀!"

"绝对不行!"爸爸大吼道。

"No way!" Dad raises his voice. "You cannot turn the garden into a public toilet!"

"But, Dad, just like I clean up after the dog, I can clean up after myself!"

"Absolutely not!" shouts Dad.

"要知道,我已经在考虑厕所的事了。我觉得厕所特傻。"安娜评论道。

"什么?这可是现代最伟大的发明之一呀!"爸爸强调说。

"你知道狗狗只是蹲下就能方便,还从来不用卫生纸吗?"安娜调皮地说。

"那是当然啦!"爸爸越来越生气了。"猫也是这样的!"

"You know, I've been thinking about the toilet. I find it rather stupid," Anna comments.

"What? It is one of the greatest inventions of modern time," insists Dad.

"Do you realise that dogs just squat, and never use toilet paper?" Anna says cheekily.

"That's obvious!" Dad is getting more irate. "And neither do cats!"

狗从来不用卫生纸

Dogs never use toilet paper.

……害得我们要挤屁股……

...made us squeeze our cheeks...

"也不知道是谁发明了抽水马桶,害得我们要使劲挤屁股,还得用这么多纸来擦。我们应该像狗和猫一样蹲下方便。"安娜坚持道。

"嗯,那你要把你的便便扔到哪儿呢?"爸爸问道。"你的家里闻起来就要像法国国王的城堡喽!当时他们不得不每几个月搬一次家,因为臭味让人无法忍受。"

"Whoever designed the toilet made us squeeze our cheeks so tightly together that we need so much paper to clean up. We should just squat like dogs and cats do," insists Anna.

"Well, where exactly do you plan to put all your mess?" asks Dad. "Your home would start smelling like the castles of the French kings! They had to move every few months because of the unbearable odours."

"难道是我们把身体制造的好东西给破坏了吗？我们用了好多年的时间才进化出可以排大小便的身体结构。那我们能不能通过往大小便里掺饮用水并添加化学物质来控制气味呢？"安娜总结道。

"安娜，我们不能回到中世纪。"爸爸指出。

"Aren't we destroying something good that our bodies make? It took millions of years to develop one exit for solids and one for liquids. And we go and mix it with drinking water and add chemicals to control the smell?" concludes Anna.

"Anna, we cannot go back to medieval times," argues Dad.

我们往大小便里掺饮用水

We go and mix it with drinking water

"那你说该怎么办呢？"

"So what do you propose?"

"是不能,但是我们不能把便便当废物简单地用饮用水冲走。我们吃地里长出来的水果和蔬菜,我们也应该把那些营养再还给土壤。"

"但这就是人们发明化肥的原因!"爸爸回应道。

"生产化肥需要许多能量!"安娜也迅速回应道。

"那你说该怎么办呢?"

"No, but we cannot treat night soil as waste simply to be flushed away with drinking water. We eat fruits and vegetables from the land, and we should get those nutrients back onto the soil."

"But that's why people invented fertilizers!" responds Dad.

"And fertilizers need a lot of energy," Anna responds quickly.

"So what do you propose?"

"我们需要理解维京人和亚马逊人是怎样用木炭把大部分粪便转化成了世界上最肥沃且最持久的土壤。"

"但是处理人类的粪便是个麻烦。"爸爸说:"而一个简单的冲水厕所就是有效的解决办法。"

"爸爸,这不难——只是我的建议有点不一样罢了。"

……这仅仅是开始!……

"We need to understand how Vikings and the Amazonian people used charcoal to turn their waste into the most fertile and longest-lasting soil in the world."

"But human waste is a problem," says Dad, "and a simple flushing toilet is the obvious solution."

"Dad, it's not difficult – what I'm proposing is just different."

... AND IT HAS ONLY JUST BEGUN!...

······ 这仅仅是开始！······

...AND IT HAS ONLY JUST BEGUN!...

Did You Know?

你知道吗?

The average toilet uses 9 litres per flush. People visit a toilet about 7 times a day, which means we use more drinking water for flushing than we do to shower and cook food.

每次冲厕的平均用水量为9升。人们每天去厕所7次,这意味着冲厕所用掉的饮用水比淋浴和做饭用掉的还要多。

The most efficient toilets only use 7 litres per person per day, cutting drinking water consumption for sanitation by a factor of 10.

最节水的厕所每人每天只用7升水,使保持卫生所消耗的饮用水减少至十分之一。

Cities spend up to one third of all their drinking water on flushing toilets, which should not use drinking water at all.

城市用于冲厕的饮用水达三分之一，因此，绝对不能再用饮用水冲厕了。

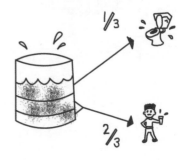

Terra preta ("dark earth") are patches of fertile soil in the Amazon made by blending charcoal, bones and manure. This turned infertile Amazonian soil to highly productive land.

亚马逊地区的特拉－普利塔（"黑土地"）是混合了木炭、骨头和粪便而形成的成片的肥沃土壤。这使得贫瘠的亚马逊土壤变成了高生产力的土地。

The Vikings created dark earth using the same ingredients as the Amazonian people, and generated soil fertility and productivity that is high by today's standards.

维京人用和亚马逊人所用的同样成分创造了黑色土壤，并且其肥力和生产力就是用今天的标准来衡量也是非常高的。

French King Louis XIII had a toilet fitted under his throne. This allowed him to grant audience to ministers and conduct business of state while attending to personal needs.

法国国王路易十三在自己的宝座下安装了一个马桶。这可以让他在解手的时候也能接见大臣和处理国家大事。

The world consumption of toilet paper reaches approximately 85 million rolls per day. The British use on average of 110 rolls of toilet paper per person per year, while people in the Baltic states use 4 times less.

世界上每天消耗的卫生纸达到了大约 8 500 万卷。一个英国人每年平均消耗 110 卷，而波罗的海沿岸国家的消耗量是英国的四分之一。

The ancient Minoan civilisation of Crete (2700–1450 BCE) was the first to use underground clay pipes with pressurised water for flushing toilets.

希腊克里特岛上古老的克里特文明（公元前 2700 年—前 1450 年）最早使用地下土制管道中的高压水来冲厕。

Think about It

想一想

Do you think we should use drinking water for flushing the toilet?

你认为我们应该用饮用水来冲厕吗?

Is human waste a problem or an opportunity?

人类的粪便是一个麻烦还是一个机会?

如果猫和狗不需要冲厕,为什么我们需要呢?

If cats and dogs do not need a flushing toilet, why do we?

When there is no rain, how would you save water?

如果没有雨水,你该如何节约水?

Do It Yourself! 自己动手!

How many times a day do you go to the toilet? And how many litres of water are used every time you flush the toilet that you have in your home? How many people are there in your family? Calculate how much drinking water your family uses by flushing every day, and then multiply the number by 365 days to get the volume of water that your family wastes every year. Ask yourself what else you could have done with that water. Make a list of uses for that water – if you can't use it in the city, where could it be used?

你每天去厕所几次？每次在家里冲厕用了多少升水？你家里有几口人？计算一下你家每天冲厕用掉了多少饮用水，然后乘上365天就能算出来你家里每年冲掉了多少水。问问你自己能用这些水做些什么，并列一张表——想想如果你在城市里没办法用这些水，这些水还能用在哪里？

TEACHER AND PARENT GUIDE

学科知识
Academic Knowledge

生物学	食物循环包括摄取、消化吸收和排便；身体通过肾脏和膀胱排出液体废物，其中富含钾元素；固体废物由细菌、纤维和脂肪组成，通过直肠排出；大便颜色可以反映健康状况；尿液含有非常少的细菌但不是无菌的。
化 学	肠道中的细菌会产生硫化氢使大便有臭味；体内钾元素过多会中毒，所以要排出去，但必须每天通过食用新鲜水果和蔬菜来补充钾离子，因为它对神经和肌肉的功能都至关重要；对植物而言，钾循环和氮循环同等重要。
物 理	钾可以溶解于水，并且是植物可吸收的现成的离子来源；小便池可以设计成不用水冲而用疏水性橡胶膜，当湿润时膜张开，水流过后就闭上，可以消除所有异味。
工程学	厕所有很多种设计：节水厕所、干厕、用涡旋装置（将尿液和粪便分开）的分离式厕所、坑式厕所、不用纸的高科技厕所，还有真空厕所。
经济学	美国波士顿在节水方面的努力使得用水量减少了43%，因此也不再需要对一条河进行分流了，还节省了大量的原来必不可少的资本投入（也减少了税收）。
伦理学	当城市饮用水严重短缺时，我们怎么能用三分之一的饮用水来冲厕呢？
历 史	古埃及已经有了用装沙子的容器做的干厕所；第一个有石头座位、扶手和冲水功能的厕所是在中国西汉王朝的陵墓中发现的；罗马人知道如何隔离人的粪便和饮用水源，并建造了公共厕所；11世纪建造的城堡都有完整的卫生间。
地 理	洛杉矶每天要进口89亿升水。
数 学	要花多少年才能收回每人每天只消耗7升水的新型厕所的投资？为了避免额外投资建造基础水利设施，我们需要节约多少水？
生活方式	75%的日本家庭的厕所都有高科技成分，比家庭电脑还要普遍。
社会学	冲水厕所被视为现代社会的象征，因此穷人们都希望拥有冲水厕所；"钱不会发臭"的说法来自罗马时期，当时皇帝用小便税平衡了预算。
心理学	希望别人来清理我们的厕所和污物是不是太傲慢了？
系统论	人们一方面不得不循环利用水，另一方面还要促进营养循环，尤其是钾循环。

教师与家长指南

情感智慧
Emotional Intelligence

爸爸

爸爸分享了他对缺水的忧虑，请他的孩子们想办法。但他被女儿的观点弄糊涂了。他问了一些问题，但是当女儿说出自己的道理时他却很快发起了脾气。女儿的想法不仅新颖，还有点奇怪。他的成见使他连考虑一下都不行。没有什么能让他满意，并且他还预见到了下一个争论又要变成一个误解的怪圈。他的思维倾向成了笑柄，却有效地结束了对话。他以自己的知识范式思考和处事，当女儿看到了使世界走上一条可持续发展道路的机会时，他看到的只有困难和问题，所以他很迷茫而且不知所措。

安娜

安娜很有创造力，准备提出新观点，虽然她知道这将会导致一定程度的不愉快。另一方面，她很坚定，当爸爸对她的观点表示强烈反对时依然坚持自己的想法。她用一个简单明了的比喻使爸爸很难和她继续争论。当遇到反对意见时，她公开反对爸爸并继续质疑他，把厕所说成是一个愚蠢的东西。安娜加强了自己的观点，说狗不需要卫生纸。当爸爸嘲笑她的观点时，她提出了生理控制方式，并明确提出她的节水概念对恢复土地营养甚至节约能源都是个机会。当爸爸要求她拿出具体方案时，安娜表达了决心，即使爸爸并不这么认为。她最后采取了和解的姿态，指出她提出的措施并不困难，只是有所不同。

艺术
The Arts

你能用卫生纸卷筒做什么呢？与其把它们扔掉，不如把这些纸板做成艺术品。可以有很多方法，但你要想出自己的创意。如果你用卷筒做个望远镜，就可以看星星了。最后，你会发现你和你的卫生纸卷筒遨游在科学的海洋里。

TEACHER AND PARENT GUIDE

思维拓展
Systems: Making the Connections

水是生命最重要的成分之一。随着世界人口增长，城市消耗的水也越来越多。当本地无法保障供水时，就必须付出高昂的代价从远处进口大量的水。为了优先满足本地用水，城市甚至把一些高耗水工业转移到都市以外的地方。市民常常忘了最重要的家庭用水就是冲厕所，占家庭用水的四分之一到三分之一。即使是节水型厕所每次也要用5升水，即一个四口之家每年要用掉5万升水。老式的标准厕所还很普遍，每个厕所每年要用掉20万升水，相当于每个家庭浪费了一个游泳池的水。对水持续增长的需求也增加了能耗。储水池和水塔在城市中广泛分布，依赖数千根管道每天24小时、每周7天不停运转，是城市最大的能量需求设施之一。即使如此，流进来的水是干净的，流出去的水却是被污染的。人体设计了一个精密的系统把尿液、血液以及消化道内的粪便分离开，从不混合。尿液富含钾元素，而粪便由细菌、纤维和脂肪组成。特殊设计的厕所可以把尿液和粪便分离开：尿液可以稀释后运到农田恢复土壤肥力，粪便可以尽快地被干燥处理以控制细菌的滋生扩散。粪便浸入水中会产生厌氧细菌，从而需要化学物质来减少水中粪便所含的细菌，并减少臭味，但也使这些水和水中的各种生物变得毫无用处。如果将粪便产生的物质干燥，就能把粪便、木炭和骨头混合用来制造表土。现在的水系统和废水处理系统破坏了其中的营养和水循环，亟需重新设计。改变厕所的标准是维持健康的水、能量和营养循环最容易、最快捷也是最有效的方式。

动手能力
Capacity to Implement

检查一下市场上所有可能的厕所系统。你会吃惊地发现除了坐便和蹲便，还有很多类型。现在选择3个你最喜欢的款式，讨论一下支持还是反对使用它们。讨论时主要注意以下几点：（1）对水的使用；（2）对大便和小便重新利用的情况；（3）产生工作机会的数量；（4）最重要的一点——容易使用且能减少气味。一旦你将众多选择减少到3个厕所系统，选择其中一个推荐给你的学校、办公室或家庭。

教师与家长指南

故事灵感来自
This Fable Is Inspired by

海克·皮佩洛
Haiko Pieplow

海克·皮佩洛是一名土壤科学专家。他在德国的罗斯托克研究植物种植和植物学，并在德国联邦环境部设立了一个有关资源效能的部门。他生活在柏林郊区的一所生物结构的太阳能房子里。2005年，他重新发现了特拉－普利塔模式，这是在亚马逊地区广泛使用的一种人类粪便和木炭的混合物，把贫瘠的土地变成了高肥力的土壤。他和乌特·朔伊布、汉斯－皮特·施密特一起出版了《特拉－普利塔——雨林地区的黑色革命》。从此，他由重新设计厕所开始，率先在德国发起了特拉－普利塔生产方式，在柏林植物园重新利用人的粪便产生高质量的表土。

图书在版编目（CIP）数据

冈特生态童书.第二辑修订版：全36册：汉英对照／
（比）冈特·鲍利著；（哥伦）凯瑟琳娜·巴赫绘；
何家振等译.—上海：上海远东出版社，2021
书名原文：Gunter's Fables
ISBN 978-7-5476-1759-5

Ⅰ.①冈… Ⅱ.①冈…②凯…③何… Ⅲ.①生态
环境－环境保护－儿童读物—汉、英 Ⅳ.①X171.1-49

中国版本图书馆CIP数据核字（2021）第213075号

著作权合同登记号图字09-2021-0823

策　　划	张　蓉
责任编辑	祁东城
封面设计	魏　来　李　廉

冈特生态童书

人人都爱的宝座

［比］冈特·鲍利　著
［哥伦］凯瑟琳娜·巴赫　绘
郭光普　译

记得要和身边的小朋友分享环保知识哦！
八喜冰淇淋祝你成为环保小使者！

Water
40

喷水管的力量
Power from a Hose

Gunter Pauli

[比]冈特·鲍利 著
[哥伦]凯瑟琳娜·巴赫 绘
郭光普 译

上海远东出版社

丛书编委会

主　任：贾　峰

副主任：何家振　闫世东　林　玉

委　员：李原原　祝真旭　牛玲娟　梁雅丽　任泽林
　　　　王　岢　陈　卫　郑循如　吴建民　彭　勇
　　　　王梦雨　戴　虹　翟致信　靳增江　孟　蝶

特别感谢以下热心人士对童书工作的支持：

匡志强　宋小华　解　东　厉　云　李　婧　陈　果
刘　丹　熊彩虹　罗淑怡　旷　婉　杨　荣　刘学振
何圣霖　廖清州　谭燕宁　韦小宏　李　杰　欧　亮
陈强林　王　征　张林霞　寿颖慧　罗　佳　傅　俊
胡海朋　白永喆　冯家宝

目录

喷水管的力量	4
你知道吗?	22
想一想	26
自己动手!	27
学科知识	28
情感智慧	29
艺术	29
思维拓展	30
动手能力	30
故事灵感来自	31

Contents

Power from a Hose	4
Did You Know?	22
Think about It	26
Do It Yourself!	27
Academic Knowledge	28
Emotional Intelligence	29
The Arts	29
Systems: Making the Connections	30
Capacity to Implement	30
This Fable Is Inspired by	31

夜里，几只条纹原海豚正沿着地中海海岸畅快地玩着。他们跟着渔船离开了海港，一只海鸥在黑暗中紧盯着他们。

A few striped dolphins are swimming along the Mediterranean coast at night. A seagull watches in the dark as they follow the fishing boats leaving the harbour.

一只海鸥在黑暗中紧盯着他们

A seagull watches in the dark

……或者鱿鱼!……

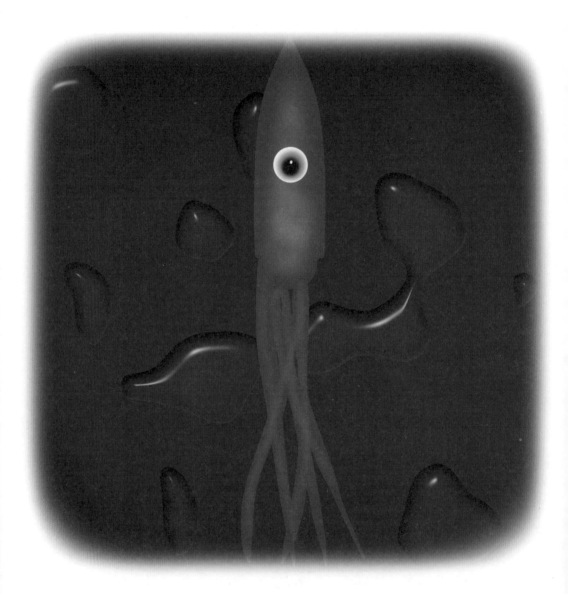

... maybe some squid!...

"这些船是去捕鱼的，美味的沙丁鱼或者鱿鱼！"海鸥大声叫道。"我们可以得到些残羹剩菜！"

"在黑暗中不可能捕鱼，"海豚回答。"没有月光，沙丁鱼是看不到食饵的。"

"These boats are going to catch tasty sardines, or maybe some squid!" cries the seagull. "There'll be leftovers for us!"

"That's impossible in the dark," responds the dolphin. "Sardines can only see their feed by moonlight."

"但这些渔民很聪明,他们围成一个圈子捕鱼,我可以抢一些鱼作为我的晚餐。"

"噢!这么说他们从我的海豚兄弟那里学会了用释放空气幕布的方式来捕获沙丁鱼群?"

"But these fishermen are smart and catch fish in a circle, and I can snap away for my dinner."

"Oh, so they've learnt from my dolphin brothers how to catch schools of sardines by releasing a curtain of air?"

……围成一个圈子捕鱼

...catch fish in a circle

打开灯光

Hold up their lanterns

"不是的，这些渔民的做法很简单。他们用船围住一个区域，然后撒网，同时打开灯光，就好像月亮出来了一样。"

"这种做法很聪明，但鱼也不傻。他们对光没有兴趣——他们只想要食物，和我们一样！"海豚答道。

"No, these fisherman keep it simple. They close off an area off with their boats, drop their nets and hold up their lanterns as if the moon is rising."

"That's clever, but fish are not stupid. They're not interested in light – they just want food, like we do!" responds the dolphin.

"耐心点,海豚!浮游生物喜欢光,而这些灯模仿月光——所以沙丁鱼成群结队地来吃等在那里的浮游生物。我呢,也就有吃的了!"

"可怜的沙丁鱼。"海豚说道。"他们看到浮游生物在灯下发光,就数以百万计地游到水面上,结果却是被捕杀,然后被装到罐子里从船上运走。"

"Be patient, Dolphin! Plankton loves light and these lamps imitate moonlight – so the sardines flock to the waiting plankton. And, I get to eat!"

"Poor sardines," says the dolphin. "They see plankton glowing in the light, so they swim to the surface by the millions, only to get caught and slaughtered. They leave the boat canned."

浮游生物喜欢光

Plankton loves light

我们海豚有时候会被网抓住

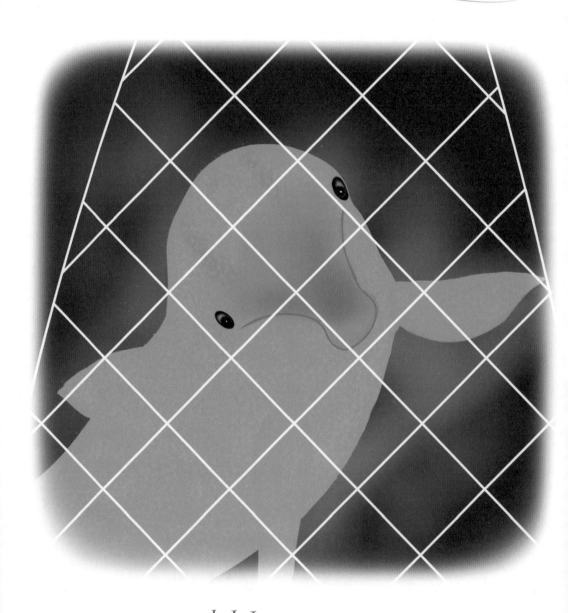

Sometimes us dolphins get caught in the nets

"所以，现在正是时候，去和渔民一起大吃一顿了！"海鸥贪婪地说。

"我不想靠得太近。我们海豚有时候会被网抓住，而且那些人类毫无同情心。"

"我知道他们非常残酷无情，但你也得承认他们用灯光引诱鱼群的做法是很聪明的。"

"So now's the time to go and feast with the fishermen!" the seagull says greedily.

"I don't want to get too close. Sometimes one of us dolphins gets caught in the nets, and these humans have no mercy."

"I know they can be ruthless, but you have to admit that their lights for tricking fish are clever."

"当然，但人们也不总是那么聪明。你看到过他们是如何尝试利用涨潮和退潮来获得能量吗？"

"看到过，这个巨大的墙挡住了高潮时的海水，然后退潮时水经管子流走，这样就可以发电了。"

"这些人居然忘了他们花园里的喷水管是怎么工作的！"

"花园里的喷水管？这和发电有什么关系呢？"

"Sure, but humans aren't always that smart. Have you seen how they're trying to make energy from the sea's ebb and flow?"

"Yes, this huge wall traps seawater in the bay at high tide, and then the water flows through pipes at low tide, making electricity."

"These people have forgotten how their garden hoses work!"

"Garden hoses? What does that have to do with the generation of power?"

……花园里的喷水管？……

...Garden hoses?

压强还是流速？

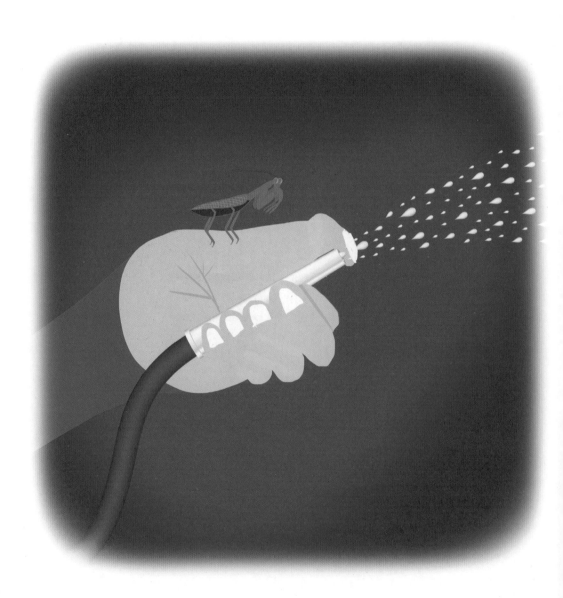

Pressure or flow?

"噢，海鸥，想想看！当管子的压强很小，水流得很慢的时候，人们是怎么做的？"

"他们用大拇指按住管嘴，即使水很少也可以喷射得很远！"

"太对了！水在小管里流的时候速度慢、压强大。"

"那用什么来产生力量呢：压强还是流速？"

"我觉得两个都需要。"

"Come on, Seagull, it's time to think. What does a human do to his garden hose when the pressure is low and the flow is slow?"

"They put their thumbs over the nozzle so that less water will spray much further!"

"Very good! Flow is slow with high pressure in a smaller pipe."

"So what is needed to create power: pressure or flow?"

"I would say both."

"你想得对，海豚！不是这个'或者'那个，而是这个'和'那个！"

"我从来不理解人们为什么只想着黑和白，支持和反对，高或低。显然，生活需要我们联合'所有和更多'已有的力量。"

"鱼和光，船和海鸥……但是你们海豚知道哪个人有足够大的拇指来按住喷水管口吗？"

"我想，要是我把鼻子放在喷水管里，你就可以得到更多的食物了！"

……这仅仅是开始！……

"You're certainly using your head, Dolphin! So it's not 'either/or', it's 'and-and'!"

"I'll never understand why people only think in black and white, for and against, high or low. It's clear that life needs us to harness 'all and more' of the available forces."

"Fish and light, boats and seagulls … But do you dolphins know anyone who has a thumb big enough to press the nozzle on a hose?"

"I think you'd get more food if I put my nose in the hose!"

... AND IT HAS ONLY JUST BEGUN!...

……这仅仅是开始！……

...AND IT HAS ONLY JUST BEGUN!...

Did You Know?

你知道吗？

Lanterns hung over the side of a boat attract plankton. Bait fish are drawn to feed on the plankton, and larger fish feed on the bait fish. The only one that does not get caught are the plankton!

悬挂在船一侧的灯光可以吸引浮游生物。诱饵鱼以浮游生物为食，诱饵鱼又是大鱼的食物。唯一不被捕获的就是浮游生物。

Fishing with lights at night is best when there is no moon. When the moon is shining, plankton feeds everywhere and is not attracted to lamps.

在没有月亮的天气用灯光捕鱼效果最好。当月光明亮时，浮游生物会四处觅食，而不会被灯光吸引。

Green or white light is best at attracting fish. Blue light is beautiful in water and red light helps with night vision, but neither attracts fish.

绿光和白光对鱼最有吸引力。蓝光照在水里非常好看，红光有助于在晚上看东西，但都不能吸引鱼。

Light changes when it moves through water. Red light is not visible beyond a depth of 3 metres; orange and yellow light is absorbed at 10 metres from the light source, where it turns grey. This works both horizontally and vertically.

光线照到水里就会发生变化。红光在 3 米以下就看不到了；橙光和黄光在离光源 10 米处被吸收，因此那里是灰暗无光的。这在水平和垂直方向上都会发生。

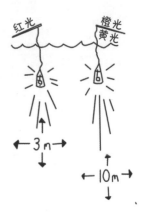

Fishing nets catch everything, including the 300,000 dolphins and porpoises that die every year entangled in nets.

渔网能捕获所有东西，每年有 30 万只海豚（包括鼠海豚）被渔网缠绕致死。

New nets save dolphins but still catch sharks and rays.

新式渔网使海豚幸免，但还是会捕捉到鲨鱼和蝠鲼。

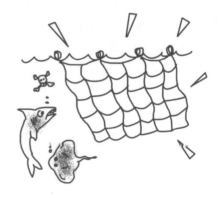

The tide, or the rise and drop of seawater levels, is caused by the combined effect of the gravity exerted by the moon, the sun and the rotation of the Earth.

潮汐，即海水的涨落，是由月亮、太阳和地球自转共同导致的引力变化引起的。

When the area through which water flows becomes smaller, and the volume of water remains the same, then the speed of water flow must increase.

当水流经过的地方变狭窄，而流量没变化，那么流动速度肯定变快了。

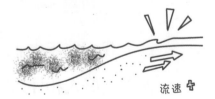

流速

Think about It

想一想

When it is dark, would you be attracted by a source of light? Or would you stay away?

黑暗中，你会被光源吸引，还是会离光源远些？

Do you agree that nets should catch fish, and not the dolphins and the sharks?

你是否同意渔网应该用来捕鱼，而不能捕海豚和鲨鱼？

What are the main advantages and disadvantages of using power from the sea?

利用海洋获取能量有什么优点和缺点？

我们能不能把人分为两类：好人和坏人？

Can we divide people in two categories: good people and bad people?

Do It Yourself!
自己动手!

Let's build a Venturi pipe. It's not difficult, but you'll need to pay attention to how it works. Find a pipe with a diameter that's big enough to fit your fist. Pour water through the pipe and watch how it flows. Then, out of stiff plastic or some other strong material, make a circle the same size as the diameter of the pipe. Make a small hole in this circle, and then fit the circle to the inside of the pipe so that it is watertight. Now pour water through your pipe, and watch how it flows. What is the difference? If you can't find a pipe and a circle, take your garden hose and turn on the water. What happens when you place your thumb over the nozzle and press it hard?

让我们来做一根文丘里管。这不难，但是你需要注意它是怎么工作的。找一根直径和你的拳头大小差不多的管子。把水灌进去，观察水是怎么流动的。用硬塑料或其他坚硬的材料做一个和管子直径相同的环。然后，在环上打一个小孔，并从管子内部固定这个环使管子不漏水。现在把水灌进管子，观察水是怎么流动的。这两种方式有什么差别吗？如果你找不到管子和环，就用你家花园的喷水管并打开水龙头。当你用大拇指使劲按住管口时会发生什么？

TEACHER AND PARENT GUIDE

学科知识
Academic Knowledge

生物学	鱼有眼睛和视网膜，可以看到聚焦的图像并侦测到移动的物体，但需要光线来区分颜色；鱼的听觉比视觉发达得多；深海鱼类只能看到黑和白，看不到色彩；潮间带是一个有着独特生物多样性的生态系统。
化　学	一个塑料瓶、水和漂白剂可以做成一个40—60瓦的光源。
物　理	绿色、蓝色和紫色的光波长较短；光线在水里发生散射是水里的悬浮小颗粒造成的，悬浮的颗粒物越多，穿过去的光线就越少；鱼类能侦测到偏振光，但人类不能；荧光主要是由人类看不到的紫外光组成的。
工程学	流过飞机机翼上表面的气流比流过下表面的快，产生了一个抬升力让飞机飞上天空；文丘里效应使燃料和空气能经过完美混合后再送入发动机。
经济学	世界上使用时间最长的潮汐发电站（在法国）运转成本最低，并且发的电比核电还便宜。
伦理学	拖网也会误捕鲨鱼、蝠鲼和小的金枪鱼，我们怎能只拯救误捕的海豚呢？我们如何才能做到根据需要去捕捞，而不是滥捕？
历　史	18世纪在英国有一个由自然哲学家和工业家组成的知识分子组织"月光社"，他们每次聚会都选在月圆的时候，这样就不用提着灯回家了；灯光捕鱼出现于公元700年的日本岐阜市。
地　理	世界上最大的潮汐发电站位于韩国的西洼（Sihwa）；最早的潮汐发电站位于法国布列塔尼地区的兰斯河上。
数　学	压强和速度的关系。
生活方式	为了满足我们的需要，人类的行为已经威胁到物种的生存并改变了生态系统；中国灯笼被全世界所接受，虽然已经不再需要用它们照明了，但它们可以在春节或农历正月十五的元宵节用来装饰城镇或房间。
社会学	停止做坏事而改成做好事是不是就够了，还是应该努力保证一切都变得更好？
心理学	当我们对人、行为和结果进行评价的时候往往想得有点绝对，但是没有什么事物是绝对好或绝对坏的。
系统论	用灯光捕鱼可以提高捕获量，缺点是许多小鱼也会被捕获，这影响了鱼类资源的可持续性。

教师与家长指南

情感智慧
Emotional Intelligence

海鸥

海鸥一想到有吃的就非常兴奋。他已经考虑到了细节，并了解渔民的捕鱼工具和方法。海鸥没有什么同情心：他首先考虑的就是自己的食物并坚持要得到自己的那一份。即使面对令人不安的事实（沙丁鱼被装罐和海豚被捕杀）时，他也把争论放在一边，重复着他的观点：用灯光捕鱼很聪明。当海豚解释说人们不总是很聪明时，海鸥还是坚持他的观点，并依然对新的事实置若罔闻。只有当受到挑战时，海鸥才开始认同海豚的观点。

海豚

海豚对海鸥只关心自己食物的自私自利的态度很反感，但没有表现出这种反感。她没有直接和海鸥冲突，而是问了一个很直接的问题来引起海鸥的注意。海豚对海鸥很有耐心，即使对海鸥的狂妄自大也很容忍。海豚向海鸥分享了自己曾离灯光太近而导致的危险，即使如此，海鸥也无动于衷。海豚没有坚持，而是换了一个话题问了几个问题，不是在情感上打动海鸥（这个策略没有成功），而是成功地用智慧吸引海鸥。她成功地把话题提高到了哲学水平，和海鸥讨论生活的原则，在这一点上双方达成了共识。

艺术
The Arts

让我们看看中国的灯笼吧！准备几张红色的A4纸和剪刀。在网络上查找所有制作灯笼的方法，然后按照说明做一个自己的灯笼。等到下一个元宵节时用这些美妙的装饰给爸妈一个惊喜！

TEACHER AND PARENT GUIDE

思维拓展
Systems: Making the Connections

　　追求产量引出了具有创造力的解决方案。灯光捕鱼已经开展了1000多年，这在早期鱼类数量充足的时候没有问题，而如今人造灯光的大量使用导致了过度捕捞。那句古老的谚语应该改为："授人以鱼，他可吃一天；而授人以渔，他将过度捕捞。"这日益成为一个严峻的事实。我们正在毁灭我们的海洋，当我们认识到情况有多严重时可能为时已晚，没有办法弥补已经造成的危害，而且问题很复杂。更严重的是拖网捕鱼，它会扫光海洋底部，不分青红皂白地捕杀海豚、鲨鱼、蝠鲼、海龟和旗鱼。捕鱼业将这些都看成废物，甚至是令人讨厌的，并把死掉的鱼和误捕的鱼倒入海洋。海豚被网缠住需要切开网才能解脱，很少能从这种折磨中幸存下来。海洋有这么多独特的资源，但是我们的管理需要更成熟和精细。尽管许多技术和商业模式已经过考验，但我们似乎还是不愿意使用这些模式。公众特别关注海豚，然而，许多其他动物也应该得到同样的待遇。我们不能仅仅减少对海洋的负面影响，还要施加积极影响：我们应该培养一种意识，即不仅要少做坏事（减少对所有生物的误捕），还要多做好事（从水的流动中获得更多能量）。我们需要多种渠道利用我们的资源：我们需要创造一套解决方案，其中没有一项是完美无缺的，但每一项都会以不同方式使事情发展得更好。

动手能力
Capacity to Implement

　　让我们来做一个莫泽灯吧！其实就是一个用回收的塑料瓶做的天花板照明灯。不需要电或铜线，是不是简单得令你吃惊！去找一个塑料瓶，最好是PET材料的。洗干净并装满水。加入氯、漂白剂或者洁厕剂以防止真菌生长。做8个这样的瓶子。如果可以，在你家、游戏房或花园棚子的天花板打8个孔，大小可以装进这8个瓶子，并用密封剂固定好。别在你爸妈的房顶打孔，除非得到他们的允许！确保这些瓶子的位置能使瓶子接收阳光，并反射到整个房间。上网找一些图片——这些图片会告诉你这些我们都能做到！

教师与家长指南

故事灵感来自
This Fable Is Inspired by

阿尔弗雷德·莫泽
Alfredo Moser

阿尔弗雷德·莫泽被称为"穷人的托马斯·爱迪生"。他出生在巴西米纳斯吉拉斯的乌贝拉巴，受过技工培训，也做过技工的工作。他很讨厌不断的停电，就寻找最简单的方法在白天为他的工作室提供照明。他认识到塑料瓶装满水后可以反射足够的光来照亮他的商店。他还加入了一点氯，以避免真菌长成绿膜阻挡光线。他把自己的发明称为"一公升的光"。他把这个发明公之于众，让所有人都可以使用，当前全世界已经建立起千万个这样的照明系统，这种系统在菲律宾做得最成功。

图书在版编目(CIP)数据

冈特生态童书.第二辑修订版:全36册:汉英对照/
(比)冈特·鲍利著;(哥伦)凯瑟琳娜·巴赫绘;
何家振等译.—上海:上海远东出版社,2021
书名原文:Gunter's Fables
ISBN 978-7-5476-1759-5

Ⅰ.①冈… Ⅱ.①冈… ②凯… ③何… Ⅲ.①生态
环境−环境保护−儿童读物—汉、英 Ⅳ.①X171.1-49

中国版本图书馆CIP数据核字(2021)第213075号

著作权合同登记号图字09-2021-0823

策　　划	张　蓉
责任编辑	祁东城
封面设计	魏　来　李　廉

冈特生态童书

喷水管的力量

[比]冈特·鲍利　著
[哥伦]凯瑟琳娜·巴赫　绘
郭光普　译

记得要和身边的小朋友分享环保知识哦!
八喜冰淇淋祝你成为环保小使者!

Water 41

巨藻之舞
The Kelp Dance

Gunter Pauli

[比] 冈特·鲍利 著
[哥伦] 凯瑟琳娜·巴赫 绘
郭光普 译

上海远东出版社

丛书编委会

主　任：贾　峰

副主任：何家振　闫世东　林　玉

委　员：李原原　祝真旭　牛玲娟　梁雅丽　任泽林
　　　　王　岢　陈　卫　郑循如　吴建民　彭　勇
　　　　王梦雨　戴　虹　翟致信　靳增江　孟　蝶

特别感谢以下热心人士对童书工作的支持：

匡志强　宋小华　解　东　厉　云　李　婧　陈　果
刘　丹　熊彩虹　罗淑怡　旷　婉　杨　荣　刘学振
何圣霖　廖清州　谭燕宁　韦小宏　李　杰　欧　亮
陈强林　王　征　张林霞　寿颖慧　罗　佳　傅　俊
胡海朋　白永喆　冯家宝

目录

巨藻之舞	4
你知道吗?	22
想一想	26
自己动手!	27
学科知识	28
情感智慧	29
艺术	29
思维拓展	30
动手能力	30
故事灵感来自	31

Contents

The Kelp Dance	4
Did You Know?	22
Think about It	26
Do It Yourself!	27
Academic Knowledge	28
Emotional Intelligence	29
The Arts	29
Systems: Making the Connections	30
Capacity to Implement	30
This Fable Is Inspired by	31

一棵庞大的巨藻坐在纳米比亚海岸边清凉的海水里，随着海浪摆动着。几头海狮穿梭在这些大海藻间，寻找着鱼吃。

"你真是一棵神奇的植物，"一头海狮有点害羞地赞美道，"你真的和竹子长得一样快吗？"

A giant kelp sits in the cold waters off the Namibian coast, moving with the waves. Some sea lions are swimming swiftly through the kelp forest, looking for fish.

"You are an amazing plant," remarks a sea lion shyly. "Is it true you grow as fast as bamboo?"

你真是一棵神奇的植物

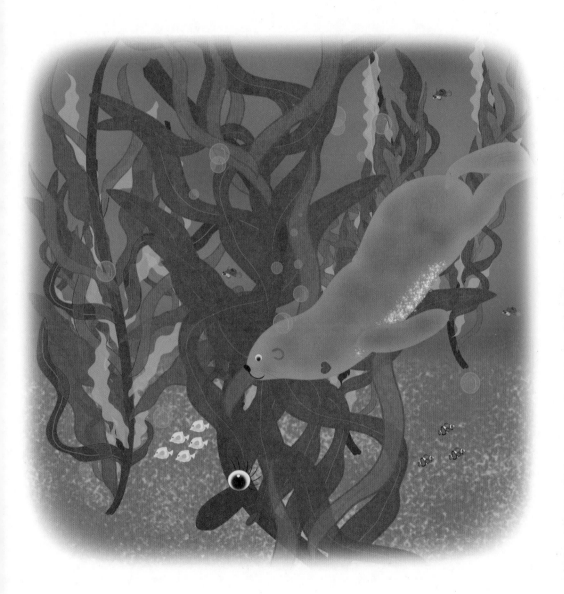

You are an amazing plant

你能长多高呢?

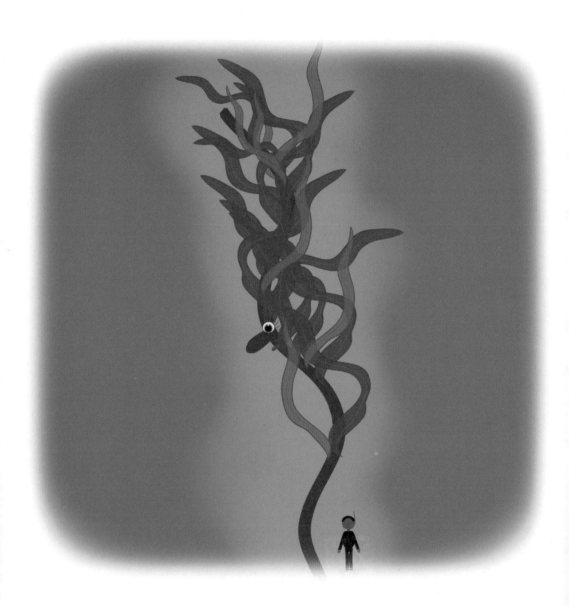

How tall will you grow?

"是的，我每天能长50厘米。"巨藻回答。

"你能长多高呢？"

"噢，我现在还是个少年，不过只要人们不过度捕捞，我可以长到50米高。"

"You are right – I can grow 50 centimetres per day," answers the kelp.

"And how tall will you grow?"

"Well, I'm still a teenager but unless people overfish, I can grow 50 metres tall."

"这可比最高的竹子还高一倍呢!"海狮回答道。"太神奇了!可你能告诉我过度捕捞与你长得高大健壮有什么关系呢?"

"哎!人类喜欢吃鱼,而鱼的食物又吃巨藻。所以如果人类捕捞太多的鱼,那些喜欢吃我们的家伙就会大吃特吃,我就长不大了。"

"That is double the size of the biggest bamboos!" replies the sea lion. "That's amazing. But tell me, what does overfishing have to do with you growing tall and healthy?"

"Well, humans like to eat fish, and fish eat stuff that eats kelp. So if people catch too many fish, then those who like to eat me will eat so much that I cannot grow."

……那些喜欢吃我们的家伙就会大吃特吃……

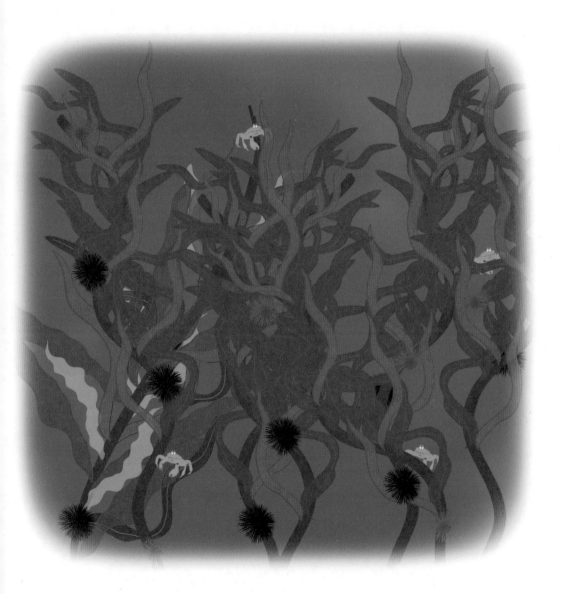

… those who like to eat me will eat so much …

我们是海洋里的森林

We are the forests of the sea

"这么说,鱼群越多、越活跃,巨藻就长得越好。这是一个实实在在的生态系统。"海狮说。

"我们能长得很大。我们是海洋里的森林,为你们和海獭提供了丰富的食物。"

"So, the more fish are alive and well, the more the kelp will grow. That is a real ecosystem," says the sea lion.

"We grow big. We are the forests of the sea, providing an abundance of food for you and the sea otters!"

"我知道你和海獭是好朋友。"

"我们沿着太平洋组成了巨型森林,帮助早期的航海家航行。我们是海洋中的巨藻高速公路,指引人们长距离航行并提供充足的食物。"

"人们是沿着巨藻高速公路旅行的吗?"

"当然,等到达彼岸他们就可以和之前一样生活。只要有巨藻的地方就有龙虾、石斑鱼和鲍鱼。我们甚至还能产生能量!"

"I know you and the sea otter are dear friends."

"Our huge forest along the Pacific Ocean helped early seafarers to navigate. We were the kelp highway of the ocean, guiding people over long distances with sure supplies of food."

"People travelled following the kelp freeway?"

"Sure, and when they got to the other side they could live on like before. Anywhere you have kelp, you also have lobsters, rockfish and abalone. We can even produce energy!"

……我们沿着太平洋组成了巨型森林……

… our huge forest along the Pacific Ocean …

……乙醇和意大利面……

ethanol and spaghetti

"巨藻能制造食物和燃料？"

"是的，我们富含矿物质和碘。如果你把我们收集在一起并放到池子里发酵，最后你就会得到大量沼气。你还能用我们制造糖，或生产乙醇给汽车作燃料。有些聪明的人还把我们做成意大利面食用。"

"你们能生产这么多好东西！为什么我们对你们了解这么少？"海狮问道。

"Kelp makes food and fuel?"

"Yes, we are full of minerals and iodine. If you harvest me and let me rot in a closed room, you will end up with plenty of biogas. You can make sugars from me and produce ethanol to fuel a car. Some smart people turn me into spaghetti and eat me."

"You make so many goodies. So why we know so little about you?" asks the sea lion.

"你知道我们有
自己的舞蹈吗，海狮？"
"舞蹈？噢，是的，我听说好像
就是你们启发了跳肚皮舞的人！"
"我们随着海洋的节奏跳舞。"

"Did you know we have our own dance, Sea Lion?"

"Dance? Well, yes, you seem to have inspired the belly dancers, from what I've heard!"

"We dance with the rhythm of the sea."

……启发了跳肚皮舞的人……

... inspired the belly dancers ...

用海啸的力量跳舞

Dance with the power of the tsunami

"那要是海啸来了怎么办呢?"
"我们会用海啸的力量跳舞!"
"怎么做到的?"

"And, what if there is a tsunami?"
"We will dance with the power of the tsunami!"
"How is that?"

"当海洋的力量太强时，我们都匍匐在海底，任由海水翻涌而过。"

"然后你们就又可以继续跳肚皮舞了，就像歌星夏奇拉教我们的一样！"

……这仅仅是开始！……

"When the power of the ocean is too strong, we all lie flat with our faces on the ground, and let it just flow over us."

"And then you can belly dance away again, just like the singer Shakira has taught us!"

... AND IT HAS ONLY JUST BEGUN!...

……这仅仅是开始！……

...AND IT HAS ONLY JUST BEGUN!...

Did You Know?
你知道吗？

Kelp can grow 50 metres tall. Its stem is flexible, allowing it to sway in any direction following the ocean currents.

巨藻能长到50米高。它的茎杆是有弹性的，能随着洋流向任何方向摇摆。

Kelp roots do not carry nutrients to the leaves. They are only an anchor. The stem is called a frond and the leaves are called blades.

巨藻的根不会为叶子供应营养，它只是一个"锚"。巨藻的茎杆称为叶状体，而叶子称为叶片。

Kelp is a popular animal feed. The leaves can photosynthesise on both sides. (The leaves on trees do it on one side only.) Kelp leaves drop off once a month, providing lots of feed.

巨藻是一种常见的动物饲料。巨藻的叶子两面都可以进行光合作用。（陆地上大树的叶子只能一面进行光合作用。）巨藻叶片每个月脱落一次，可以提供大量饲料。

Kelp forests are extremely rich and diverse, supporting an abundance of shellfish, marine mammals, seabirds and seaweeds. There are kelp forests along both the Pacific Rim and Southern Africa.

巨藻森林具有极大的丰富性和多样性，支持着大量的贝类、海洋哺乳动物、海鸟和海藻的生活。环太平洋和南非都有巨藻森林。

Kelp forests reduce wave power, offer grip and support for boats, and create an abundance of food. These forests may have facilitated the emigration of people from Asia to the Americas over the North Pacific Coast. The kelp ecosystem is therefore known as the "kelp highway".

巨藻森林减弱了海浪的力量，为船只提供固定物和支持，还能提供大量食物。这些森林很可能曾帮助亚洲人穿过太平洋北部海岸迁徙到美洲。巨藻生态系统因此被称为"巨藻高速公路"。

The giant bamboo (Guadua angustifolia) is only half the size of the giant kelp (Macrocystis pyrifera). Giant bamboo can grow up to 90cm per day, and giant kelp up to 50cm. Bamboo dies when it flowers, and lives up to 70 years; kelp can live up to 100 years.

一种叫瓜多竹的巨竹也只有巨藻的一半高。巨竹每天能生长90厘米，而巨藻每天只长50厘米。竹子开花后就会枯死，可以活70年，而巨藻能活100多年。

Kelp has the highest concentration of iodine of any food. Kelp noodles are made from kelp, sodium alginate and water. It is free from gluten, fat, cholesterol, protein and sugar.

在所有食物中，巨藻含碘量最高。巨藻面条就是用巨藻、海藻酸钠和水做成的，不含麸质、脂肪、胆固醇、蛋白质和糖。

Giant kelp is called the Sequoia of the Seas. It moves graciously with the flow of the water as if dancing.

巨藻被称为海洋中的红杉。它随着海水优雅地摆动，就像是在跳舞。

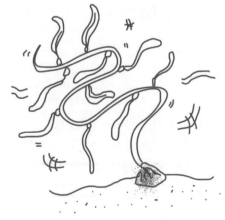

Think about It
想一想

Kelp relies on a healthy population of fish to thrive, but does safeguarding fish populations necessarily translate into more kelp?

巨藻依赖合理的鱼类数量而茁壮生长，但是鱼群越多是不是就必然意味着更多的巨藻呢？

Is it important to be the greatest, the fastest, the oldest or the best?

成为最大、最快、最老或最好真的很重要吗？

当面对一股势不可挡的强大力量时，你会怎么办呢？

What would you do if faced with an overwhelming force you could not stop?

Which music inspires you to dance?

什么音乐会让你跳起舞来？

Do It Yourself! 自己动手!

Go to your local shops. First pass by a supermarket and ask if there are any seaweeds for sale. Most likely the answer will be "No". Then ask which products contain seaweed. Most likely you will get an "I don't know" answer. The answers you might receive in Japan will be different from those received in France (Brittany) or Spain (Galicia). Check ice cream, microwave meals and frozen-food labels yourself. You will likely recognise the word "carrageenan", or perhaps a general description like "emulsifier", but you might not have known that these are made from seaweed. Now stop at the local health-food store. Ask if there are any seaweeds for sale or if it is included in any of the products. You are likely to be listening to the answer for the next 10 minutes. Seaweed could be everywhere.

去当地的商店看看。先去超市问问是否有海藻卖，答案很可能是"没有"。然后问下什么产品含有海藻，最可能的答案是"我不知道"。你在日本得到的答案可能与在法国布列塔尼和西班牙加利西亚的不同。检查一下冰淇淋、微波食品和冷冻食品标签，你可能会发现"卡拉胶"的字样，或者"乳化剂"之类的一般性描述，但是你可能不知道这些东西都是从海藻中提取出来的。现在到当地的健康食品商店看看，问问是否有海藻卖，或者是不是有什么产品含有海藻。你很可能会听到10分钟都说不完的答案。海藻无处不在。

TEACHER AND PARENT GUIDE

学科知识
Academic Knowledge

生物学	巨藻富含碘，碘会促进甲状腺素的合成，调节食物代谢，同时也解毒；自闭症与缺碘有关；营养、能量和生长是按照生态系统中的一套制衡机制运行的；巨藻是一种褐色水藻。
化 学	用海藻做的纤维能用来缝合伤口；海藻中的钙离子和体液中的钠离子可以使氧分渗透到细胞里，促进组织生长。
物 理	冲击和断续的运动，正弦运动；海藻酸钠可增强黏度并用作乳化剂；巨藻依靠浮力漂浮在水中。
工程学	与其对抗地球和海洋的运动，不如随着这些节奏一起运动，这是针对自然力量的最好防御方式；海藻纤维做的毛巾已经于1993年开始供应市场；海藻纤维泳装于2000年出现，2012年又出现了淡藻纤维做的婴儿服装。
经济学	一吨干的巨藻能生产200—250千克海藻酸钠，可以用来制成200千克纤维；用巨藻生产的产品有100多种。
伦理学	如果一个生态系统内所有组成部分的功能都能发挥到最好，这个生态系统就能繁荣发展，所以一种生物不能过度开发利用另一种生物：生态系统需要地球上的所有成员相互协调，以达到一个动态平衡。
历 史	舞蹈作为一种宗教仪式出现在古埃及和古希腊。
地 理	大型巨藻森林位于太平洋中；巨藻森林沿着非洲海岸分布，从南非到纳米比亚，沿途都受本格拉寒流的影响。
数 学	浮力的计算；海啸波的数学特征。
生活方式	海白菜和裙带菜是最常见的食用褐藻；舞蹈是一种情绪表达方式，也是一种治疗方法。
社会学	肚皮舞和西班牙的弗拉明戈舞很相似。
心理学	即使你觉得你是最好的，承认某个人可能在某些方面比你好也是一种谦逊的态度。
系统论	我们太依赖于土地上的产品了，我们唯一利用的海洋资源就是鱼，而在海藻森林中却蕴含着巨大的未开发的生物量，只要我们控制捕鱼量，就能让这些生物量可持续地提供100多种产品。

教师与家长指南

情感智慧
Emotional Intelligence

海狮

海狮很羡慕巨藻，并害羞地赞美她。她想了解更多，主动提出了很多问题。她很专心，学得很快，理解了有关生态系统新的信息及其重要性。随着理解加深，海狮认识到还有很多知识她还不了解。她听到了新的理论（巨藻高速公路），这些一鳞半爪的信息呈现给她的东西比她之前想象的还要多。巨藻有很强的节奏感和灵活性，还知道如何应对海啸，这让海狮感到惊奇。最后，海狮赞美了巨藻，并将巨藻比作当代最著名的音乐家和（肚皮舞）舞蹈家之一的夏奇拉，以此表达对巨藻的仰慕。

巨藻

巨藻很镇静，但很了解自己的能力、时代和位置。巨藻很享受海狮的陪伴，花时间回应和解答她的困惑，包括对其他因素（鱼）的依赖和自己无法控制的威胁（过度捕捞）。巨藻还顺便提到了自己的重要历史价值。巨藻没有狂妄自大，她恰如其分地分享了她的潜力和才能。然而，巨藻很有自知之明，并意识到了自己可能遇到的事情和自己的局限性。她的舞蹈让海狮大吃一惊，她还告诉海狮自己应对海啸的神秘能力，这也是她们的友谊中互相信任的表现。

艺术
The Arts

该上舞蹈课了！让我们用阿梅利亚·特拉平（Amelia Terrapin）编排的舞蹈，在运动中学习科学。怎么做呢？相关网站介绍说你需要学习基本的舞蹈动作。然后和朋友们讨论：当海啸来的时候你们怎么跳舞？在小组面前试一下，一起商量出一种最好的方式，通过舞蹈来表现"海啸"。现在尝试表演做生物饮料以及和鱼做朋友。每次都要想一下怎么用舞蹈动作来"说话"，而不能真的说话。阿梅利亚做这件事已经很多年了，所以她是个伟大的教练。

TEACHER AND PARENT GUIDE

思维拓展
Systems: Making the Connections

巨藻是一种卓越的高产生态系统：很少有人深刻认识到它是世界上生长最快、用途最多的生物之一。巨藻不仅在海洋森林生态系统中是个关键物种，还是可持续产业的基础。它应用于医疗、土壤、食品、纺织和能源等领域的数百种产品中；而且我们只要收获就行，从来不用种植它们。你能想象出如果庄稼只需收获而不用种植，也不用管理，这能节约多少钱和时间吗？巨藻提供的纺织材料可以用咸水加工处理，比起棉花节约了很多淡水。但是棉花纤维只能来源于种子，而巨藻中的海藻酸钠占整个植株生物量的25%。尽管可以从大多数巨藻（和其他海藻）中获取的产品都有明确的记载，并在科学上得到证明，甚至在世界范围都有生产，然而没有一种产品成为主流产品被使用。旅游业创造了新的探险活动，比如潜水到有防护的大白鲨洞穴中；但是，还没有旅游者去观看同样精彩壮观的巨藻森林，以及其中的鱼群和海獭。主要原因可能是不了解：人们不知道巨藻森林里有如此多姿多彩的环境。由于过度捕鱼，巨藻森林正在急剧消失，甚至红树林也不得不为养殖业（主要是养虾）和旅游业腾出空间。我们不仅仅要知道巨藻控制海浪和应对最猛烈的海啸的生存能力，还要知道它保护着海洋森林中的所有生命。人们还不了解巨藻对生态系统的这些贡献，但未来的岁月会证明巨藻森林和热带雨林一样重要。

动手能力
Capacity to Implement

海藻能为未来提供很多食物、医疗用品、纺织品、肥料和清洁产品，随着需求增长，供应也将越来越快。在你生活的地方找一下有谁为工业生产提供海藻原料。如果现在还没有，想象一下你自己变身海藻产业背后的推动力。你会注重什么？哪种产品最有意义？你会做原材料的供应商吗，或者促进现有产品的销售？想象一下你如何把海藻产业变成你的谋生之道。

教师与家长指南

故事灵感来自
This Fable Is Inspired by

夏奇拉·伊莎贝尔·迈巴拉克·里波尔
Shakira Isabel Mebarak Ripoll

夏奇拉出生在哥伦比亚西北部的港口城市巴兰基利亚，并在那里长大。她父母分别是哥伦比亚人和黎巴嫩人。她在学校的时候就开始表演。夏奇拉把肚皮舞再度介绍给更多的观众并鼓舞了数百万人和她一样跳舞。她一直是销量最好的艺术家之一，她的唱片在全世界销售了1300多万张。她在2010年国际足联世界杯的主题歌曲《非洲时刻》是历来最受欢迎的音乐视频之一。夏奇拉通过赤足基金会建立了多所有舞蹈演播室的学校，以帮助哥伦比亚的贫困孩子。

图书在版编目(CIP)数据

冈特生态童书.第二辑修订版：全36册：汉英对照 /
(比)冈特·鲍利著；(哥伦)凯瑟琳娜·巴赫绘；
何家振等译. —上海：上海远东出版社，2021
书名原文：Gunter's Fables
ISBN 978-7-5476-1759-5

Ⅰ.①冈… Ⅱ.①冈…②凯…③何… Ⅲ.①生态
环境-环境保护-儿童读物—汉、英 Ⅳ.①X171.1-49

中国版本图书馆CIP数据核字(2021)第213075号

著作权合同登记号图字09-2021-0823

策　　划　张　蓉
责任编辑　祁东城
封面设计　魏　来　李　廉

冈特生态童书
巨藻之舞
[比]冈特·鲍利　著
[哥伦]凯瑟琳娜·巴赫　绘
郭光普　译

记得要和身边的小朋友分享环保知识哦！
八喜冰淇淋祝你成为环保小使者！

Food 44

蛆的口水

Maggot Spit

Gunter Pauli

[比]冈特·鲍利 著
[哥伦]凯瑟琳娜·巴赫 绘
李原原 田烁 译

上海远东出版社

丛书编委会

主　任：贾　峰

副主任：何家振　闫世东　林　玉

委　员：李原原　祝真旭　牛玲娟　梁雅丽　任泽林
　　　　王　岢　陈　卫　郑循如　吴建民　彭　勇
　　　　王梦雨　戴　虹　翟致信　靳增江　孟　蝶

特别感谢以下热心人士对童书工作的支持：

匡志强　宋小华　解　东　厉　云　李　婧　陈　果
刘　丹　熊彩虹　罗淑怡　旷　婉　杨　荣　刘学振
何圣霖　廖清州　谭燕宁　韦小宏　李　杰　欧　亮
陈强林　王　征　张林霞　寿颖慧　罗　佳　傅　俊
胡海朋　白永喆　冯家宝

目录

蛆的口水	4
你知道吗?	22
想一想	26
自己动手!	27
学科知识	28
情感智慧	29
艺术	29
思维拓展	30
动手能力	30
故事灵感来自	31

Contents

Maggot Spit	4
Did You Know?	22
Think about It	26
Do It Yourself!	27
Academic Knowledge	28
Emotional Intelligence	29
The Arts	29
Systems: Making the Connections	30
Capacity to Implement	30
This Fable Is Inspired by	31

一群鹌鹑正在四处觅食，那块地方满是苍蝇。
"我们正在享用最美味的午餐。"一只鹌鹑咯咯笑着说。

A covey of quails is foraging around in an area infested with flies.

"We are having one of the greatest luncheons ever," giggles one quail.

最美味的午餐

Greatest luncheons ever

苍蝇就像是食物工厂

Flies are like food factories

6

"没错,可奇怪的是,人类似乎不喜欢苍蝇。"

"是呀,但凡人类自己不懂的,他们就不喜欢。其实,苍蝇就像是食物工厂。"

"同意,对于我们鹌鹑来说的确如此,但人类从来没想过吃蛆吧。"

"嗯,人类很喜欢吃我们小小的、营养丰富的鹌鹑蛋。可当他们看到蛆在垃圾里爬来爬去,就赶紧屏住呼吸,甚至闭上眼睛。"

"Indeed, it is so strange that people simply don't like flies."

"Well, whatever people do not understand, they don't like. Flies are like food factories."

"Yes, for us that is true, but people could never imagine eating maggots."

"Well, they do like our tiny and super nutritious eggs. But when they see maggots crawling in waste, they pull up their noses and even shut their eyes."

"哈,这些胖乎乎的小蛆虫可是我很长时间以来吃过的最美味的东西了。我们还是动作快点吧,要不它们又要变成苍蝇了。说实话,苍蝇可没这么好吃。"

"我同意,但是这些小小的苍蝇卵一旦变成蛆后,它们制造蛋白质的速度可是地球上最快的。你知道吗,只要食物充足,1千克蝇卵只需3天,就可以变成300多千克富含蛋白质的美味呢!"

"Oh, these chubby maggots are simply the tastiest I have had in a long time. We'd better hurry or they will turn into flies. And to be honest, flies simply are not as tasty."

"I agree, but there is nothing on earth that makes protein faster than these tiny fly eggs once they turn into maggots. Can you imagine, providing they have enough food, one kilogram of fly eggs turns into more than 300 kilograms of wonderful, protein-rich food within just three days!"

只需3天，富含蛋白质的美味

Protein-rich food within just three days

非常特殊的口水

Very special saliva

"我相信,在这方面没有谁能做得更好,即便是藻类、蘑菇或细菌,都不能和蛆相提并论。你知道蛆有一种非常特殊的口水吗?"

"他们的口水能有什么特殊的呢?"

"No one can do better, I believe not even algae, mushrooms or bacteria match this performance! Did you know these maggots have very special saliva?"

"What is so special about their spit?"

"它能帮助治愈伤口。"
"真的吗？怎么做到的？"
"嗯，它不但能清除伤口上死去的组织，还能促进新的细胞生长。"

"It helps heal wounds."
"Really? How does that work?"
"Well, it helps cells to grow while it cleans the wound of dead tissue."

它能帮助治愈伤口。

It helps heal wounds.

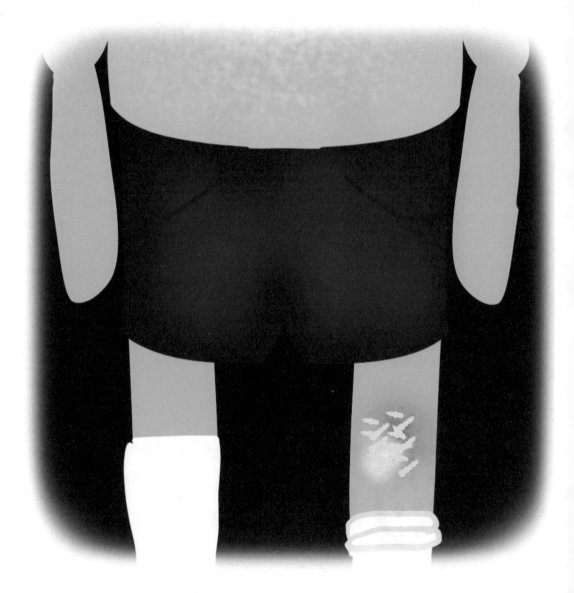

"听起来确实不错,但是你愿意让蛆在你身上爬来爬去吗?"

"如果没有其他选择,我也许愿意吧。假如伤口不能愈合,那么医生可能不得不给我截肢;这意味着我要失去一只脚或一个翅膀。"

"他们为什么不把口水挤出来呢?"

"That sounds great, but would you like maggots crawling all over your body?"

"Perhaps, if I had no other choice. If a wound does not heal, the doctor may have to amputate; that means losing a foot or a limb."

"Why don't they milk the spit?"

"嗯，提取口水，听起来很有挑战性哦！"

"嘿，如果在海边，你的头扎进水里，想想，会发生什么？"

"Hmmm, extracting spit – now that sounds like a challenge."

"Hey, what happens to you when your head goes under water when you are at the beach?"

听起来很有挑战性哦

Now that sounds like a challenge

啊！你知道我一定会吐出来！

Ugh! You know I will throw up.

"没什么呀，我会很小心地闭着嘴，屏住呼吸。"

"但是，如果一个巨浪袭来，把你整个儿冲得上上下下，灌进了很多咸咸的海水，这时你会怎么办？"

"啊！你知道我一定会吐出来！"

"Nothing. I take care to close my beak and breathe out through my nose."

"But if a big wave surprises you, turns you upside down and you gulp a lot of salt water, then what?"

"Ugh! You know I will throw up."

"这正是我想的。所以，我们可以把蛆放进盐水中，然后撇去它们吐出的口水。虽然这听起来有些倒胃口，但是的确有助于那些有伤口的人减轻痛苦。"

"你说对了，人类应该意识到，仅靠时间是不能愈合所有伤口的，而蛆虫却能提供一些帮助。"

……这仅仅是开始！……

"That's what I thought. So let us put the maggots in salt water and then skim their spit. It may not sound appetising, but it will certainly provide relief for people with open wounds."

"You are right, people should realise that time alone does not heal all wounds, and that a maggot can certainly offer some help."

... AND IT HAS ONLY JUST BEGUN!...

……这仅仅是开始!

...AND IT HAS ONLY JUST BEGUN!...

Did You Know?
你知道吗？

Quails have been bred domestically for over 4,000 years and it is thought that the Chinese quail is the ancestor of all breeds of quails. The commercial variety that is most popular originated in Japan.

鹌鹑的人工驯养已经有4 000多年历史了，有人认为中国的鹌鹑是所有鹌鹑种类的祖先。那些最受欢迎的商业化品种则起源于日本。

Quails are migratory birds travelling from Africa to Europe and glide using air currents. There are also mountain quails that migrate up and down the mountains on foot.

鹌鹑是一种候鸟，它们在非洲与欧洲之间迁徙，会利用气流来滑翔。还有一类鹌鹑是在山顶与山脚之间迁徙。

When the female has too many eggs and cannot keep them all warm, the male will join her on the nest. The chicks can walk and eat immediately after hatching.

如果雌鸟有很多蛋要孵，以至于不能让所有的蛋保持适宜的温度，那么雄鸟就会帮着雌鸟在巢中孵卵。雏鸟在孵化后过不多久就会行走和吃东西了。

The Aborigines of Australia, the Mayans in the Andes, and the Italians (during war in the 15th century) used maggots to treat wounds.

澳大利亚土著人、安第斯山的玛雅人以及15世纪经历战乱的意大利人都曾利用蛆来治愈伤口。

Maggots feeding off food scraps and abattoir-waste can produce the same amount of protein in 10 days that a pig generates in six months.

蛆仅靠吃食物残渣在 10 天中所生产的蛋白质总量，与一头猪半年内的产量是一样的。

The rearing of maggots is highly productive with an annual output of 1.2 tons of black soldier fly larvae per square metre.

养蛆是非常高产的，每平方米年均可生产 1.2 吨黑蝇幼虫。

Fly larvae have a good appetite: one kilogram of eggs will convert into more than 300 kilograms of protein in less than a week, provided that there is enough food.

苍蝇幼虫胃口很好，如果食物充足，那么1千克卵可以在不到一周的时间内转化为300千克的蛋白质。

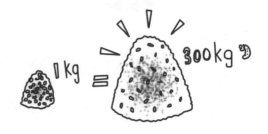

The most costly "vomit" is a substance called ambergris, produced by sperm whales. This substance is used in perfume. It is valued at more than US$ 20,000 per kilogram. This "exquisite" matter is usually discovered by dogs on the beach.

最名贵的"呕吐物"是由抹香鲸形成的一种叫作龙涎香的物质。这种物质被用于制作香水，它的价格高达每千克2万余美元。这种宝贝非常容易被海滩边的小狗闻到。

Think about It
想一想

Would you rather take antibiotics, or would you prefer maggots cleaning your wounds?

你是愿意用抗生素，还是愿意让蛆来清理你的伤口？

你认为缓解疼痛的良方是时间，还是止疼药？

Is time a good remedy for pain, or would painkillers be the best option?

Do you think that flies are good friends to human beings?

你认为苍蝇是人类的好朋友吗？

你认为哪个价值更高，是屠宰场的废弃物，还是栖息于废弃物上的蛆的口水？

What has the greatest value, the waste from the abattoir or the spit of the maggot farmed on the waste?

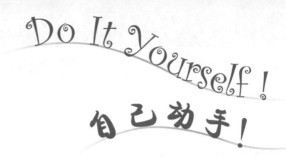

Do It Yourself!
自己动手!

How much protein do you eat? Make a list of all you eat during a normal day. List food in terms of protein, carbohydrates and fats. Now look at what maggots eat, if their main source of nutrition is abattoir waste. Then verify how much of the food you eat created how much waste. And then ask the question: How much waste do the maggots leave behind?

你要食用多少蛋白质呢？把你一天吃过的东西列个清单。将这些食物按照蛋白质、碳水化合物和脂肪来分类。现在，看看蛆都吃什么，它们的主要营养来源都是废弃物。然后，算一下你吃的食物产生了多少废弃物。接下来的问题是，蛆又留下了多少废弃物呢？

TEACHER AND PARENT GUIDE

学科知识
Academic Knowledge

生物学	蛆在治疗过程中所扮演的角色是吞噬细菌；蛆的唾液腺中含有变形杆菌，能产生抑制细菌的物质；苍蝇幼虫能够杀死链球菌，链球菌对于抗生素而言具有很强的抗药性。
化 学	蛆释放减少坏死组织的酶；生物手术的应用随着抗生素的发现而减少，现在细菌变得越来越具有抗药性，苍蝇幼虫治疗法又重新引入；幼虫能够产生碳酸钙来改变细菌繁殖环境；活蛆奶酪是撒丁传统食物，由羊奶制成，里面有活的苍蝇幼虫，能释放强酸来分解奶酪的脂肪，进而做出低脂肪的软奶酪。
物 理	蛆的蠕动刺激抗菌物的释放；基于比重不同来进行物质分离（如在盐水中分离呕吐物，在呕吐物中分离唾液）。
工程学	蛆虫疗法又称生物手术；设计一个口袋让蛆虫施展功效，避免让活的蛆虫直接在伤口上爬动。
经济学	通过比较蛆虫疗法的成本、并不见效的抗生素疗法的成本以及为病人截肢所产生的医疗、社会、经济代价，来估算一下蛆虫疗法的竞争力；饲料成本占畜牧业食品生产成本的60%—70%；替代效应的重要性在于，它不仅能更有效地利用有限的资源，而且还能产生多重效益增加当地经济收入。
伦理学	关于动物权利的争论：是将蛆碾碎来提取活性成分，还是让蛆活着，通过盐水催吐以提取唾液，最后让蛆成为鸡或鹌鹑的食物。
历 史	古埃及人为了食用鹌鹑肉和鹌鹑蛋而饲养鹌鹑；几个世纪以来澳大利亚土著部落一直沿用蛆虫疗法，安第斯山脉的玛雅人也是如此。
地 理	鹌鹑从非洲南部到欧洲的迁徙路径。
数 学	在将产业集群的利润率与核心产业对比时电子数据表格的局限性；多重效益的产生为当地经济催生了更多重的效益。
生活方式	可以通过改善生活方式（如戒烟）和改变饮食习惯（如选择无糖食物）来使伤口更好地愈合；鹌鹑蛋要比鸡蛋营养丰富得多。
社会学	社群中的成员有可能会几乎同时呕吐；秘鲁亚马逊流域的死藤水可导致剧烈呕吐，用于清除胃肠中的寄生虫。
心理学	时间不能治愈所有伤口；我们为什么不喜欢我们不了解的事物；蛆虫治疗法最主要的障碍是心理因素，这种障碍同时存在于患者和医务人员中。
系统论	多元价值的创造并非都能被量化，如卫生条件的改善和减少毒素的摄入。

教师与家长指南

情感智慧
Emotional Intelligence

鹌鹑

鹌鹑对自己的食物很满意，不理解人类为何不喜欢这种营养丰富的食物。人类因自己不了解蛆的价值而不喜欢蛆，而鹌鹑却并非这样无知，因此，鹌鹑对人类没有太多认同。鹌鹑认为人类有些无知，还不愿意更新自己的知识。鹌鹑有时间观念，享用大自然给他们的充足食物。他们表达了对苍蝇幼虫生产能力的钦佩，并认为他们是世界上最棒的。鹌鹑的对话深入到了超越食物本身的层面，比如关于蛆对医疗卫生方面作用的讨论，这些充满探索的交流覆盖了广泛的议题。总之，这些对话展现了他们的处世哲学和反思性思维，坚持超越自我世界，认为有些东西虽然不被认可，却是很有用的。

艺术
The Arts

图表为我们理解复杂数据提供了简单的方式。自己制作一张图表吧，数据是屠宰场的废物量，蝇卵和蛆的数量，转化为鹌鹑饲料、人类医疗用品、鱼饲料的数量，海洋中被捕获用于人工繁殖的鱼苗的数量。看看你可以使用哪些不同类型的图表来展现以上数据（如饼图、条形图、柱状图），然后开动脑筋吧！

TEACHER AND PARENT GUIDE

思维拓展
Systems: Making the Connections

　　蛆是很多鱼类和鸟类的理想食物，人们却不屑一顾，还担心苍蝇在我们的食物中产卵。然而，蛆是生物链中十分重要的环节，蛆有能力将腐败物的废渣转化成50%—60%的蛋白质和25%的脂肪，这是多么理想的饲料。现在的许多动物饲料都是由大豆和鱼粉组成，可是，大豆仅含有35%的蛋白质，鱼粉的蛋白质含量相对多一些。我们是在将一种食物转化成另一种质量更低的食物，其营养含量远不如最原始的饲料，这也解释了人们为什么要忍受饥饿。现在，我们不仅要关心食物链中损失的营养，还要关心我们可持续发展实践中损失的更好的机遇，比如养殖蛆虫。昆虫饲养行业的兴起是一大进步，但我们仍然过于聚焦于单一的行业。那些将蛆虫制作成食物的企业没有看到蛆虫唾液的价值，医疗卫生行业也没有看到提供更健康饮食的机遇。创造成功的关键是实现多赢。创造经济、社会、环境共赢的能力将最终实现可持续的健康发展和食品供应，甚至能为增强土壤肥力提供坚实保障。

动手能力
Capacity to Implement

　　做一次蛆虫营销大师吧！你知道患者和医护人员都不喜欢蛆，人们也不喜欢接近苍蝇，看到蛆在腐烂的尸体上爬来爬去也会感到恶心。那么，现在你就有责任为蛆虫开展营销工作啦！你要将蛆虫以人类朋友的形象展示出来，告诉人们蛆虫可以为提升我们的生命质量做出很大的贡献。因此，你要想出能够帮助人们克服一知半解和厌恶心理的方法，找到兴趣点，甚至可以让人们变得喜欢苍蝇、蝇卵和蛆虫。不要太严肃，记得要幽默些哦！

教师与家长指南

故事灵感来自
This Fable Is Inspired by

戈弗雷·扎木略
Godfrey Nzamujo

戈弗雷·扎木略出生在尼日利亚北部城市卡诺，后被送到美国加利福尼亚学习。他是一位成功的学者，拥有农学、经济学和信息技术学位。他的理想是改善非洲人民的生活。于是，他放弃了美国的事业返回非洲，在贝宁建立了宋海组织。他在非洲的事业开始于1985年，此前尼日利亚拒绝为他提供可利用的土地。他将自己的全部精力投入到建立一个生产中心，在那里，赤贫及没有接受教育的人都有生产粮食、找到工作的机会。他让一名铁匠制造生产工具，训练农民育种、沤肥。正是在这种充分获取一切可利用之物效益的探索中，扎木略将蛆虫养殖整合为更有价值的产业链：从位于波多诺伏的第一个生产中心，扩大到帕罗库、萨瓦罗、肯维基三个地区。宋海组织每年培训300名非洲人，教他们如何生产粮食和获得工作，通过利用现有之物来脱贫致富。

图书在版编目(CIP)数据

冈特生态童书.第二辑修订版:全36册:汉英对照/
(比)冈特·鲍利著;(哥伦)凯瑟琳娜·巴赫绘;
何家振等译.—上海:上海远东出版社,2021
书名原文:Gunter's Fables
ISBN 978-7-5476-1759-5

Ⅰ.①冈… Ⅱ.①冈…②凯…③何… Ⅲ.①生态
环境-环境保护-儿童读物—汉、英 Ⅳ.①X171.1-49

中国版本图书馆CIP数据核字(2021)第213075号

著作权合同登记号图字09-2021-0823

策　　划　张　蓉
责任编辑　程云琦
封面设计　魏　来　李　廉

冈特生态童书
蛆的口水
[比]冈特·鲍利　著
[哥伦]凯瑟琳娜·巴赫　绘
李原原　田　烁　译

记得要和身边的小朋友分享环保知识哦!
八喜冰淇淋祝你成为环保小使者!

Food
43

海洋农夫

Farmers of the Sea

Gunter Pauli

［比］冈特·鲍利 著
［哥伦］凯瑟琳娜·巴赫 绘
李原原 田烁 译

上海远东出版社

丛书编委会

主　任：贾　峰

副主任：何家振　闫世东　林　玉

委　员：李原原　祝真旭　牛玲娟　梁雅丽　任泽林
　　　　王　岢　陈　卫　郑循如　吴建民　彭　勇
　　　　王梦雨　戴　虹　翟致信　靳增江　孟　蝶

特别感谢以下热心人士对童书工作的支持：

匡志强　宋小华　解　东　厉　云　李　婧　陈　果
刘　丹　熊彩虹　罗淑怡　旷　婉　杨　荣　刘学振
何圣霖　廖清州　谭燕宁　韦小宏　李　杰　欧　亮
陈强林　王　征　张林霞　寿颖慧　罗　佳　傅　俊
胡海朋　白永喆　冯家宝

目录

海洋农夫	4
你知道吗？	22
想一想	26
自己动手！	27
学科知识	28
情感智慧	29
艺术	29
思维拓展	30
动手能力	30
故事灵感来自	31

Contents

Farmers of the Sea	4
Did You Know?	22
Think about It	26
Do It Yourself!	27
Academic Knowledge	28
Emotional Intelligence	29
The Arts	29
Systems: Making the Connections	30
Capacity to Implement	30
This Fable Is Inspired by	31

一位老人坐在夏威夷北岸的鱼塘旁。他注视着一群正在冲浪的孩子们。"这是一片肥沃的土地。"他说。

"您真是生活在梦想世界里,爷爷。"一位年轻的冲浪者回答道,"您要知道,夏威夷的食物和燃料全部都依赖进口。嗯,差不多全部。"

An elder sits near a fishpond on Hawaii's North Shore.

He is watching some children on their surfboards.

"This is the land of plenty," he says. "You are living in dream world, Grandpa," responds a young surfer. "You know Hawaii imports all its food and fuel. Well, nearly all."

注视着一群正在冲浪的孩子们

Watching some children on their surfboards

在波利尼西亚附近海域航行

Sailing the currents of Polynesia

"要知道，我们曾经拥有100多万人口，而且完全可以自给自足。"

"那是您还在波利尼西亚附近海域航行的时代了，依靠鸟儿、月亮和星星指引航向，是吧？"年轻人嘲笑道。"我们已经听过您的故事了，我们非常喜欢您的故事，可时代不同了啊。是时候向前看了。"

"To think that we were once more than a million people, and that we had all we needed right here."

"And that was when you were sailing the currents of Polynesia, guided by the birds, the moon and the stars, right?" teases the youngster. "We have heard your stories and we love them, but that was then and this is now. It is time to look to the future."

"拥有从过去经验获得的智慧，才能更好地期待未来。"爷爷说。

"您不是真的想让我们划着独木舟去钓鱼吧，爷爷？我们现在可以从国外进口任何我们想吃的鱼。"

"One often has a better view on the future when you have the wisdom of the past," says the grandfather.

"You don't really want us to get into canoes and go fish, do you? We are getting all the fish we need from overseas."

拥有从过去经验获得的智慧，才能更好地期待未来

Better view on the future when you have the wisdom of the past

我们在海里筑起石头墙

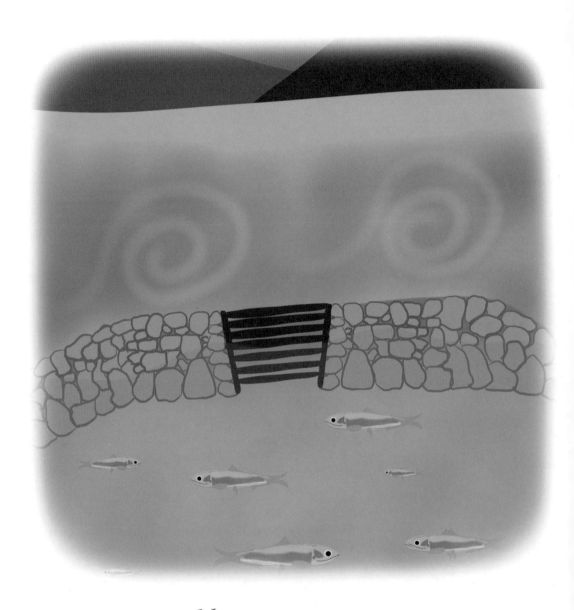

We build stone walls into the sea

"这就是我们要重新思考这个问题的原因,不能一直依赖进口,要看看我们已经拥有了什么!以前,我们并不仅仅依靠出海打鱼,我们还有鱼塘,就像眼前这个。"

"您怎么能从这儿钓到鱼呢,爷爷?"

"我们在海里筑起石头墙,然后用木制的闸门将那些随着涨潮带来的鱼困住。"

"This is why we need to rethink this and instead of importing all the time look at what we already have! Back in the days we did not rely solely on boats to fish, we had fishponds, like this one right here."

"How did you get the fish in there, Grandpa?"

"We built stone walls into the sea and trapped any fish that came in with the incoming tide, using sluice gates made of wood."

"听起来您说的鱼塘建得像个大教堂一样结实。"

"事实确实如此。这个鱼塘已经有600多年的历史了。如果能够得到妥善维护,它可以一直在这里,提供大量的鱼。"

"那么这些鱼吃什么呢?"

"Sounds like your fishpond was built like a cathedral."

"In fact, it was. This pond is over 600 years old and if well maintained will last forever, providing an abundance of fish."

"And what do the fish eat?"

这个鱼塘已经有600多年的历史了

This pond is over 600 years old

牛才吃草呢,爷爷。鱼可不吃草。

Cows graze, Grandpa Fish don't eat grass.

"我们让它们吃草。"

"牛才吃草呢,爷爷。鱼可不吃草。"

"不是这样的,聪明的年轻人!你应该跟爷爷多待一些时间,多学点知识,比如说知道我们种植海藻来喂鱼,就像农场工人种植青草来喂牛一样。"

"We let them graze."

"Cows graze, Grandpa. Fish don't eat grass."

"Not so, my bright young lad! You should spend more time with Grandpa and learn more, like the fact that we grow algae for our fish – like a rancher grows grass for his cattle."

"这听起来好得让人难以置信。"

"现在甚至更好。你知道我们在山里和平原上种植芋头吗?"

"知道。但是这得用淡水种植,而淡水对海鱼来说一点儿都不好。"

"That sounds too good to be true."

"It gets even better. You know we farm taro in the mountains and on the plains?"

"Yes, but that is with fresh water that is no good for sea fish."

我们在山里种植芋头

We farm taro in the mountains

鱼儿因此生长得非常喜人

Fish flourish on it

"从芋头种植地洛岛山上流下的水富含营养元素。我们将这些水引入鱼塘，鱼儿因此生长得非常喜人。"

"我明白了！您将海岸边鱼塘的水产养殖与土地种植相结合。一切都能各尽其用，一点儿都不浪费。一切都顺其自然，充分利用地心引力和土壤的力量。太棒了！"

"The water flowing downhill from the lo'l, our taro field, is rich in nutrients. We let it flow into our ponds and the fish flourish on it."

"I see! You are combining aquaculture, in your ponds on the shore, with farming on land. Everything gets used and nothing goes to waste. Everything flows naturally, using gravity and the power of the earth. That is so cool!"

"这就是我们过去用来为本地人提供食物的方法。我们现在应该还是这样做。"

"现在我们谈论的才是真正的创新！过去的智慧可以改变未来——让土地、海洋和人们做到最好。"

……这仅仅是开始！……

"That is how we used to take care of feeding all our people and we should do so again!"

"Now we are talking about real innovation! Using wisdom from the past that can change the future – allowing the land, the ocean and the people to be their best."

... AND IT HAS ONLY JUST BEGUN!...

……这仅仅是开始！……

...AND IT HAS ONLY JUST BEGUN!...

Did You Know?
你知道吗?

Hawaiians were some of the first fish farmers. Other cultures have existed as hunter-gatherers, feeding millions of people for aeons, without any imports.

夏威夷人是最早的渔业养殖者之一。其他文明则在长达亿万年的时间里,利用狩猎和采集,无需进口任何食物而养活了几百万人。

Surfing started in Hawaii and has evolved into a $10 billion business.

冲浪运动起源于夏威夷,现在已经发展成一项收入达到 100 亿美元的产业。

Hawaiians found their way around the Pacific Ocean by celestial navigation for thousands of years before the compass was invented.

在罗盘发明之前的几千年里，夏威夷人通过天文观察为他们在太平洋上航海指引方向。

Hawaii has the longest erupting volcano in modern history. It has been active since 1983.

夏威夷拥有现代史上喷发最久的火山，这座火山从1983年开始一直喷发到现在。

The island of Kauai in Hawaii is the wettest spot on earth, with an annual rainfall of 11,700 mm. That is 10 metres and 170 centimetres per year.

夏威夷的考艾岛是地球上最湿的地方，年降雨量达到 11 700 毫米，也就是每年降雨达 11.7 米。

Hawaii has the highest landmass in the world. When what is above sea level is added to what is below, it comes to 11,000 metres.

如果把海平面以上部分的高度和海平面以下部分的高度相加的话，夏威夷是世界上最高的陆地，其高度达到 11 000 米。

Hawaii is one of the few states getting bigger all the time. This is due to volcanic eruptions adding landmass.

夏威夷是美国为数不多的面积不断扩大的几个州之一。这是因为火山喷发在不断增加陆地面积。

The native language of Hawaii has one of the shortest alphabets in the h, k, l, m, n, p, and w, as well as a glottal stop called an 'okina.

夏威夷本地方言是世界上拥有最短的字母表的语言之一，只有h、k、l、m、n、p和w，以及被称作"冲音"的声门闭塞音。

Think about It
想一想

How much of the food you eat is produced locally?

你吃的食物中有多少是本地产的?

当你抬头看夜晚的星空时,你能分辨出南北吗?

When looking at the night sky, can you tell where north and south is?

Are there any cultures today that build houses, fishponds or animal pens that would last for 600 years?

现在,还有哪些文明可以建造能够使用600年之久的房子、鱼塘或牲畜棚圈呢?

火山是破坏了人类的生计还是开创了新的未来?

Are volcanoes destroying the livelihood of people or paving the way for a new future?

Do It Yourself!
自己动手!

Set up a small fish tank, fill it with water and add a few fish and water plants. It is easy to set up one by yourself, but you can also buy an aqua farm kit on the Internet from certain sites. Take a close look at what is going on in your tank. Discover how fish and water plants feed off each other and how they take in additional energy from the air. See how the waste from the fish serves to feed the plants and how the plants keep the water clean.

建造一个小鱼缸，装满水，再放上几条鱼和水草。你自己就可以非常容易地建造一个小鱼缸，但是你也可以在网上购买水产农场养殖工具。仔细观察鱼缸里发生的变化。找出鱼和水草是如何互相提供养料，以及它们如何从空气中吸收额外的能量。观察鱼的排泄物如何为水草提供养料，而水草如何保持鱼缸里的水质清洁。

TEACHER AND PARENT GUIDE

学科知识
Academic Knowledge

生物学	从以水生植物为食到以底泥动物为食的鱼类营养层级；哪种海洋动物以吃海藻为生；芋头的营养价值，多年生植物芋头的球茎也被称作"象耳"；从高山到大海的营养物质层级。
化 学	传统的夏威夷农场严禁使用农业化学品，因为在沼泽地里耕种时，化学品的使用会迅速污染河流和海洋。
物 理	冲浪板的几何形状；双重船壳独木舟的几何结构；运用天文导航的古代科学；潮汐涨落；地心引力的力量。
工程学	如何建造能够存在几百年的建筑物；水闸的建造和作用；用线钓鱼（垂钓）和用网捕鱼的对比；水产养殖生产系统的建造。
经济学	建筑物分25到40年折旧；建造可持续500年的建筑物所带来的资金积累；昂贵的新技术创新与能够利用现有资源以低成本生产的创新之间的对比。
伦理学	如果每天都可以从本地获得新鲜食物，为什么还要进口食物呢？
历 史	世界的中心一度被认为位于大西洋附近，现在又变为位于最大的水体——太平洋了；为什么人类要建造寺庙、金字塔和大教堂来抵挡时间的流逝？
地 理	群岛的定义；地球上最大的群岛在哪儿？
数 学	加法与乘法的对比：如果加上零，则总数不变；在乘法中，如果乘以零，则结果为零；增加一种因素的效果：比如你在上游地区增加一种有毒化学品，那么下游的一切都会受到影响。
生活方式	体育运动和体育锻炼的重要性；把耕种、钓鱼、修建一类的体力劳动当作生活经历。
社会学	为了展望更美好的未来而向以往的经历学习的能力；认识并欣赏已经拥有的，而不是到别处去寻找新的更好的东西。
心理学	鼓励人们做到最好；评价一个人要看他已具备的品质，而不是你希望他应具备的品质；"如果你评价一条鱼是看它爬山的能力，那么这条鱼一辈子都是失败的"（爱因斯坦）；两代人之间互相学习传承的重要性；讲故事在学习中的作用。
系统论	营养物质从高山流到山谷，生产食物、创造更多的营养物质，这些营养物质在顺流而下的途中被捕获，从而生成一连串的能量和从蔬菜到肉类的各种各样的食物，将高地、低地和海洋联系起来。

教师与家长指南

情感智慧
Emotional Intelligence

冲浪者

年轻的冲浪者是一位具有批判精神的现实主义者，自信地分享他的观点，比如什么东西已经过时了，以及他为什么认为他的爷爷与现实脱节。以前他听过爷爷的故事，他表现出对爷爷的爱意和对爷爷的智慧的敬意，但他还是坚信爷爷所说的岛上居民能够自给自足的时代已经过去了。他还没有准备好回到过去，像他的祖先那样生活。尽管他已经心有成见，但还是愿意创造一定的空间学习新事物，开阔眼界，更进一步地了解本民族的传统和文化。他有信心向爷爷提问。当他意识到将传统体制重新运作非常容易时，他感到非常庆幸。

爷爷

老人是一位乐观主义者，并不接受被人看作是梦想家。虽然他的孙子提出了很多负面的事实，但他仍然坚持以传统观念将祖先的土地看作是一片富饶的土地。他致力于让他的孙子看到自己所看到的机会。因为爷爷以前讲过这个故事，他知道不能将自己的观点强加在别人身上，而是应该通过摆事实、讲道理来引导孙子自己去发现。他宽容地接受批评，所提供的信息战胜了孙子的质疑，并将孙子转变成一位共同建造更好未来的同盟者。

艺术
The Arts

买一些芋头，如果在你家附近的商店里买不到，可以买些其他块茎类植物，比如土豆、地瓜或山药，因为这些作物总是长着最奇怪的形状。选择一些长得奇怪形状的作物。仔细观察每一个作物，想象一下你可以用火柴棍、坚果或小鹅卵石添加一些特征进行创作。你的芋头可能会变成你以前从未见过的龙或怪兽哦！

TEACHER AND PARENT GUIDE

思维拓展
Systems: Making the Connections

世界上所有的文明都曾经在水、食物和能源方面自给自足。这种状况随着工业化改变了。人类开始物物交换，用燃料交换食物。他们还开始建造大规模的水源分配系统以增加用水。大规模生产食物并用船运送到世界各地，这种方式正变得越来越昂贵。可是人们还是愿意付钱。进口食品意味着进口水。因为生产每吨粮食需要1 000吨水，所以购买100万吨粮食就相当于购买10亿吨水。在过去几十年里，我们一直寻求规模经济和批量生产，相信这样会降低成本，提高产量。但如果将美国种植的农作物与巴西种植的同一种农作物相比较，我们会发现，传统种植方式重复利用水和营养物质，能够降低生产成本，因为水可以至少重复使用三次，而且一种农作物产生的废物可以成为另一种农作物的营养物质。传统的方式可以利用所有可用资源产生更多价值。这与"现代"农业仅着眼于最终产物的价值而忽视任何剩余物或副产品的做法完全不同。将土地种植与海水养殖整合的夏威夷方式十分"成功"，这不仅体现在能够提供健康均衡的饮食，而且体现在能增加食物种类和扩大季节性生鲜农产品的生产规模。世界上并没有一个标准的系统，却有一种通用的设计框架，可以设计出一种模式来实现植物、海藻和鱼的多样化共存，最终找到海洋中最理想的关系。地心引力的自然力量让水、物质和营养物质从高处流向低处，从而增加产量，却几乎没有产生多少废物，而潮汐则为在循环往复的生命周期中产生更多生命提供所必需的营养物质和氧气。

动手能力
Capacity to Implement

让我们算一笔生意账：假设你打算以当前国际市场的价格从国外进口鱼、小麦和燃料。研究一下你需要付出的资金。然后模拟夏威夷的应用模式，计算一下：每公顷能够生产多少千克芋头？鱼塘里可以养殖多少千克鱼？你需要雇佣多少人？估算一下需要投入多少劳动力、精力、食物和水，基于这些统计作个比较，并得到结论：哪种生产系统能够生产更多的食物？哪种系统所需要的成本最低？

教师与家长指南

故事灵感来自
This Fable Is Inspired by

阿洛哈·麦可古菲
Aloha McGuffie

阿洛哈·麦可古菲是位夏威夷本地居民，她是夏威夷玛诺阿遗产中心的管理人。她和丈夫马克·麦可古菲共同居住在那里。该中心的3.5英亩土地上种植着波利尼西亚本土植物和一些引进作物。那里还坐落着一座夏威夷本地寺庙。在远离故土的企业界里闯荡25年后，阿洛哈回到夏威夷，在那里她重新认识了夏威夷的本土传统文化，这种传统文化的基础是夏威夷的核心价值观——"去热爱和关心你生活与成长的地方以及那里的人们"。阿洛哈练习传统合唱和舞蹈，包括草裙舞，并与年轻一代分享她对传统方式的了解，向他们介绍并解释将高山和大海联系起来的奇妙的土地管理系统。

图书在版编目(CIP)数据

冈特生态童书.第二辑修订版:全36册:汉英对照/
(比)冈特·鲍利著;(哥伦)凯瑟琳娜·巴赫绘;
何家振等译.—上海:上海远东出版社,2021
书名原文:Gunter's Fables
ISBN 978-7-5476-1759-5

Ⅰ.①冈… Ⅱ.①冈… ②凯… ③何… Ⅲ.①生态
环境-环境保护-儿童读物—汉、英 Ⅳ.①X171.1-49

中国版本图书馆CIP数据核字(2021)第213075号
著作权合同登记号图字09-2021-0823

策　　划	张　蓉
责任编辑	程云琦
封面设计	魏　来　李　廉

冈特生态童书
海洋农夫
[比]冈特·鲍利　著
[哥伦]凯瑟琳娜·巴赫　绘
李原原　田　烁　译

记得要和身边的小朋友分享环保知识哦!
八喜冰淇淋祝你成为环保小使者!

Food
42

捕鱼不用网

Fishing without Nets

Gunter Pauli

[比] 冈特·鲍利 著
[哥伦] 凯瑟琳娜·巴赫 绘
李原原 田烁 译

上海远东出版社

丛书编委会

主　任：贾　峰
副主任：何家振　闫世东　林　玉
委　员：李原原　祝真旭　牛玲娟　梁雅丽　任泽林
　　　　王　岢　陈　卫　郑循如　吴建民　彭　勇
　　　　王梦雨　戴　虹　翟致信　靳增江　孟　蝶

特别感谢以下热心人士对童书工作的支持：

匡志强　宋小华　解　东　厉　云　李　婧　陈　果
刘　丹　熊彩虹　罗淑怡　旷　婉　杨　荣　刘学振
何圣霖　廖清州　谭燕宁　韦小宏　李　杰　欧　亮
陈强林　王　征　张林霞　寿颖慧　罗　佳　傅　俊
胡海朋　白永喆　冯家宝

目录

捕鱼不用网	4
你知道吗？	22
想一想	26
自己动手！	27
学科知识	28
情感智慧	29
艺术	29
思维拓展	30
动手能力	30
故事灵感来自	31

Contents

Fishing without Nets	4
Did You Know?	22
Think about It	26
Do It Yourself!	27
Academic Knowledge	28
Emotional Intelligence	29
The Arts	29
Systems: Making the Connections	30
Capacity to Implement	30
This Fable Is Inspired by	31

一只海豚宝宝正在玩着水泡环,一头鲸看见后问道:

"你是从你妈妈那儿学来的吗?"

A baby dolphin is playing with a water ring. A whale watches and asks:

"Did you learn that from your mom?"

海豚宝宝玩着水泡环

Baby dolphin plays with a water ring

连我祖母都会呢!

Even my grandma knows how to do it!

"是呀!"海豚回答,"连我祖母都会呢,就像是带着呼啦圈跳舞一样。"

"Oh yes!" responds the dolphin, "Even my grandma still knows how to do this. It's like dancing with the people's hula-hoop."

"是的，我见过人类是怎样扭动身体来旋转呼啦圈的。"

"这项运动能促进身体健康，让心脏跳得更快，还能燃烧脂肪。"鲸说，"说起脂肪，为了后面的长途跋涉，是该吃点东西了。我要一头扎下去，用我的气泡口袋抓一些磷虾。"

"Yes, I have seen how people move their bodies to keep the hoop going."

"It keeps one fit, makes the heart beat faster and burns fat." says the whale, "Talking about fat, it's time to get some food for the long trip ahead. I will go down and catch some krill in a pocket of air."

人类扭动身体

People move their bodies

水中的气泡口袋

Pockets of air in the water

"什么？"海豚问道，"你在水中用气泡做口袋？"

"它看上去更像是一个装满了空气的大袋子。"鲸微笑着说。

"难道你要用这些气泡来捕鱼吗？"海豚问。

"What?" says the dolphin, "Do you make pockets of air in the water?"

"It may look more like a big bag of air." smiles the whale.

"And these air bubbles, do you use them for fishing?" asks the dolphin.

"没错！我们会不停地旋转，让这些气泡连成环状，然后，我们最爱的食物——磷虾，就会被装进由一个个微小气泡组成的袋子里。这样，我们只需要一边游一边张着嘴巴，就能吃到虾了。不知不觉间，我们的肚子就鼓起来了！"

"Exactly. We swirl around and around, blowing bubbles in a loop, so that krill, our favourite food, is caught in an envelope of tiny air bubbles. From there it's as easy as swimming up with our mouths open. And before we know it our tummies are full!"

在水中旋转……

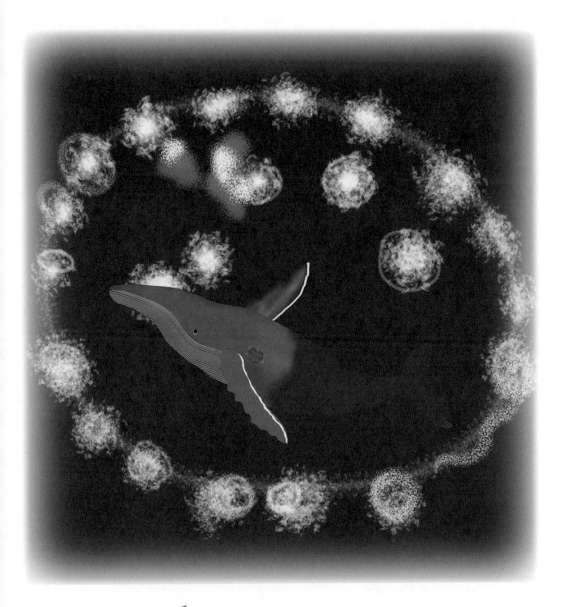

Swirl around in the water …

一个团队来产生很多气泡

Team to create lots of bubbles

"这听起来和我们的做法差不多啊。"海豚说,"但是,我们需要一整个团队才能产生很多气泡来捕鱼。"

"我们可以在一个气泡里捕获超过100千克的食物。"鲸说。

"这太不可思议了!我们的目标只是那些随着气泡上浮的小鱼们。然后,我们整个家族,还有所有的伙伴们,只需要把鱼吞进自己的嘴里就可以了。"

"That sounds just like what we do," says the dolphin, "but we need a whole team to create a lot of bubbles to catch fish."

"We can catch more than a hundred kilograms in one air bubble," says the whale.

"Now that is impressive! We only want the small fish that float to the top with our bubbles. Then my whole family and all our friends simply scoop the fish into our mouths."

"这听起来也很简单啊!"鲸说。

"是的,但是我们要注意别捉到那些体重大的鱼妈妈们,她们的肚子里装满了鱼卵,会孵化出更多小鱼,那可是我们下个季节的食物来源。"

"That sounds very easy as well!" says the whale.

"Yes, but we have to be careful not to catch the heavier mommy fish that are full of eggs, as they will make more fish for all of us next season."

注意别捉到鱼妈妈

Be careful not to catch the mommy fish

永远不会被气泡捕捉到

Never get caught in our air bubbles

"这个做法很聪明。鱼妈妈年纪越大,她肚子里的鱼卵也就越多。"

"而且,因为鱼妈妈的身体更重,所以永远不会被我们的气泡捕捉到。"

"That is very clever. The older the mommy fish, the more eggs she will have."

"And because she is heavier she will never get caught in our air bubbles."

"要是人类也能学会不用网就能捕鱼该有多好啊!还有,如果他们能放掉那些鱼妈妈就更好了,这样,他们就可以避免过度捕捞、破坏海床了。"鲸叹气道。

……这仅仅是开始!……

"If only humans would learn how to fish without nets! And if they always released the mommy fish, they would prevent overfishing and avoid destroying the sea bed." sighs the whale.

... AND IT HAS ONLY JUST BEGUN!...

······这仅仅是开始！······

AND IT HAS ONLY JUST BEGUN!

Did You Know?
你知道吗？

𝒜 female sturgeon can lay 200,000 to 500,000 eggs. A salmon only produces 2,500 to 7,500 eggs, while the lemon sole can lay 600, 000 eggs.

一条雌性鲟鱼的产卵量为 20 万—50 万枚。一条三文鱼的产卵量仅为 2 500—7 500 枚，而一条檬鲽鱼的产卵量却高达 60 万枚。

𝒯he sunfish is the champion: It has up to 300 million eggs in its ovaries.

翻车鱼是产卵冠军：它卵巢中的卵子数量多达 3 亿枚。

The sunfish is the heaviest fish, weighing up to 1,000 kg. It can be as tall as it is long, so it looks circular in shape, like the sun. The sunfish loves to eat jellyfish, but as these have few nutrients they need to eat tons of them. In Japan and Korea the sunfish is a delicacy, but in Europe catching it is prohibited.

翻车鱼是最重的鱼，体重足有1000千克。它的身高和体长差不多，因此外形看起来是圆形的，就像太阳一样。翻车鱼喜欢吃水母，但因为水母的营养含量并不高，所以需要吃几吨的水母才行。在日本和韩国，翻车鱼是美味佳肴，但在欧洲，翻车鱼是禁止捕捞的。

Sturgeon only need two eggs from each female to survive spawning to maintain its shoals.

每条雌性鲟鱼只需产两颗鱼卵，就足以维持其种群的数量。

Fish farming is an alternative to fishing in the wild. Unfortunately, the main feed for farmed fish is fish caught at sea.

人工养殖是替代野外捕捞的一个选择。然而，不幸的是，人工养殖的主要饲料却是海洋中的野生鱼类。

People started using fishing nets about 40,000 years ago. Techniques for catching fish include gathering, spearing, netting, angling and trapping. The beaver will also trap fish. The bear catches fish with its paws.

大约4万年前，人们开始用网来捕鱼。捕鱼的技术包括围、叉、网、钓、诱。海狸也会诱捕，熊则用掌来抓鱼。

The worldwide fish catch from the ocean has been stable at just under 100 million tons per year. However, most fish reserves are under stress and there is insufficient natural supply to support the human appetite.

每年，全球范围内的海洋捕鱼量都稳定在1亿吨以内。然而，大部分鱼类的保有量都面临着巨大压力，如果仅仅依靠自然供给，是很难满足人类的胃口的。

尽管科学家们只探索了1%的海洋深度，仍有数以百万计的水生物种尚待发现。然而目前，我们已知的鱼类品种已约有32 000种，这比其他所有的脊椎动物加起来还要多。

Even though scientists have only explored 1% of the ocean depths, millions of aquatic species are yet to be discovered. Today we know of about 32,000 different species of fish. This is more than all kinds of vertebrates combined.

Think about It
想一想

Would you ever kill a fish that has eggs and is about to have little ones? How is it that fishermen have never thought about this?

你会杀掉一条肚子里有鱼卵的鱼吗，尤其是在这种鱼已经濒临灭绝的情况下？为什么那些捕鱼者就从未考虑过这些问题呢？

制造出足够大的气泡来捕鱼有多简单呢？

How easy is it to make bubbles that are big enough to catch fish?

What might a dolphin think about fishing nets left in the sea?

对于海洋里遗留的渔网，海豚会怎样想？

鲸和海豚用气泡捕鱼有多久了？我们可以向它们学习吗？

For how long have whales and dolphins been fishing with air bubbles? Can we learn from them?

Do It Yourself!
自己动手!

Take a fish bowl and fill it with water and some small fish. Make small holes in a plastic tube and lead this down to the bottom of the fish bowl, making sure some of it lies at the bottom of the bowl. Or if you do not have a tube with holes, look for some big drinking straws. Blow air through it. What happens to the fish?

准备一个鱼缸,给它加满水,放入几条小鱼。然后,找一根塑料管,在塑料管上扎几个小洞,把它插入水中,让一部分塑料管接触到鱼缸底部。如果没有塑料管,可以用平时喝饮料的粗吸管来代替。向吸管中吹气,观察鱼缸中的小鱼会发生什么变化。

TEACHER AND PARENT GUIDE

学科知识
Academic Knowledge

生物学	生殖系统（海鱼类在海床上产卵）；过度捕捞导致生态失衡；远洋深海鱼冬眠的影响；磷虾是食物链的组成部分；鱼类和肉类的区别；鱼油和动物脂肪的区别。
化学	水中的溶解氧；渔网使用的聚合物材料的寿命；含有ω-3脂肪酸的深海鱼油对人类健康的重要性。
物理	气泡是如何使鱼漂浮到水面上的；帆运用了怎样的几何学原理来提升航海效率；帆在推动或拉动轮船前行时的设计有何差异；计算流体动力学。
工程学	传统渔船与双体船的设计对比；如何利用洋流来为船增加动力；固定帆与移动帆的区别；美洲杯比赛推动了帆船设计的革新；如何利用压缩机来制冰；通盘考虑外观、风速、船速、倾角以及洋流的帆船模型；风洞测试。
经济学	中间人的角色；在船上加工鱼和在陆地上加工鱼的对比；海洋运输的燃料成本；燃料补助及经济效益；将鱼的排泄物转化为动物饲料；植物蛋白转化为鱼肉蛋白和植物蛋白转化为动物蛋白的对比。
伦理学	对处在产卵期的鱼类或其他物种的杀害；尽管鱼类存量已近枯竭，仍然没有停止捕鱼。
历史	捕鱼技术的进步；航海技术的革新；埃及呼啦圈舞的起源。
地理	85%的鱼类存量都面临下降的危险；哪些地方有充足的鱼类资源。
数学	如何基于预期寿命、生育率、婴儿存活率计算出稳定的人口数量，以及如何将该计算方法应用于计算鱼类的存量；如何计算补充鱼类存量所需的年限；计算鲸制作气泡口袋转了多少圈，制造了多少水泡环。
生活方式	鱼卵成为商品，没有人将食用鱼卵与鱼类存量枯竭联系在一起，这种生活方式可持续吗？吃鱼与吃肉的对比。
社会学	代际学习的重要性。
心理学	如何通过团队协作来实现成功。
系统论	衡量健康食物标准的变化（鱼取代了肉，养殖鱼取代野生鱼）是怎样造成鱼类存量锐减的。

教师与家长指南

情感智慧
Emotional Intelligence

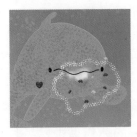

海豚宝宝

海豚宝宝意识到了自己在家庭中的角色,而且在与鲸的对话中,他也展示出了自我意识。他的交流方式富有同情心,他向自己的家庭成员比如祖母,还有鲸都表达了尊敬之意。他有着旺盛的求知欲,不停地问问题,知道自己的知识是有限的,并渴望学到更多的知识。海豚已经和其他海洋生命建立了关系,有着超前的思维,敢于分享自己关于未来渔业的想法。

鲸

成年鲸愿意花时间和年幼的海豚在一起聊天,显示出了他与海洋中其他哺乳动物的共鸣。他耐心地阐述观点并回答数不尽的问题。他用简单的语言阐述了具有一定理解难度的问题,让不是专家的海豚宝宝也可以轻松领会。对话后来发展为对海豚坚持维护海洋生态可持续发展的赞赏,并为海豚宝宝提供了反思的话题,在鼓励梦想的同时并没有脱离实际。鲸给那些肩负管理职责却没有意识到可持续发展重要性的人类敲响了警钟,同时展现出了一种处世哲学。

艺术
The Arts

准备一个装有水的鱼缸,还有一些塑料管。将塑料管卷成环状,放到鱼缸底部。关掉灯,坐在黑暗中,在鱼缸后方放置一个黄色光源和一个蓝色光源,将光源打开,一幅生动的水景画就会呈现在你的面前。

TEACHER AND PARENT GUIDE

思维拓展
Systems: Making the Connections

人类的过度捕捞导致85%的鱼类资源濒临枯竭，关于定量控制的争论也从未停止。科学分析数据明确显示，过度捕捞现象十分严重，我们必须要限制捕捞。然而，不幸的事实是，半数以上的捕捞鱼并未供人类消费，而是转变为其他动物的饲料，这些动物包括鸡、猪，甚至还有其他鱼类。这是一种非常低效的蛋白质转换方式。出海捕鱼，捕获后再将鱼冷冻起来，这一过程也增加了能源消耗成本。其实，大部分的燃料需求是可以削减的。总之，不管是从经济角度还是生态角度来看，目前的捕鱼业都是不可持续的。然而，几乎没有人意识到，在整个捕鱼业中最不可持续的其实是捕捞那些带卵的雌鱼。更糟糕的是，对鱼卵的消费（如鱼子酱）还经常被看作是享用美食。如果有选择地进行捕捞，并革新捕鱼技术（如气泡捕鱼技术、雌鱼识别技术等），那么鱼类资源很快会得到恢复。是时候告别那些古老的捕鱼方法，向鲸和海豚学习捕鱼技术了。如果不捕捞雌鱼，或捕获雌鱼后将其放生，那么捕鱼业还是可以实现可持续发展的。

动手能力
Capacity to Implement

在全世界范围内，渔业都是高度管制的。浏览一下你所在国家关于渔业的法律法规，同时也研究一下一些与之相关的国际协议。在熟悉了相关规则与实践之后，起草一份《产卵期鱼类权利宣言》。然后看看你的同学、朋友和家庭成员中能有多少人支持你的行动。你认为你能动员1000个人来支持你的宣言吗？

教师与家长指南

故事灵感来自
This Fable Is Inspired by

艾瑞克·勒克雷
Eric le Queré

艾瑞克·勒克雷的职业生涯始于法国海军，这给了他周游世界的机会。他学习了渔业贸易，并越来越关心与水资源及能源消耗息息相关的工人低收入问题。后来，他在摩洛哥开始了渔业生意，并吸引了大批工程师来重新思考整个捕鱼业的概念设计：从船体设计到船上加工鱼以及如何节约水资源和能源。他发现了一个新兴工业的创意平台，这个平台不仅能为渔民带来更多收入，而且最重要的是，建立一个最终能保持鱼类种类多样性以及海洋生产力的渔业体系。而这个平台的灵感，正是来自海洋中最聪明的哺乳动物。

图书在版编目(CIP)数据

冈特生态童书.第二辑修订版:全36册:汉英对照/
(比)冈特·鲍利著;(哥伦)凯瑟琳娜·巴赫绘;
何家振等译.—上海:上海远东出版社,2021
书名原文:Gunter's Fables
ISBN 978-7-5476-1759-5

Ⅰ.①冈… Ⅱ.①冈…②凯…③何… Ⅲ.①生态
环境-环境保护-儿童读物—汉、英 Ⅳ.①X171.1-49

中国版本图书馆CIP数据核字(2021)第213075号

著作权合同登记号图字09-2021-0823

策　　划　张　蓉
责任编辑　程云琦
封面设计　魏　来 李　廉

冈特生态童书
捕鱼不用网
[比]冈特·鲍利　著
[哥伦]凯瑟琳娜·巴赫　绘
李原原　田　烁　译

记得要和身边的小朋友分享环保知识哦!
八喜冰淇淋祝你成为环保小使者!

Food 45
带刺的荨麻
Stinging Nettles

Gunter Pauli

［比］冈特·鲍利 著
［哥伦］凯瑟琳娜·巴赫 绘
李原原 田烁 译

上海远东出版社

丛书编委会

主　任：贾　峰

副主任：何家振　闫世东　林　玉

委　员：李原原　祝真旭　牛玲娟　梁雅丽　任泽林
　　　　王　岢　陈　卫　郑循如　吴建民　彭　勇
　　　　王梦雨　戴　虹　翟致信　靳增江　孟　蝶

特别感谢以下热心人士对童书工作的支持：

匡志强　宋小华　解　东　厉　云　李　婧　陈　果
刘　丹　熊彩虹　罗淑怡　旷　婉　杨　荣　刘学振
何圣霖　廖清州　谭燕宁　韦小宏　李　杰　欧　亮
陈强林　王　征　张林霞　寿颖慧　罗　佳　傅　俊
胡海朋　白永喆　冯家宝

目录

带刺的荨麻	4
你知道吗？	22
想一想	26
自己动手！	27
学科知识	28
情感智慧	29
艺术	29
思维拓展	30
动手能力	30
故事灵感来自	31

Contents

Stinging Nettles	4
Did You Know?	22
Think about It	26
Do It Yourself!	27
Academic Knowledge	28
Emotional Intelligence	29
The Arts	29
Systems: Making the Connections	30
Capacity to Implement	30
This Fable Is Inspired by	31

一只瓢虫带着她的孩子穿越森林。他们从一株植物飞到另一株上,吃着植物上的蚜虫。瓢虫一抬头,刚好看到一只小山羊伸着舌头,馋馋地盯着那些茂盛的绿色荨麻。

"孩子,你饿了吗?"瓢虫问。

A ladybird guides her young through the forest. They move from one plant to the next, feeding on aphids. She looks up and sees a young goat with his tongue sticking out, hungrily looking at the lush green nettles.

"Are you hungry, kid?" she asks.

"孩子,你饿了吗?"

"Are you hungry, kid?"

从来不知道被荨麻刺扎到会这么难受

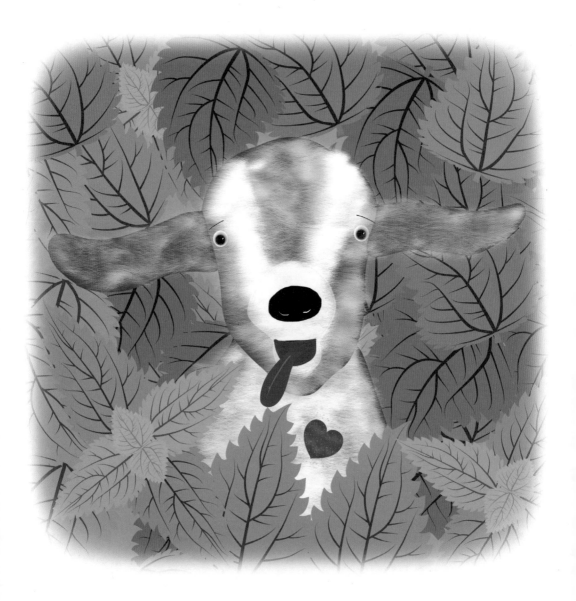

Never knew nettles sting so badly

"我实在是太饿了。我想吃这些荨麻,可是舌头却被荨麻的刺扎得瘙痒难熬,你看,我舌头上到处都是荨麻刺扎出来的小红泡。"山羊哀叹。

"你妈妈没有告诉过你要小心一些吗?"

"她当然说过啊,但我从来不知道被荨麻刺扎到会这么难受。"

"I am so very hungry. I tried these nettles and now my tongue is itching, and it has lots of red bumps all over," laments the goat.

"Your mom never told you to be careful?"

"Of course she did, but I never knew nettles could sting so badly."

"荨麻是我们这些昆虫最好的避难所了,因为荨麻能保护我们不被奶牛或山羊吃掉。你的舌头被扎一次,以后就不会再来这里吃了。"瓢虫笑着说。

"但是,我妈妈告诉过我,这些荨麻有助于治疗疾病,比如关节炎和风湿病。"

"Well, nettle fields are a great refuge for us insects, as there is little danger that we will end up in a cow or a goat's belly. Your tongue will only hurt once. After that you'll never come back for more," laughs the ladybird.

"But Mom did tell me these stings are good for treating health problems, like arthritis and rheumatism."

……我们昆虫最好的避难所……

... great refuge for us insects ...

妈妈告诉我荨麻富含铁

Mom told me, nettles are rich in iron

"你妈妈说的很对。但是,第一,你这么年轻,不会得这些病的;第二,我从来没听说过舌头会得风湿病。"瓢虫咯咯地笑着。

"妈妈还告诉我,荨麻富含铁,铁是我身体里造血必需的元素。"

"She is absolutely right. But, firstly, you are too young to suffer from these illnesses, and secondly, I don't know of any animal with rheumatism of the tongue," chuckles the ladybird.

"Mom also told me nettles are rich in iron, and that is what I need to make lots of blood."

"你有一个非常有智慧的妈妈。"瓢虫点头说道,"你为什么不做荨麻汤喝呢?这样你就不用生吃它们了。"

"我还没有学会怎样做汤呢。汤好像是人类的食物,不是我们动物的。"

"You have a very wise mother," says the ladybird, nodding. Why don't you make a soup of the nettles, instead of eating them raw?"

"I haven't learned how to make soup yet. It seems more like something people eat, not animals."

做荨麻汤

Make soup of the nettles

他们的衣服不扎人吗?

Didn't their clothes sting?

"人类利用荨麻有很长的历史了,"瓢虫认真地说,"他们的王室都曾经穿过荨麻做的衣服呢。"

"啊,那他们的衣服不扎人吗?"小山羊惊讶地睁大了眼睛。

"People have been using nettles throughout history," muses the ladybird. "Their royals even used to be dressed in it."

"But didn't their clothes sting?" The kid's eyes go wide with surprise.

"当然不！他们会先煮荨麻，然后用荨麻纤维来制作漂亮的服饰。荨麻制成的衣服非常结实，过一百多年都不会烂。"

"好神奇啊！"小山羊喊道，"那就是说，爷爷买了一件荨麻制成的衣服，他的孙子的孙子的孙子都还可以穿咯！不过，我猜等他的后代穿的时候，他的衣服一定太过时了。"

"No, of course not. They would first boil it and then use the fibres to make beautiful dresses and cloaks. The clothes were so strong, they would last for a hundred years."

"That's amazing!" cries the kid. "So the clothes a granddad bought would still be dressing his grand-grand-grandchildren. But I'm afraid his clothing must have been out of fashion by the time they got to wear it."

用荨麻纤维来制作漂亮的服饰

Fibres to make beautiful dresses and cloaks

为什么人们要穿纯棉的衣服

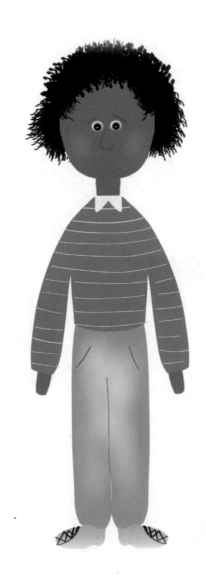

Why do people dress up in cotton

瓢虫笑了。"是的，过去，祖辈留给下一代的衣服是一种特殊的礼物，就像日本妇女那样，她们现在仍然将自己的和服留传给后代，那些和服有几百年的历史了。和服是永远不会过时的。"

"真的吗？那为什么人们要穿纯棉的衣服或石油产品制成的衣服呢？石油这种资源可是不可持续的。"

The ladybird smiles. "Yes, in the old days receiving clothing was a special gift from one generation to the next, just like Japanese women who still hand down their kimonos, hundreds of years old. Kimonos never go out of fashion."

"Really? So why do people dress themselves up in cotton, or in clothing made from petroleum, a resource that can't last?"

"嗯，聪明的小朋友，原谅他们不懂得珍惜现有资源的行为吧！我可不会用我的荨麻来做任何交换的。荨麻不仅给我提供食物和避难所，而且还是一个很好的邻居呢，荨麻丛里到处是漫天飞舞的蝴蝶。"
……这仅仅是开始！……

"Well, my clever young friend, they should be forgiven for not appreciating what they have. I will not trade my nettle field for anything. Not only does it give me food and shelter, it is also a wonderful neighbourhood to live in, one so full of butterflies."
... AND IT HAS ONLY JUST BEGUN!...

……这仅仅是开始!……

...AND IT HAS ONLY JUST BEGUN! ...

Did You Know? 你知道吗?

Nettles have been used for centuries in natural medicine to reduce inflammation and treat allergies.

人类使用荨麻作为缓解炎症、治疗过敏的天然药物已经有几个世纪的历史了。

Nettles taste like spinach. They can be brewed as a tea, blended as a pesto and made into a purée. Nettle soup is common in eastern Europe. In the UK and the Netherlands it is a popular ingredient in cheese making.

荨麻的味道像菠菜。荨麻可以用于制茶，也可以混合成香蒜酱，做成果泥食用。荨麻汤在东欧国家很常见。在英国和荷兰，荨麻经常添加用于制作奶酪。

Nettles have been used for clothing, tablecloths, bedsheets, sandbags, rucksacks and even parachutes. The national dress (gho) of Bhutanese men is made from nettle.

荨麻已经用于制作衣服、桌布、床单、沙袋、帆布包，甚至降落伞。不丹男子的传统服饰（帼）也是荨麻制品。

荨麻是多年生植物，不需要灌溉或杀虫剂。它们生命力顽强，靠自然生长就可收获。没必要像种植庄稼那样来种植荨麻。人们由于对荨麻的历史和用途缺乏认识，而错误地将其当作野草。

Nettles are perennials and need no irrigation or pesticides. They grow abundantly and can be harvested from Nature; there is no need to plant it as a crop. People are ignorant about the history and the use of nettles and therefore classify this plant as weed.

The nettle's sting was developed as a defence against grazing animals. With the rare exception of hungry goats and sheep, grazers will leave the nettles alone, creating a protected area for insects and plants alike.

荨麻的刺是为了抵御食草动物而慢慢进化出来的。除了偶尔有些饥饿的山羊和绵羊之外，食草动物一般都不会去碰荨麻，这就为那些昆虫和类似的植物留下了生存空间。

Ladybirds are beneficial predators living on plants while feeding on mites and aphids. During its lifetime of 3 to 6 weeks each ladybug eats approximately 5,000 aphids.

瓢虫是寄生在植物上的食肉动物，以螨虫和蚜虫为食。瓢虫的寿命为3—6个星期，在整个生命周期中，一只瓢虫大约要吃掉5 000只蚜虫。

Cotton has long replaced nettles as a source of fibre for clothing. Cotton now covers 2.5% of the world's cultivated land. It however uses 16% of the world's agricultural chemicals.

长久以来，棉花取代了荨麻，用于制作衣服。棉花的种植面积已经达到了全世界耕地的 2.5%，使用了全世界 16% 的农药。

Ladybirds are also called ladybugs, but they are neither bugs nor birds. In fact, they are a type of beetle.

瓢虫的英文名为 ladybird 或 ladybug，但瓢虫既不是鸟（bird），也不是臭虫（bug）。事实上，瓢虫是甲虫的一个种类。

Think about It
想一想

Do you think that nettles, even though they sting, play an important role in our lives?

你认为扎人的荨麻在我们的生活中扮演着重要的角色吗?

如果让你穿上荨麻制成的牛仔裤,你感觉怎样?

How would you like to be dressed in jeans made from nettles?

Would you enjoy living in a world without fashion, where you inherit your clothes from your grandparents?

你愿意生活在没有时尚的世界吗?在那里,你穿的是祖辈留下来的衣服。

你如何看待给其他物种以错误的命名?比方说,ladybird 事实上既不是女士也不是小鸟。

What do you think of a language that calls something a name it is not, for instance a ladybird, which is neither a bird nor a lady?

Do It Yourself!
自己动手!

Look for a patch of nettles. Around Europe and North America you will find many. Take the time to make sure that there are no snakes or spiders. Now put on some gloves and pick a bag of nettles. Look under the leaves for ladybirds, moths, butterflies, larvae and aphids. See if you count at least ten different forms of life. When you get home wash the nettles and use them to make a soup. Do you need a recipe? Ask you mom to treat the nettles like spinach and let us know what it tastes like!

寻找一片荨麻。在欧洲和北美,你可以找到很多荨麻。要确保那里没有蛇或蜘蛛。然后戴上手套,摘一袋荨麻。仔细观察荨麻叶子上的瓢虫、飞蛾、蝴蝶、昆虫的幼虫和蚜虫等。看看你能不能数出10种以上不同的生物。回家后,冲洗一下荨麻,然后把它们做成汤。需要食谱吗?问问妈妈如何烹饪菠菜的方法吧,然后告诉大家荨麻的味道哦!

TEACHER AND PARENT GUIDE

学科知识
Academic Knowledge

生物学	荨麻是蝴蝶、蛾、毛毛虫等幼虫的食物来源；野生或家养的大型牲畜不吃荨麻，因为荨麻带刺；瓢虫的自卫机制包括释放化学物质和保护色；如果食物短缺，瓢虫会吃掉自己软体的同胞；传统的农耕技术会在庄稼中留出一小块地方供昆虫来生存繁衍，以保障庄稼免了虫害。
化学	荨麻富含维生素A、维生素C、铁、钾，锰和钙，是治疗风湿病的药物；荨麻多产说明土壤肥沃含氮；荨麻富含三种色素：叶绿素，叶黄素和胡萝卜素，可以用来制作冰淇淋或用于烘焙食品。
物理	荨麻是中空的，里面充满了气体，为荨麻衣服提供了天然的保温效果；夏天时，纤维会聚集得很紧，保温效果降低。
工程学	荨麻纤维是白色丝状的，最长可达50毫米，比亚麻更加柔顺、细滑，是制作纺织物的理想原料；荨麻是一种天然的阻燃剂，可被用作石棉和墙面覆盖物的替代物。
经济学	棉花取代荨麻是由于其纤维丰富，但棉花种植产生了很多意想不到的后果，如大量的灌溉需求、农药的使用、农作物单一种植和转基因等问题；进口物品的选择标准不是质量和用途，而是时尚与消费者偏好。
伦理学	荨麻或许不被归为有机产品，却是在与生态和谐共生的环境中生长的。
历史	世界上第一个荨麻织坊要追溯到青铜器时代的丹麦；伊丽莎白女王一世睡在荨麻床单上；拿破仑的军队穿着荨麻制成的衣服；第一次世界大战时，因为棉花受到限制，德国军队也是穿着荨麻衣服。
地理	荨麻多见于欧洲、北美和亚洲，现在绝大多数的荨麻供给源于尼泊尔。
数学	荨麻与棉花之间的资源效率对比计算；棉花需要农药种植，依赖进口，流行周期短，工人收入低；荨麻保存时间长，不需农药，可以世代流传。
生活方式	用"100%纯荨麻"或其他可再生、储量充足的纤维取代"100%纯（有机）棉"；欧洲传统用餐习惯喜欢将汤作为第一道菜，但现在不这样了。
社会学	瓢虫被叫作ladybird始于中世纪，那时候，农民向圣母玛利亚祈祷，寻求保护庄稼不受害虫侵扰；后来，一种保护庄稼的益虫出现了，人们就给它起名"ladybird"，以纪念圣母玛利亚。
心理学	妈妈的天性是保护，希望自己的孩子远离危险；享受父母与孩子共同进行实时实地的学习的机会。
系统论	荨麻促进生物多样性，保持土壤平衡。

教师与家长指南

情感智慧
Emotional Intelligence

瓢虫

瓢虫是一个细心的观察者，她注意到了小山羊非常饿。她忙碌而又精神集中，并没有将全部注意力从其食物——蚜虫上转移过来。瓢虫想知道是谁在教小山羊这些知识，想到应该是小山羊的妈妈，但是瓢虫也表达了这样的观点，那就是最好的老师也许应该是自己的亲身实践。而且，瓢虫为小山羊提供了时间和实地学习的经验，以及更广阔的视角——荨麻不仅仅是食物。瓢虫细致深入地讲述了关于生活方式的内容，分享了她所生活的生态系统中的作用方式。瓢虫表达了自己对小山羊情感上的认同。她将学习提升到了哲学层面，那就是我们要学会满足于当前自己所拥有的资源。

小山羊

小山羊饿极了，以至于不能克制自己的情绪。他露出了一副很痛苦的表情。他做好了冒险尝试吃一次营养丰富的荨麻的准备，却换来了瘙痒和不适。他坚信自己已经从妈妈那里学到了从健康到食物方面的大量知识，并且知道如何运用这些知识。小山羊意识到自己能力有限（比如做荨麻汤），但我们要知道小山羊还是虚心好学的。他求知欲很强而且很信任瓢虫。他想知道如何能简单应对现代时尚。小山羊还提出了更深刻的问题：人们为什么总在两种坏东西中做选择呢？这也为瓢虫阐述自己的处世哲学创造了氛围。

艺术
The Arts

荨麻富含黄绿到黄这一色调的色素。摘下一些荨麻叶子，放在水中煮沸。从锅里捞出煮过的荨麻，然后让水蒸发掉，这样就能收集到颜料了。当锅里的水蒸发到仅剩1厘米深时，你的颜料就收集好了，然后就可以作画啦！水蒸发得越多，颜色就越深。你可以把颜料混合到香草冰淇淋中，做出黄色或绿色的冰淇淋。

TEACHER AND PARENT GUIDE

思维拓展
Systems: Making the Connections

　　大自然不是孤立的系统，而是由很多相互联系的系统组成的一个复杂体系。生命是相互交织的网络。荨麻是牧场的组成部分，通常生长在牧草旺盛的地带。荨麻将这片牧草变成一个生命力非常旺盛的地带。荨麻不仅为蝴蝶、蛾等小生物提供生存之所，而且还通过展开协作来保护它的生殖繁衍不被饥饿的蚜虫所影响。这也解释了荨麻何以发展成为瓢虫的避难所。高产的荨麻贡献着养料、纤维、彩色颜料、健康产品等甚至更多的东西，却仍然被称作杂草。人们已经忘记了，500年以前，几乎每一个身份显赫的人都是身穿荨麻制品，睡在荨麻制成的床单上。人们还忘记了荨麻提供给我们的充足营养。非常高效、高产的荨麻生态系统，免费为我们提供了许多产品与服务。如果我们还认识不到自己本地现有资源的重要作用，不关心我们周围的生存环境，那么我们就还要依赖进口，而且搞不清这些舶来物是否对人体有害，是否危害着环境。我们的无知导致了很多机遇的丧失，比如促进本地就业、提高经济收入、提升能源利用效率等。种植荨麻不需农药，不需灌溉；加工荨麻也只需加工棉花或化纤产品所需原料的一小部分。是时候重新认识并利用我们周围现有的资源了。荨麻生命力旺盛，即便是经常收割，也能很快再长出来。我们需要增强发现并利用现有资源的能力，扭转资源长期短缺的困境，满足每个人的生存必需。

动手能力
Capacity to Implement

　　针织物不仅用于制作衣服，也用来制作书包、鞋、床品以及墙面装饰物。现在，我们要利用一种特殊纤维制成的针织物来设计一款书包。这种纤维在大自然中非常常见，不需人工种植，却可以轻松收获。你的第一个任务是找到这种纤维，第二个任务是想出把它从纤维变成纤维织物的办法，比如机器加工或针织的办法，然后做成书包售卖你的产品，让那些帮助你实现梦想的人健康幸福地生活。现在，你有行动方案了吗？

教师与家长指南

故事灵感来自
This Fable Is Inspired by

西比拉·索隆多
Sybilla Sorondo Myelzwynska

1963年，西比拉·索隆多出生于美国纽约，17岁时获得在伊夫圣罗兰设计工作室工作的机会。她致力于设计独特服饰，创立自己的品牌。在历经20多年成功的职业生涯后，西比拉决定休息一下，并出售了自己商业中的大部分收益。在思考了自己对地球、社会的责任后，她决心建立针织物自由联盟，旨在促进纺织业对可再生能源的利用以及确保原材料质量。西比拉的愿望不仅是制作漂亮舒适的衣服，还要让衣服背后承载一个美丽的故事。西比拉承诺，要消除服饰面料的有毒成分，寻找转基因棉花的替代品，减少对农药的依赖。她还承诺要改善生产流程，为工人创造更好的工作条件。西比拉要提供更多充满关爱、生态友好的纺织品。值得一提的是，西比拉烹制的荨麻汤非常美味，这些荨麻是她家周围牧场里种植的。

图书在版编目(CIP)数据

冈特生态童书.第二辑修订版:全36册:汉英对照 /
(比)冈特·鲍利著;(哥伦)凯瑟琳娜·巴赫绘;
何家振等译.—上海:上海远东出版社,2021
书名原文:Gunter's Fables
ISBN 978-7-5476-1759-5

Ⅰ.①冈… Ⅱ.①冈… ②凯… ③何… Ⅲ.①生态
环境-环境保护-儿童读物—汉、英 Ⅳ.①X171.1-49

中国版本图书馆CIP数据核字(2021)第213075号

著作权合同登记号图字09-2021-0823

策　　划　张　蓉
责任编辑　程云琦
封面设计　魏　来　李　廉

冈特生态童书
带刺的荨麻
[比]冈特·鲍利　著
[哥伦]凯瑟琳娜·巴赫　绘
李原原　田　烁　译

记得要和身边的小朋友分享环保知识哦!
八喜冰淇淋祝你成为环保小使者!

Food 46

都市农场

Farming in the City

Gunter Pauli

[比] 冈特·鲍利 著
[哥伦] 凯瑟琳娜·巴赫 绘
李原原 田烁 译

上海远东出版社

丛书编委会

主　任：贾　峰

副主任：何家振　闫世东　林　玉

委　员：李原原　祝真旭　牛玲娟　梁雅丽　任泽林
　　　　王　岢　陈　卫　郑循如　吴建民　彭　勇
　　　　王梦雨　戴　虹　翟致信　靳增江　孟　蝶

特别感谢以下热心人士对童书工作的支持：

匡志强　宋小华　解　东　厉　云　李　婧　陈　果
刘　丹　熊彩虹　罗淑怡　旷　婉　杨　荣　刘学振
何圣霖　廖清州　谭燕宁　韦小宏　李　杰　欧　亮
陈强林　王　征　张林霞　寿颖慧　罗　佳　傅　俊
胡海朋　白永喆　冯家宝

目录

都市农场	4
你知道吗?	22
想一想	26
自己动手!	27
学科知识	28
情感智慧	29
艺术	29
思维拓展	30
动手能力	30
故事灵感来自	31

Contents

Farming in the City	4
Did You Know?	22
Think about It	26
Do It Yourself!	27
Academic Knowledge	28
Emotional Intelligence	29
The Arts	29
Systems: Making the Connections	30
Capacity to Implement	30
This Fable Is Inspired by	31

在一座大城市中，一个草莓和一支芦笋正在寻找栖身的地方。草莓建议到市中心去找找看。

In a big city, a strawberry and an asparagus are out looking for a place to settle and grow. The strawberry suggests they look downtown.

寻找栖身的地方

A place to settle and grow

几乎没有土壤

Hardly any soil

"你不会是疯了吧!"芦笋拒绝了草莓的主意,"你知道的,市中心除了混凝土和肮脏的空气之外,什么都没有。另外,那里几乎没有土壤。"

"可是,你总归是需要水的,对吗?"草莓问道。

"You are out of your mind!" The asparagus balks at the idea. "You know there's nothing downtown but concrete and dirty air. And there's hardly any soil."

"Well, you need water, right?" asks the strawberry.

"那当然，我们植物如果没有水就不能生存了。但是，在市中心，雨水中缺乏矿物质，而且直接从排水管流进了下水道，然后就进入大海了。"

"那水几乎流不进大海里。"草莓回答道，"但是，我可以保证，在市中心的屋顶上能够喝到全部我们所需要的水。"

"Of course, we plants cannot survive without water. But downtown, the rainwater is lacking minerals, and it flows from gutters straight into the sewage, and then into the sea."

"It hardly gets to the sea," responds the strawberry. "But I can guarantee that we can get all the water we need from the roofs downtown."

在屋顶上喝到我们所需要的水

Get water we need from the roofs

你难道从来没有见过温室吗?

Have you never seen a greenhouse?

"好吧,不过假如你是在那些天寒地冻的北方城市,比如蒙特利尔或者北京,你要怎样熬过冬天啊?"

"你难道从来没有见过温室吗?"

"OK, but if you're looking at these cold and freezing northern cities like Montreal or Beijing, how will you get through winter?"

"Have you never seen a greenhouse?"

"当然见过。"芦笋说,"在温室里,农民让你在水中生长,让你温暖地过冬。"

"我们走吧!在城市里为自己建造一座温室,然后在泡沫玻璃上安家。"草莓说道。

"Well, yes," says the asparagus. "That's where the humans make you grow in water and keep you warm over the winter."

"Let's go and build ourselves a glass house in the city, and then let's settle down to grow on foam glass," says the strawberry.

我们走吧！在城市里为自己建造一座温室

Lets go and build ourselves a glass house in the city

谁来为我们买单啊?

Who will pay for all this?

"有干净的泡沫玻璃和暖和的温室当然好啊,"芦笋喃喃道,"可是,谁来为我们买单啊?"

"嗯,首先,总有一群人喜欢吃新鲜的芦笋和草莓。你要知道,如果我们覆盖了整个屋顶,我们可以同时为1000个人提供食物!"

"听上去不错,可这还是不够为我们的房屋买单啊。"

"It's all very well having clean foam glass and warm greenhouses," the asparagus grumbles, "but who's going to pay for all of this?"

"Well, first of all there are humans who will love to eat us fresh. You know, if we cover a whole roof, we could offer food for 1,000 people all the time!"

"That sounds great, but that won't be enough to pay for our house."

"不要急着下结论。我们寻求的是双赢,而不是单方面得到好处。"草莓说道。

"但是,超市得支出更多的成本来包装、冷藏、运输我们。这样,我们就成输家了!没有人愿意买我们!"

"超市和市中心建筑物的主人都会是赢家。"

"Don't jump to conclusions. We should be looking for a win-win, not a win-lose," says the strawberry.

"But the supermarkets spend so much money packaging, cooling and shipping us. So we will lose! And they won't like us!"

"They could be winners along with the owners of these downtown buildings."

双赢，而不是单方面得到好处

Win-win, not a win-lose

农场在他们的房顶上

Farm on their roofs

"人们怎么会希望有座农场在他们的房顶上呢？"

"因为你和我可以让房屋冬暖夏凉，而他们还能靠健康食品来赚钱。"

"这样的话，我们帮助他们节约能源和资金，又给他们创造了在市中心就可以吃到新鲜蔬菜和沙拉的机会。有道理。我同意。"

"Why would they want to have a farm on their roofs?"

"Because you and I keep the roof cool in the summer and warm in the winter, while making money on healthy food."

"So we help them save energy and money, and give them a chance to eat the freshest veggies and salads in town. That makes sense. I buy that."

"如果我们在屋顶上快乐生长,开心得想唱歌了,那会怎么样呢?"

"希望人们能加入我们齐声合唱!"

……这仅仅是开始!……

"And what happens if we're growing so happily up there that we want to sing a song?"

"Hopefully people will come and join the chorus!"

... AND IT HAS ONLY JUST BEGUN!...

……这仅仅是开始!……

... AND IT HAS ONLY JUST BEGUN! ...

Did You Know?
你知道吗？

By 2010, more than 50% of the people in the world lived in cities, and it is expected to increase to 75% within the next generation.

截至2010年，世界上50%的人口生活在城市中，预计到下一代会增加到75%。

Cities do not produce food. Worse, the biological waste produced in cities does not return to the land, thus losing nutrients to produce food in the future.

城市不会生产食物。更糟糕的是，城市中产生的有机垃圾并没有回到土壤中，从而流失了可以在未来生产食物的养分。

Farming does not have to be done on soil. You can farm on humid air (aeroponics) or on water (hydroponics).

农业不一定非要在土壤中进行，在潮湿的空气中（气栽法）或水中（水耕法）都可以进行。

If all commercial flat roofs were used for farming, then a city could feed 25% of its inhabitants, and would employ 12 people for every 1,000m² farm.

如果所有的商业建筑平顶都用于农业种植，那么一个城市就可以解决25%的居民的温饱问题，每1 000平方米的农场还可以解决12个人的就业。

Farms on the roof save packaging and transport, re-use rainwater, and save energy by keeping buildings cool in summer and warm in winter.

屋顶农场节省了包装和运输环节，实现了雨水的再利用，还可以让建筑物冬暖夏凉，节约能源。

Urban farms do not have to be limited to rooftops; vertical farming could dramatically increase output, provide shade and increase air quality.

城市农场并不仅限于屋顶；垂直耕种可以大量增加产量，提供阴凉，改善空气质量。

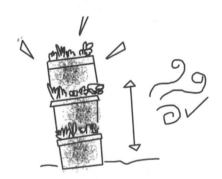

Storing and transporting vegetables over long distances increases the risk of molds, yeast and bacteria, requiring packaging and chemical controls, thus increasing cost and pollution.

长距离储存、运输蔬菜增加了滋生霉菌、酵母菌和细菌的风险，对包装和化学制剂的需求又增加了经济成本和污染。

Foam glass is made from recycled glass bottles. It weighs 10 times less than soil and can be recycled forever. It is ideal for rooftop farming.

泡沫玻璃由回收的玻璃瓶制作而成。它比土壤轻10倍，而且可以永久地循环使用，是屋顶农业的理想材料。

Think about It
想一想

Would you rather eat vegetables from a faraway farm, or from a rooftop around the corner from your home?

你更愿意吃哪一种蔬菜,是从远方农场运来的,还是自家附近屋顶上种植的?

如果是来自附近的食物,你希望它被包装得像超市里的那样吗?

When food comes from around the corner, do you want it packaged the same way as supermarkets do?

Should we use rainwater in cities, or does it belong to the sea?

我们应该利用城市中的雨水,还是就让它回归大海呢?

你给草莓的建议是什么,是住在城市里,还是回到农田中?

What would you recommend to the strawberry: Live in the city, or grow on a field?

Do It Yourself! 自己动手!

Have you ever farmed strawberries? It's easy! Go and buy some seeds, a small pot and some soil, and then plant the seed. One important lesson: water only from the bottom since overhead water could rot the crown of the plant and damage the fruit. Make certain that the room is not too warm, since a hot room (>16°C) will inhibit flowering. The tricky part will be to pollinate the flowers by hand, using a soft paintbrush. You will have to learn to recognise the difference between a male and a female flower – once you have identified them, gently brush around the male flower, and swirl the pollen onto the female flower. As the strawberries grow, put some straw on the bottom so they remain clean. After the harvest, cut the old leaves and expect the strawberry to give you fruit for the next three years.

你种过草莓吗?非常简单!买一些草莓种子、一个小花盆,再准备一些土,然后在花盆里撒上种子。重要的知识:只能在底部浇水,因为如果从上面浇水的话,水会让花冠腐烂,也就弄坏了果实。别让房间温度太高,超过16℃就会抑制开花。比较难的环节是给花授粉,可以用一个软毛的画笔来操作。你要学会如何区分花朵的雄蕊和雌蕊,找到花的雄蕊后,轻轻地用刷子刷一下它,然后将刷子上沾到的花粉再刷到雌蕊上。随着草莓逐渐生长,在它们的底部放一些稻草来保持清洁。收获后,将老叶子剪掉,这样以后三年中,还会长出草莓。

TEACHER AND PARENT GUIDE

学科知识
Academic Knowledge

生物学	许多植物更喜欢通过根部吸收水分；从顶部浇水有可能会损伤植物和果实。空气污染影响光合作用，因而阻碍植物正常生长。温室可以用来建造室内植物花园，保持生物多样性。植物也会表达感情，植物开花与发芽的模式主要受绿光影响；光照频率影响生物的数量和营养量；高波段的红光可以增加西红柿的产量，增加菠菜中维生素C的含量。
化 学	雨水中缺乏矿物质，如果要让植物生长，还需再给雨水添加养料。
物 理	温室通过控制植物根叶之间的温度差来控制水的渗透；温室的LED灯可为植物生长提供光源；老式廉价的钠元素灯是做不到的，而且还会以热量转化的形式损失能源。
工程学	温室可以保护庄稼免受寒冷、炎热、沙尘和虫害的侵扰；在根茎加热系统中，碳纤维保持着根叶之间的温度差异，优化渗透效果；一些温室只需12V直流电便可运行。
经济学	研究并对比本地原产食物的成本与外地生产的食物成本（包装、加工、冷藏、运输、仓库保管）；利用商业建筑屋顶空间需要另外的建筑结构方面的投资，但是也可以带来收入，节约能源。
伦理学	如果城市从土地中获取了营养，比如水果、蔬菜，那么城市就要反哺土壤，以便为子孙后代留下可持续发展的空间；重新补给土壤不会导致土壤退化，而合成化肥则做不到这一点。
历 史	罗马帝国皇帝提比略全年都有黄瓜吃，园丁将黄瓜种植在马车上，白天推到室外，晚上再推进室内；在法国，温室的术语为"橘园"，因为它最初是用于保护橘子度过寒冷的冬天。
地 理	荷兰是世界上温室密度最高的国家；温室单位面积产量最高的欧洲国家是西班牙，世界上面积最大的温室在中国。
数 学	"食物里程"用于计算食物收获之地与餐桌之地之间的距离；计算一下你所买的食物的食物里程吧。
生活方式	尽管与时令不符，人们还是希望全年都能吃到新鲜的水果，这种对反季节水果的需求加剧了南北半球之间的食品贸易。
社会学	在英国维多利亚时期，拥有一个玻璃温室种植花草和非应季水果是许多家庭社会地位的象征。
心理学	自力更生可以给人自信，并激发应变能力；仇外心理是指对外来、陌生或不解的事物的不明原因的恐惧。
系统论	城市食物的供给不仅可以通过利用建筑空间来改善生活、节约能源，还可以通过改进包装与运输来实现。

教师与家长指南

情感智慧
Emotional Intelligence

草莓

草莓提出了一个新奇的冒险想法：去城市中生存。当芦笋对这个建议提出反对意见时，草莓并没有变得保守，而是以一个中立的问题加以回应，所持立场也十分坚定。当芦笋提出了关于寒冷的保留意见时，草莓以另一个问题加以回应。草莓分享了关于玻璃和泡沫玻璃的信息，这些知识是芦笋熟知的，因此更容易达成一致。然而，还有成本的问题没有说清楚。草莓基于事实给予了回答，并给出了策略规划：共赢，共赢的主体也包括那些持反对意见的人和那些起初注定要失败的人。草莓消除了芦笋的顾虑，因为实施这个提议能让所有人获利。当草莓发现芦笋已经被说服后，她露出了喜悦的情绪，提议一起唱首歌，这也是达成共识的表现。

芦笋

芦笋非常自信，起初他与草莓的关系很紧张。这使得芦笋在第一次听到去城市生活的提议时表现得很直接，不知收敛，没有意识到对草莓的冒犯。芦笋草率地回答草莓的问题，并坚持自己消极的态度。当芦笋不得不接受草莓的逻辑后，他又不断提出新的否定的观点。当芦笋和草莓在水、温室的态度上基本达成一致后，芦笋开始担心谁将承担成本，而且意识到仅靠食物的收入是不足以平衡收支的。共赢的策略规划并没有立即让芦笋接受，只有当芦笋明白了提议的逻辑后，他才重新变得自信起来；他建议让每个参与者都来庆祝这个新方案。

艺术
The Arts

下载一些城市的照片，找一些商业中心的鸟瞰图。把这些图片打印在一张大纸上，然后在房顶画上花园和温室。结果大有不同吧！绘画能帮你设计一些可能出现的场景。现在，你只需做出一个选择：让它变成现实！

TEACHER AND PARENT GUIDE

思维拓展
Systems: Making the Connections

　　城市总有一天会变成世界75%人口的栖息之地。如果我们继续沿用今天的生产和消费模式，那么城市贫民区将会越来越多，贫穷与饥饿会四处蔓延，导致出现不安全感与仇外情绪。今天，我们在城市以外的区域制造食物，并依赖交通来运输这些食物。城市规划者已经将城市划分为居民区、商业区和工业区，这也迫使人们和物品不得不总是在城市中来往穿梭，引发了大量交通问题。城市农业为重新整合这些高耗能、高成本的行业提供了机遇。由于城市中没有廉价的土地，所以要开发其他空间：商业和工业建筑的屋顶正在招手呼唤呢。食物的成本不能简单地比较计算，如比较屋顶种植1千克草莓的成本和农业化种植、运输、储藏1千克草莓的成本。为了理解这个提议的优势，通过透明的方式来比较价值是非常重要的。业主基础设施的成本会因能源的节约而被分摊，但最重要的额外价值是建筑有了增加额外收入的能力，这转化成了财产的账面价值或资本收益。再进一步，城市农业还有很多潜在的吸引力，比如增加游客量（额外收入），这对于购物中心来说尤其可贵，增加了业主的原始价值。对环境的红利则是，将黑漆漆的柏油屋顶（满足绝缘和防水的需求）变成了绿色屋顶（植物在此释放水和氧气，吸收二氧化碳），进而将建筑物变得冬暖夏凉。如果这些好处不限于一处建筑物，而是扩展到整个区域，那么城市会节约很多能源，扭转当地气候变化趋势。建筑物屋顶上生产的农产品也可以用于当地饮食消费。

动手能力
Capacity to Implement

　　你所在的城市中有多少空屋顶呢？找找那些工厂、配送中心还有购物中心的广阔屋顶，你一定会为这么大面积的宽敞空地而感到惊讶。（并不是所有屋顶都可用，因为有些屋顶不够坚硬，有些没有便利的通道。）策划一个方案，将10%的空地变成农场。你可以生产多少食物？你可以创造多少个就业岗位？你会成立一个团队来开展这项工作吗？如果你觉得自己人年轻了，问问自己这个问题吧：你觉得自己多大年纪才可以种植草莓？

教师与家长指南

故事灵感来自
This Fable Is Inspired by

穆罕默德·哈格
Mohamed Hage

穆罕默德·哈格是一位屡创商业奇迹的年轻企业家。鲁法农场是他在2002年创立的第三家公司。他最早的公司——赛普拉传媒是许多加拿大企业的E-mail服务商。他在激发员工工作热情方面很有天赋，而且在联络不同领域专家方面有很多诀窍。鲁法农场的创意源于哈格童年时的一段经历，那时新鲜的食物就产自贝鲁特城外的村庄中。他和劳伦·拉斯迈尔、叶海亚·巴德兰组成了团队。劳伦·拉斯迈尔是一位经过专业培训的生物化学家，他采用生物方法来预防虫害，调节微气候，进行作物轮作和农作物品种选择。叶海亚·巴德兰是一名专业的建筑工程师，为鲁法农场的成功提供了必要的专业技术。

图书在版编目(CIP)数据

冈特生态童书.第二辑修订版:全36册:汉英对照 /
(比)冈特·鲍利著;(哥伦)凯瑟琳娜·巴赫绘;
何家振等译.—上海:上海远东出版社,2021
书名原文:Gunter's Fables
ISBN 978-7-5476-1759-5

Ⅰ.①冈… Ⅱ.①冈…②凯…③何… Ⅲ.①生态
环境-环境保护-儿童读物—汉、英 Ⅳ.①X171.1-49

中国版本图书馆CIP数据核字(2021)第213075号

著作权合同登记号图字09-2021-0823

策　　划	张　蓉
责任编辑	程云琦
封面设计	魏　来　李　廉

冈特生态童书
都市农场
[比]冈特·鲍利　　著
[哥伦]凯瑟琳娜·巴赫　绘
李原原　田　烁　译

记得要和身边的小朋友分享环保知识哦!
八喜冰淇淋祝你成为环保小使者!

Housing 47

另一面的阳光
Sun on Both Sides

Gunter Pauli

[比]冈特·鲍利 著
[哥伦]凯瑟琳娜·巴赫 绘
唐继荣 译

上海远东出版社

丛书编委会

主　任：贾　峰
副主任：何家振　闫世东　林　玉
委　员：李原原　祝真旭　牛玲娟　梁雅丽　任泽林
　　　　王　岢　陈　卫　郑循如　吴建民　彭　勇
　　　　王梦雨　戴　虹　翟致信　靳增江　孟　蝶

特别感谢以下热心人士对童书工作的支持：

匡志强　宋小华　解　东　厉　云　李　婧　陈　果
刘　丹　熊彩虹　罗淑怡　旷　婉　杨　荣　刘学振
何圣霖　廖清州　谭燕宁　韦小宏　李　杰　欧　亮
陈强林　王　征　张林霞　寿颖慧　罗　佳　傅　俊
胡海朋　白永喆　冯家宝

目录

另一面的阳光	4
你知道吗？	22
想一想	26
自己动手！	27
学科知识	28
情感智慧	29
艺术	29
思维拓展	30
动手能力	30
故事灵感来自	31

Contents

Sun on Both Sides	4
Did You Know?	22
Think about It	26
Do It Yourself!	27
Academic Knowledge	28
Emotional Intelligence	29
The Arts	29
Systems: Making the Connections	30
Capacity to Implement	30
This Fable Is Inspired by	31

一株老鹳草正在南非桌山的阳光下享受美好的一天。一条毛毛虫从老鹳草的叶片上爬过，在其中一片叶子下方安顿下来，想打个盹。叶片下方是纳凉的好地方，能避开火辣辣的阳光，周围的鲜花也让这里成为一片美丽的休憩之处。

A geranium is enjoying a day in the sun on Table Mountain when a caterpillar crawls over one of its leaves and settles on the underside for a nap. The cool spot in the shade provides cover against the hot sun and the bright flowers make it a beautiful place for a rest.

正在阳光下享受美好的一天

Enjoying a day in the sun

不用多久，太阳不仅会照到叶子顶面，还会照到叶子底面

Soon the sun will be shining below as well as on top!

"要是你认为待在那里又安全又凉爽,那你可得小心了!不用多久,太阳不仅会照到叶子顶面,还会照到叶子底面。"老鹳草提醒道。

"那不可能!"毛毛虫回答道。"只要太阳照在叶片上方,那么叶片底面总会有阴凉地儿。这是显而易见的。"

"If you think you can stay safe and cool there, beware, because soon the sun will be shining below as well as on top!" remarks the geranium.

"That cannot ever happen!" replies the caterpillar. "If the sun shines on top, then there will always be shade below. That is quite obvious."

"看上去似乎的确如此,但我身体下方的地上有一滩水,水会将阳光向上反射。一旦被反射的光线照到我的叶片底面,你就得享受一份'温暖惬意'了。这就像人们使用的新型太阳能电池板一样。"

"享受?我失去了安全又凉爽的休憩处,怎么还能享受啊?"

"That may seem true, but there is a pool of water on the ground below me, and the water will reflect the sunlight back up. If the reflected light hits the bottom of my leaf you could enjoy some lovely warmth. It works just like with these new solar panels people use."

"Enjoy? How could I enjoy losing my safe and cool resting spot?"

水会将阳光向上反射

Water will reflect the sunlight back up

太阳能板只有顶面能捕获太阳能

Solar panels capture solar power only on top

"当然，你可能会不太高兴，是吧？或许你得再找别处休息了。但你相信吗，人们安放在屋顶用来捕获太阳能的所有太阳能板，只有顶面能照到阳光，底面从来接收不到阳光照射？"

"Well, you would not be too happy, would you now? You may have to find another resting spot. But can you believe that all these solar panels that people has been putting on their roofs to capture solar power only receive sun on top, never below?"

"是因为人们希望模拟自然吗？你们植物也只是利用其中一面的光能，对吧？"

"这与是否向大自然学习无关，更多地在于如何利用现有的资源。如果两个表面都能利用好，就能获得双倍的能量。"

"Is that because they want to imitate nature? You plants only use the sun on one side, don't you?"

"This has nothing to do with learning from nature. This has more to do with using what you have. By using both surfaces, you can get double the power."

如果两个表面都能利用好，就能获得双倍的能量

By using both surfaces, you can get double the power

我们应该尽可能多地利用现有的资源

We have to do more with what we have

"你是建议让这些用沙子做的太阳能板的两面都得到阳光的照射吗?"

"没错!如果你能同时从太阳能板的顶面和底面获得能源,为什么只利用顶面呢?我们应该尽可能多地利用现有的资源。"

"So are you suggesting that these solar panels made from sand should have the sun shining on both sides?"

"Absolutely! If you could make energy on top and below, why would you only use the top? We have to do more with what we have."

"瞧，我喜欢阴凉地儿。没有阴凉地儿，就太热啦，甚至感觉要烧起来！而且我听说，如果太阳能电池变得过热，它们就不能正常工作了。"

"如果你热过头了，会怎样做呢？"

"Look, I like the shade. If there is no shade then it gets too hot. You could even burn up! And I have heard that if solar cells get too hot, they don't work so well."

"What do you do when it gets too hot?"

如果你热过头了,会怎样做呢?

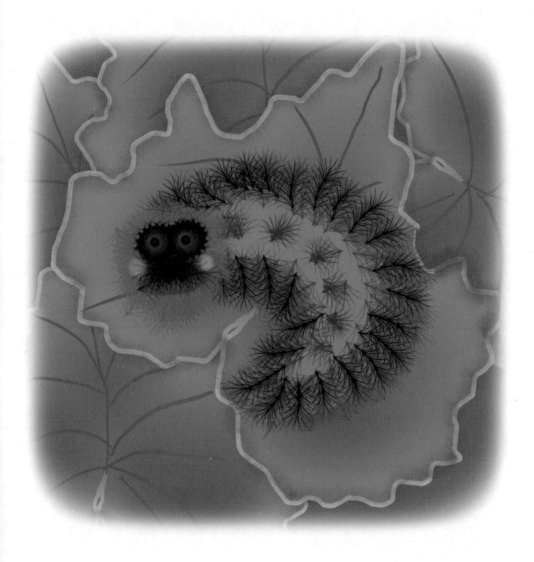

What do you do when it gets too hot?

凉爽又清新的身体

A cool and fresh body

"我？嗯，我喜欢去冲个澡。"

"对呀！如果太热，你会用水来降温。然后，你会得到什么？"

"凉爽又清新的身体，不是吗？"

"Me? Well, I like to take a shower."

"Exactly! When it is too hot, you use water to cool down, and then what do you have?"

"A cool and fresh body?"

"就这些吗？"老鹳草追问道。

"哦，我懂了！这样的话，房子就能获得更多能源，同时还有热水。这热水还可能很干净！这将是未来的生活方式。"

……这仅仅是开始！……

"Is that all?" wonders the geranium.

"Oh, now I get it. A house can have more energy and have hot water. It can even be clean, hot water! That is the way to live in the future."

... AND IT HAS ONLY JUST BEGUN!...

……这仅仅是开始！……

... AND IT HAS ONLY JUST BEGUN! ...

Did You Know?
你知道吗?

Traditional solar panels only use the power of the sun on one side, and never on both sides.

传统的太阳能板只有一面能利用太阳能，从未两面同时利用。

Optics will allow the sun to shine directly on to the top of the panel, and shine in a concentrated way on the bottom.

光学设备将使得太阳不仅能直接照射太阳能板的顶面，而且能以某种方式将光线聚焦于它的底面。

If solar cells are kept at 50° C while exposed to the sun, they generate the most electricity.

暴露于阳光下的太阳能电池，当它们的温度保持在 50℃ 时，发电效率最高。

A car engine gets hot through combustion, but thanks to water-cooling, the engine keeps on working.

汽车发动机由于燃烧而变热，但水冷设备让它保持运转。

Tiny water pipes inside a photovoltaic panel maintain the panel at a constant temperature, while generating hot water.

光伏电池板内的微小水管让电池板保持恒定温度，同时产生热水。

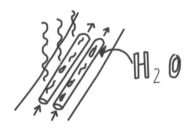

Solar energy streams like a direct current (DC), while our electric network operates on the alternate current (AC) standard.

太阳能像直流电一样流动，而我们的电力网络系统则是按交流电的标准运行的。

*C*hargers for computers, toys and games do not only charge, they also convert the electricity from AC (the power of the grid) to DC (the power system of solar panels).

计算机、玩具和游戏机的充电器不仅仅用于充电，还将电流从电网系统中的交流电转换为像太阳能板系统一样的直流电。

A solar panel made of sand generates more energy (thermal) from heating and cooling water than from the electricity (power) it generates.

用沙子做成的太阳能板，通过加热和冷却水产生的热能比它发电产生的能量还要多。

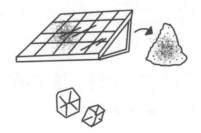

Think about It

How competitive would solar panels be if they generated double the amount of energy, some as electricity and some as heat?

如果太阳能板产生的能量可以翻倍，其中电能和热能各占一部分，它的竞争力会如何？

How might engineers look at making photovoltaic cells thinner and lighter to save energy, while thicker cells generate double the revenue?

工程师们会怎么看这个问题，是将光伏电池变得更薄更轻来节约能源，还是将太阳能电池变厚来获得双倍的能量？

Does specialisation, with one set of solar experts for electricity and another set of solar experts for hot water, have the effect that people end up buying two systems instead of one?

随着专业化程度的加深，太阳能专家中有的专注于发电，有的专注于产生热水，这是否会导致人们最终购买两套太阳能电池板系统，而不是一套？

What are energy efficient ways to cool down when it is hot?

处于高热状态时，怎样降温才是节约能源的？

Do It Yourself!

自己动手!

Take a piece of black metal (if you can't find black metal, then paint it black). It should be at least 10 cm long and 2 cm wide, but the bigger the better. Then find some aluminium foil that will reflect the sun from an open space above, onto the bottom of the black metal plate. Find out, through trial and error, firstly the right angle and secondly the best curve of the aluminium foil so that the sunlight is reflected onto the bottom surface of the panel for the longest possible time. You should design the foil so that the sun is concentrated on the bottom surface for at least four hours. It may be a good idea to work in four or five teams and share your experiences to find the best solution faster. It may not be easy, but it will be fun!

找一块黑色金属（如果没有，就把其他金属涂成黑色）。尺寸应不小于10厘米长、2厘米宽，越大越好。找些能向上反射阳光的铝箔，使它们能在开阔地方将阳光反射到黑色金属板的底面。通过反复试验，首先找到正确的角度，其次是铝箔的最佳弧度，从而令阳光反射到黑色金属板底面的时间最长。设计好铝箔后，确保阳光能集中照射到黑金属板底部至少4小时。不妨分成4—5组分别进行试验，大家分享经验，从而更快找到最佳解决方案。这可能不太容易，但会很有趣。

TEACHER AND PARENT GUIDE

学科知识
Academic Knowledge

生物学	南非凡波斯植被的生物圈；历史上曾将老鹳草与天竺葵类植物的名称混淆。
化 学	如何将沙子（硅）转化为石英（二氧化硅）；用盐酸和三氯硅烷来提纯石英；与铅有关的健康危险。
物 理	光伏电池发电的理想温度；通过紫外线和加热来净化水；水中的热传递；用热泵从冷源产热，或从热源制冷。
工程学	如何把同一个装备中的热水生产与能源生产相结合；将基于某类工业标准（汽车）上的技术和创新集群应用到另一类工业（太阳能）上；如何将剧毒和能源密集的太阳能电池生产过程转变为可持续的过程。
经济学	生产力计划；用你现有的资源生产更多的东西；关注核心竞争力，要把那些显而易见但在中心范围之外的机遇创造性地转化成核心业务。
伦理学	做得不那么糟糕，仍然是糟糕；若有机会做得更好却拒绝去尝试，那是真的很糟糕；如果一个微小的改变就让它们变得非常具有竞争力，那么还能认为有必要对它进行补贴吗？
历 史	镜子最早于公元前8000年在土耳其使用；希腊人知道凹凸镜的特性；亚历山大·贝克勒尔是首位观察到光伏效应的人（1839年）；1894年，首个太阳能电池专利被授予梅尔文·西弗里。
地 理	南非桌山和好望角；地球哪个部分接收到最多的阳光照射？北极的太阳能产业会有竞争力吗？
数 学	几何光学；反射与折射。
生活方式	我们接受了只利用太阳能板一面的标准做法，从不质疑它的效率；对于单个的光伏电池，我们除了希望获得更多的电能输出之外，也没有强制要求它表现出其他方面的性能。
社会学	社会对可再生能源的接受较慢，似乎需要政府干预。
心理学	心理性近视：我们被培养成只关注我们自己希望和需要的事物，对其他任何不符合我们固有逻辑的观点却缺乏思考的能力和胸怀。
系统论	我们还处于脱离实际的学术象牙塔中，只进行着非常集中但相互隔绝的科学研究。

教师与家长指南

情感智慧
Emotional Intelligence

毛毛虫

毛毛虫知道自己希望得到什么——只有阴凉而已,他不相信任何有可能威胁到自己的安乐窝的事物。他只依据自身的利益来考虑问题,对其他现实情况视而不见。老鹳草迫使毛毛虫去观察自己的降温行为。这有助于激励毛毛虫去聆听、思考,并超越过去的思维定势。毛毛虫通过亲身体会,理解了老鹳草一直在设法分享的逻辑和思想。由于获得了新的见解,他现在不仅被说服了,甚至变得很热心。

老鹳草

老鹳草处于一种静坐不动的生活状态。他知道自己无法移动,因此花时间观察周围的各种现实情况。他对毛毛虫抱有同情心,警告毛毛虫不久将失去阴凉。当毛毛虫表示难以置信时,老鹳草花时间解释了为什么毛毛虫所处的位置不能长时间保持阴凉。老鹳草扩展了这一讨论主题,进行了更深入的独到反思。他将讨论引向了更广泛的话题,这种思考方式很有借鉴意义,甚至还提出一些称得上人生观的核心原则。老鹳草随后从解释转为提问,质疑毛毛虫的行为。这种方式不仅吸引了毛毛虫的全部注意力,而且帮助老鹳草引导毛毛虫获取新信息,最终让毛毛虫形成了对事物新的积极态度。

艺术
The Arts

找到一个你想用作手绘目标的物体,先以它为模特绘制一幅简单的素描。然后,通过凸透镜观察这个物体,并将在镜子中看到的景象画下来。再通过凹透镜观察同一个物体,并将镜子中看到的景象画下来。最后将三幅画放在一起,你会观察到什么现象?

TEACHER AND PARENT GUIDE

思维拓展
Systems: Making the Connections

太阳每天都在发光。即便云层很厚，仍然可以看到太阳光。据说只需到达地球土壤和水体的太阳总能量的1%，就可以满足我们的全部能源需求。然而，不可能把整个地球都盖上太阳能设备，所以我们需要用最有效率的方法来处理此事。当我们应用光学技术将太阳光汇聚到太阳能板底部时，若能把微小的水管置于太阳能板内部来给光伏电池降温，那么为了容纳这些小管，这种太阳能板会比正常的太阳能板更厚。从发电的角度来说，这将增加原材料的成本。但另一方面，这种太阳能电池"三明治"中的小管，却能够以非常低的成本产生热水。这种综合考虑热水和电力的技术，将从太阳光中捕获更多的能量，而额外的材料成本却远低于新增的能量价值。因为高热可以杀死水中的细菌，所以水一旦被加热，同时也会被净化。为了在太阳能电池底部产生聚焦能量的镜面效果，在太阳能板和反射装置的底部之间有20厘米的空隙，这个空间充满了空气。如果将太阳能系统用作建筑的实际屋顶，而不是放置在屋顶上，这个空间就可以当作隔热层，帮人们省去购买保温或御寒的特殊绝缘材料的花费。实际上，最初的太阳能电池已经逐渐发展成一个提供许多好处的系统，包括电能、热水、用作屋顶、隔热和净化水质。有了可再生能源，房主能节约资金去进行其他投资，也能确保我们的生活品质更高、更健康。

动手能力
Capacity to Implement

你们学校使用太阳能吗？你们学校使用任何形式的可再生能源吗？为什么不请求约见校长，询问给建筑加热和降温需要多少能源，以及全年的照明需要多少电力？或许校长会告诉你用于加热、降温和照明的总预算。如果学校所有建筑的屋顶都装上光伏电池做成的太阳能板，估算这将产生多少电力。然后，如果所有可利用的屋顶空间都装上太阳能板，将产生多少电力和热量。学校是否有足够的屋顶空间来进行太阳能利用，使得学校可以不依赖外部电网？请列出你的论据，系统阐述你的看法，然后看能否说服校长。

教师与家长指南

故事灵感来自
This Fable Is Inspired by

斯特凡·拉松
Stefan Larsson

斯特凡·拉松学习的专业是计算机科学和电气工程。他曾经在太阳能和生物燃料领域从事过研究工作,并成为瑞典达拉纳技术高中工业管理学部的助理教授。他职业生涯的很大一部分时间用于在瑞典和德国的工业能源巨头大瀑布电力公司(Vattenfall)研究聚光太阳能。当大瀑布电力公司决定不把太阳能作为其未来战略的一部分后,他离开了该公司。然后,他与尼古拉斯·斯腾隆德一起创办了太阳能战士公司(Solarus),目标是研发、生产和销售聚光太阳能技术,通过将电能和热能结合在一起,使太阳能更具竞争力。他将第一个试点项目放在了自己的祖国瑞典。他意识到,如果太阳能在靠近北极圈的地方都具有竞争力,那么它将在地球任何地方都具有竞争力。拉松教授是瑞典聚光太阳能研究基金会的董事会成员。

图书在版编目（CIP）数据

冈特生态童书.第二辑修订版：全36册：汉英对照／
（比）冈特·鲍利著；（哥伦）凯瑟琳娜·巴赫绘；
何家振等译.—上海：上海远东出版社，2021
书名原文：Gunter's Fables
ISBN 978-7-5476-1759-5

Ⅰ.①冈… Ⅱ.①冈… ②凯… ③何… Ⅲ.①生态
环境－环境保护－儿童读物—汉、英 Ⅳ.①X171.1-49

中国版本图书馆CIP数据核字（2021）第213075号
著作权合同登记号图字09-2021-0823

策　　划	张　蓉
责任编辑	祁东城
封面设计	魏　来　李　廉

冈特生态童书

另一面的阳光

［比］冈特·鲍利　著
［哥伦］凯瑟琳娜·巴赫　绘
唐继荣　译

记得要和身边的小朋友分享环保知识哦！
八喜冰淇淋祝你成为环保小使者！

用身体发光

Light My Fire

Gunter Pauli

[比]冈特·鲍利 著
[哥伦]凯瑟琳娜·巴赫 绘
唐继荣 译

上海远东出版社

丛书编委会

主　任：贾　峰
副主任：何家振　闫世东　林　玉
委　员：李原原　祝真旭　牛玲娟　梁雅丽　任泽林
　　　　王　岢　陈　卫　郑循如　吴建民　彭　勇
　　　　王梦雨　戴　虹　翟致信　靳增江　孟　蝶

特别感谢以下热心人士对童书工作的支持：

匡志强　宋小华　解　东　厉　云　李　婧　陈　果
刘　丹　熊彩虹　罗淑怡　旷　婉　杨　荣　刘学振
何圣霖　廖清州　谭燕宁　韦小宏　李　杰　欧　亮
陈强林　王　征　张林霞　寿颖慧　罗　佳　傅　俊
胡海朋　白永喆　冯家宝

目录

用身体发光	4
你知道吗?	22
想一想	26
自己动手!	27
学科知识	28
情感智慧	29
艺术	29
思维拓展	30
动手能力	30
故事灵感来自	31

Contents

Light My Fire	4
Did You Know?	22
Think about It	26
Do It Yourself!	27
Academic Knowledge	28
Emotional Intelligence	29
The Arts	29
Systems: Making the Connections	30
Capacity to Implement	30
This Fable Is Inspired by	31

一只萤火虫沿着河流飞行，寻找自己的伙伴。这是个漆黑一片的夜晚，他的尾巴明亮地一闪一闪。一只正在寻找美餐的青蛙抬起头来，惊喜地看到这么多的萤火虫。可有上百只呢！

"你们成群结队是为了自卫吗？"青蛙好奇地问。

"不是的！我们正在为篝火舞会照明。"萤火虫回答道。

A firefly is looking for his friends along the river. It is a dark night and his tail is flashing brightly. A frog, on the hunt for a good meal, looks up and is amazed to see so many fireflies. There are hundreds of them!

"Are you gathering an army to defend yourselves?" wonders the frog.

"No, we are making light for my fire party," responds the firefly.

为篝火舞会照明

Making light for my fire party

"舞会？我还从来不知道虫子会开舞会呢！"

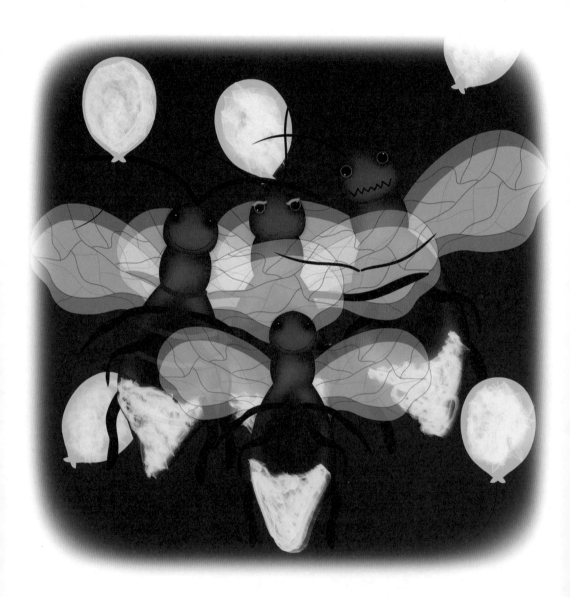

"A party? I never knew that flies party!"

"舞会？我还从来不知道虫子会开舞会呢！"

"你只要瞅瞅，就会发现我们遍布全世界。我们萤火虫大家族的一些成员已经学会如何合拍地闪耀他们的光芒啦！"

"但是，你们是如何发光的呢？"

"A party? I never knew that flies party!"

"Look and you'll find us all over the world. Some members of our family have learnt how to flash their lights in harmony!"

"But how do you flash your lights?"

"要发光,我们需要镁元素。"

"镁元素是不是那种植物用来制造叶绿素的金属?利用那个神奇的配方,就能把太阳能变成能量和食物?"青蛙问道。

"嗯,你现在知道我们从哪里获得能量补给了吧!植物的花蜜里有很多好东西,包括这种神奇的镁元素。"

"For that we need magnesium."

"Isn't magnesium the metal that plants use to make chlorophyll? Use that magical formula to turn the power of the sun into energy and food?" asks the frog.

"Well, now you know where we get our supplies from. Plant nectar is full of goodies, including this magical magnesium."

植物的花蜜里有很多好东西

Plant nectar is full of goodies

9

我们还得有氧气

We also need oxygen

"你们发光需要的就只是一种金属吗?"

"不是的！在自然界，没有什么事情只靠一样东西！我们还得有氧气。"

"只要人们不砍伐树木，不搞单一种植毁掉我藏身的灌木丛，那么在森林中氧气用都用不完呢！"青蛙发表了自己的意见。

"And that is all you need to make light - one metal?"

"No, in nature nothing relies on only one thing! We also need oxygen."

"That is abundant here in the forest, as long as people don't start cutting down the trees and planting monocultures without any undergrowth for me to hide," comments the frog.

"我听说,人们已经利用这个神奇的配方,让花儿在夜里发光。要是用来做路灯就太酷了。"

青蛙赞同萤火虫的说法。"人们是时候学会如何利用你们的发光方式了。他们可是花了很大代价来保证路灯能亮起来呢。"

"I hear people have used this magic formula to make flowers light up and glow at night. Pretty cool streetlights."

The frog agrees. "It's about time people learnt how to use your way of making light. They are spending a fortune on keeping street lights on."

花了很大代价来保证路灯能亮起来

spending a fortune on street lights

……捕杀鲸类就为了得到鲸油……

... killing whales to have oil ...

"你知道吗？人们有一阵在全世界捕杀鲸类，就为了得到鲸油来点亮路灯。我们也得小心，免得所有的同伴也惨遭毒手！"萤火虫发出警告。

"人们更应该学会你们的发光方式！"

"You know that people once went around the world killing whales to have oil to power their street lights? We have to be careful that we don't all get killed too!" warns the firefly.

"People should rather learn to copy the way you do it!"

"我们比那些叫作LED的小灯看上去更加光亮吗？"

"哦，你是指那些发光二极管吗？"青蛙问道。

"It seems that we shine brighter than those little lights called LED?"

"Oh, you mean those light-emitting diodes?" asks the frog.

发光二极管

Light-emitting diodes

光线穿过空气的速度要快多啦

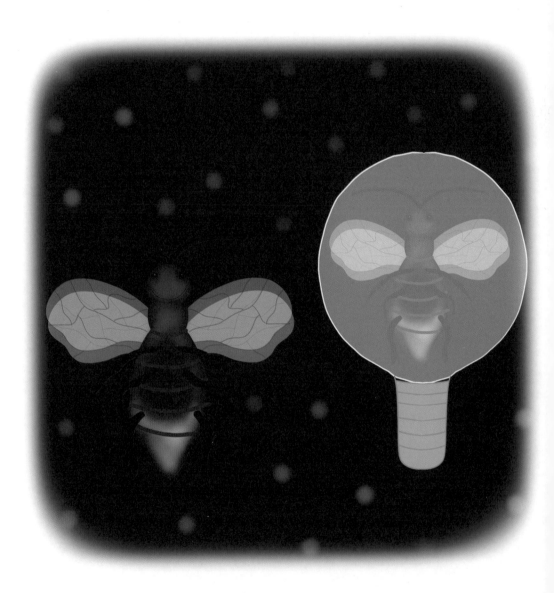

Light travels faster through air

"当然是啦!光线穿过空气的速度比穿过身体或玻璃要快多啦。我们早就解决这个问题了,人类却还有很多光学方面的知识需要学习。"

"那么能不能告诉我,你们是怎样解决这个问题的,通过改变物理定律吗?"

"Of course! Light travels faster through air than through a body or glass. We solved that a long time ago, but people still have a lot to learn about optics."

"So tell me, how did you solve it, by changing the laws of physics?"

"不，我们没那么聪明，但我们更巧妙地利用物理学，还有特别是几何学：我们身上鳞片的形状和构成形式使得亮光更加耀眼。"

"这就是你们的味道尝起来这样苦的缘故吗？"青蛙轻笑着说。"老实说，我也就真正饿得受不了才不得不吃你们！"

……这仅仅是开始！……

"No, we are not that smart, but we make clever use of physics, especially geometry: the shape and the form of the scales on our bodies makes lights shine more brightly."

"Is that why you taste so bitter when I eat you? To be honest," chuckles the frog, "I would have to be very hungry indeed to go for you!"

... AND IT HAS ONLY JUST BEGUN!...

……这仅仅是开始！……

... AND IT HAS ONLY JUST BEGUN! ...

Did You Know?

你知道吗？

People create light using incandescence. A filament inside a lamp gets hot and emits light.

人们用白炽灯照明。灯泡里面的灯丝变热并发出光亮。

Fireflies create light through luminescence, mixing chemicals with oxygen.

萤火虫用冷光照明，这是通过混合化学物质与氧气实现的。

Luminescence neither requires nor generates heat and therefore it is known as 'cold light'.

冷光之所以被称为冷光，是因为它既不需要也不会产生热。

We do not yet know exactly how glowworms and fireflies switch their light on and off.

我们还不太清楚萤火虫及其他发光虫是如何开启和关闭它们的发光系统的。

Cold light is produced by many species all over the planet, from fungi to jellyfish, squid, krill and even fish.

许多物种都可以发出冷光，它们遍布全球，包括真菌、水母、鱿鱼、磷虾甚至鱼类。

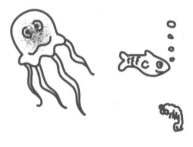

Hot or cold, light comes from one source: an excited electron that moves up and down on the atomic orbit. When it settles back down it releases a photon, a tiny light.

光无论冷热，都出自一个源头：被激发的电子在原子轨道向上和向下移动。当电子从高能量轨道跃迁回低能量轨道并稳定下来后，便释放一个光子，也就是一个微小的光点。

𝒜 single-cell dinoflagellate of half a millimetre long can send a light signal 5 metres away, the equivalent of a 2-metre tall person being able to communicate over a distance of 20 kilometres.

只有半毫米长的单细胞甲藻（腰鞭毛虫）能发出远达 5 米的光，相当于 2 米高的人能与 20 千米外的人进行交流。

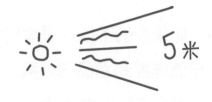

𝒜 decrease in bioluminescence can be correlated to relative levels of toxicity. This innovative method eliminates the need to use live animals in laboratories to test the potential danger of new products for people.

生物发光现象的减少与环境中毒性的相对水平相关。利用这种创新原理，就不需要在实验室用动物活体来检测新产品对人们的潜在危险性。

Think about It

While humans use electric power to produce light, animals and fungi produce light-emitting chemicals they find in their food and in the air or water. Which do you prefer?

人们利用电能制造光，而动物和真菌则利用它们在食物、空气或水中发现的可发光化学物质发光，你更喜欢哪一种？

如果我们不能理解萤火虫和水母怎样控制它们的光照，我们怎么能真正了解自然呢？

If we do not understand how fireflies and jellyfish turn their lights on and off, what do we really know about Nature?

Scientists used to test the toxicity of soap or lipstick on the eyes of rabbits. Aren't you sad that some manufactures of soap and cosmetics still do?

过去科学家在兔子的眼睛上测试肥皂或口红的毒性。一些肥皂和化妆品的生产商现在仍旧这样做，对此你不会难过吗？

有些物质的味道是苦的，这是否意味着它对你有益或者有害？

When something tastes bitter, does it indicate that it is bad for you or good for you?

Do It Yourself !

自己动手！

Get a flashlight. Make sure it is not too bright and does not hurt your eyes when you look into it. First shine the light on a sheet of white paper. Take a photograph. Now shine the light through a pane of clean, clear glass onto a sheet of white paper. Photograph this. Now take a fish tank, made of transparent glass, and fill it with water. Now shine the light, with the same intensity, through the glass and the water onto the white sheet of paper. Take another photograph. Compare the three photographs and discuss the differences with your friends. What do we learn?

拿一个手电筒，确保它的光线不太亮，你看它的时候不会感到刺眼。首先，将手电筒照射到一张白纸上，拍一张效果照片。然后，将手电筒的光穿过一块干净透明的玻璃照射到一张白纸上，再拍张照片。现在，把一个用透明玻璃制成的鱼缸装满水，让同样强度的光穿过玻璃和水照射到白纸上，也把效果拍摄下来。对比这三张照片，和你的朋友讨论各张照片有什么不同。我们学到了什么知识？

TEACHER AND PARENT GUIDE

学科知识
Academic Knowledge

生物学	生活中光的作用：可以是一种防御机制，也是传播孢子、交流、寻觅食物和识别天敌的一种手段；萤光素酶可以催化发光反应；多数发光物种被抓住后，会失去它们的发光能力；萤火虫的成虫与幼虫都发光；深海的主要光源是水母、鱿鱼、虾等动物而不是太阳；动物和发光菌之间的共生关系；在一个科的鲨鱼类和42个科的其他鱼类中发现了生物发光现象。
化学	镁在冷光中的作用体现在以镁为核心的叶绿素，氧气在普通发光中的作用体现在以铁为核心的血液；物种改变冷光颜色的能力。
物理	冷光开闭之谜；光穿过玻璃或空气传播；白炽灯光与生物发光的不同；海水吸收红色、橙色和黄色的阳光，散射紫色光线，所以蓝绿光是唯一到达海洋微光区的光线；生物发光与荧光的不同；生物发光是海洋中交流的主要形式；动物可以开启和关闭生物荧光，调节荧光强度、颜色甚至光的角度分布；声波信号能在水中有效地发送，但是没有方向性，生物荧光则更精确和可控制；发光二极管的原理。
工程学	将冷光化学系统转变为分析系统，用来检测环境污染物；一些有毒成分也可用于医疗诊断、农业和食品安全保证。
经济学	引入地方经济发展的概念，促使本来要离开某地的资金，现在能用在本地。
伦理学	在测试产品的毒性时用萤火虫的酶就能获得可比较的结果，为什么还要用动物活体呢？
历史	鲸油曾经被用作燃油路灯的燃油来源，导致鲸类在全世界范围内受到大规模屠杀。
地理	深度从200米延伸到1 000米的海洋微光区，处于海洋中有阳光照射的区域（即真光带）和"午夜"不透光的区域（即无光带）之间，是大多数发光生物的家园。
数学	用于生物发光成像的算法。
生活方式	虽然我们享受欣赏自然的乐趣，但我们并未真正从了解自然进步到向自然学习。
社会学	通过聚会或参军等社会活动发展对家庭或社群的归宿感。
心理学	晚上路灯亮起时的安全感。
系统论	光是自然产生的，没有任何碳排放或有毒残留物。

教师与家长指南

情感智慧
Emotional Intelligence

青 蛙

青蛙观察着周围的环境。他得出了自己的结论,但一说出来很快就碰壁了。起初他并不信任萤火虫的超前逻辑,但他改变了以自我为中心的态度,转而通过提问试图了解更多。他眼界开阔,把看似分散的各个方面与生态系统(涉及氧气和森林)的巨大挑战联系起来。他勇于冒险,对一些奇思妙想持开放态度,并把这些想法拿出来讨论,虽然他根本不知道它们是否行得通。他向萤火虫展示了情感共鸣,意识到一旦人类以某种方式模仿萤火虫的照明系统可能带来的死亡危胁。不过,他承认自己也是萤火虫的潜在威胁,尤其当他非常饥饿时更是如此。

萤火虫

萤火虫无所畏惧地与他的一个天敌交谈。青蛙是萤火虫的潜在天敌,但萤火虫并不焦虑,还花时间向青蛙解释冷光如何工作。萤火虫确保青蛙理解其中的化学原理,但自己却不求任何回报。他思想开放,有意愿分享其家族在过去的岁月里积累的智慧,并力争让对方更好地了解冷光的特性。萤火虫以尊重的态度倾听青蛙的创意想法,并迅速形成一种坚定的主张,决不接受人类可能带来的危险,即外出抓捕他们来制造路灯。萤火虫的自我意识非常强,他还提示人类应该在哪些方面开展研究。

艺术
The Arts

拍一张用传统方式照明的街道。设想现在有各种形式的冷光源。画一条同样的街道,通过移除街灯的电杆、电线和电箱来进行所有必要的调整,营造新的环境。请记住,仅有萤火虫和真菌可不够,你需要确保灯光的自然提供者有天然的食物基地,并将它融入你的艺术设计中。所有这些成分的组合将提供一种全新的街景视野。让你的想象力随意驰骋吧!

TEACHER AND PARENT GUIDE

思维拓展
Systems: Making the Connections

　　照明已成为生活的一部分。我们完全不能想象家中不能随意开关照明灯光是怎样的情形。政府已经以高成本承担了为夜间提供道路照明的责任，这被视作应对犯罪和为市民提供安全感的重要一步。

　　发电主要依赖化石燃料。虽然工程师们一直在寻找可再生能源和更有效的能源系统，但相关产业却一直局限于设计新的电灯。通过数十年的发展，照明系统已从白炽灯演变为日光灯（CFL，即紧凑型荧光灯）和发光二极管（LED）灯。一个白炽灯泡在其使用寿命期内消耗的电费超过它最初售价的5—10倍。LED灯不使用灯丝，使用寿命是白炽灯的10倍，预计LED领域将会出现无数发明创造。

　　我们该从一套全新的照明系统获得灵感、受到启发！这套系统在海洋中占统治地位，并被陆地上的真菌和昆虫成功采用。它的原材料随处可得，能就地生产，废物很少而且是有机的。由于电网最主要成本之一的铜电缆中的铜需要采矿和加工，因而新的照明系统比电网更好。自然的冷光系统不需要任何导线或变压器，唯一的挑战是科学家必须解决如何开启和关闭冷光。对于自然界中冷光怎样开启和关闭，以及怎样根据意愿转变冷光的颜色和聚光还缺乏一定程度的了解，表明我们对自然界的真正运作方式还是懂得太少。或许我们已经能破解基因组和绘制DNA图谱，但我们仍然缺乏某些理解大自然的高效运作所需的基础知识。

动手能力
Capacity to Implement

　　建设和运营一套公共路灯系统，需要投入多少？首先要铺设电缆，然后是生产和树立灯杆，并确保它们能耐受20年的腐蚀。在此之后是安装路灯，并安排人员进行管理，以便在日落后开启路灯，日出时关闭。试着估计上述各项的开支，以及需要多少人参与。现在的问题是：谁来为路灯系统的电费买单？现在再设想你拥有通过化学方式产生的冷光。不要担心相关的工程问题，因为还没有可用的技术；你只是关注新路灯系统的设计，该系统需要每月为路灯照明补充"食物"。你还需要建设一个生产所需化合物的工厂。哪一项建议将提供更多的工作岗位？哪一项建议将耗费更多？举行一个趣味竞赛吧！

教师与家长指南

故事灵感来自
This Fable Is Inspired by

露西娅·阿特霍图
Lucia Atehortua

阿特霍图博士从哥伦比亚的安蒂奥基亚大学毕业并获得自然科学学位后,前往纽约城市大学学习并获得博士学位。她进行了40年的植物学和生物技术研发,尤其在藻类、真菌和植物方面建树颇丰。她最初专注于对自然进行观察,后来转向利用在自然产品中发现的生物技术来设计制造系统。她的专长之一是蝎尾蕉属植物,这些植物能从其巨大的花朵中提取荧光素酶类化合物并将其用于道路照明。阿特霍图博士的工作已经得到广泛认可,荣获数十项奖励;她的许多发明已经成功地获得了专利。她现在仍然在故乡哥伦比亚麦德林市生活和工作。

图书在版编目（CIP）数据

冈特生态童书.第二辑修订版：全36册：汉英对照 /
（比）冈特·鲍利著；（哥伦）凯瑟琳娜·巴赫绘；
何家振等译.—上海：上海远东出版社，2021
书名原文：Gunter's Fables
ISBN 978-7-5476-1759-5

Ⅰ.①冈… Ⅱ.①冈… ②凯… ③何… Ⅲ.①生态
环境–环境保护–儿童读物—汉、英 Ⅳ.①X171.1-49

中国版本图书馆CIP数据核字（2021）第213075号

著作权合同登记号图字09-2021-0823

策　　划	张　蓉
责任编辑	祁东城
封面设计	魏　来　李　廉

冈特生态童书
用身体发光
［比］冈特·鲍利　著
［哥伦］凯瑟琳娜·巴赫　绘
唐继荣　译

记得要和身边的小朋友分享环保知识哦！
八喜冰淇淋祝你成为环保小使者！

Housing 49

水晶宫殿

A Crystal Palace

Gunter Pauli

[比] 冈特·鲍利 著
[哥伦] 凯瑟琳娜·巴赫 绘
唐继荣 译

上海远东出版社

丛书编委会

主　任：贾　峰
副主任：何家振　闫世东　林　玉
委　员：李原原　祝真旭　牛玲娟　梁雅丽　任泽林
　　　　王　岢　陈　卫　郑循如　吴建民　彭　勇
　　　　王梦雨　戴　虹　翟致信　靳增江　孟　蝶

特别感谢以下热心人士对童书工作的支持：

匡志强　宋小华　解　东　厉　云　李　婧　陈　果
刘　丹　熊彩虹　罗淑怡　旷　婉　杨　荣　刘学振
何圣霖　廖清州　谭燕宁　韦小宏　李　杰　欧　亮
陈强林　王　征　张林霞　寿颖慧　罗　佳　傅　俊
胡海朋　白永喆　冯家宝

目录

水晶宫殿	4
你知道吗？	22
想一想	26
自己动手！	27
学科知识	28
情感智慧	29
艺术	29
思维拓展	30
动手能力	30
故事灵感来自	31

Contents

A Crystal Palace	4
Did You Know?	22
Think about It	26
Do It Yourself!	27
Academic Knowledge	28
Emotional Intelligence	29
The Arts	29
Systems: Making the Connections	30
Capacity to Implement	30
This Fable Is Inspired by	31

一群海鸥正奋力飞过一座垃圾山。而一辆接一辆的卡车正从城市运来更多的垃圾。

"太令人吃惊了，人类都扔掉了些什么呀！"一只海鸥粗声粗气地说。"那么多东西的终点都不应该在这里。更糟的是，海洋中鱼类和鸟类会把塑料错当作食物。"

"可不是！瞧瞧这成千上万的塑料瓶子。"另一只海鸥回答。

A flock of seagulls is working its way through a mountain of waste. Truck after truck delivers more waste from the city.

"It is astounding what people throw away," squawks one. "There is so much that should never end up here. And it is even worse, at sea, where fish and birds make the mistake of taking plastics for food."

"Yes, just look at these millions of plastic bottles," says another.

太令人吃惊了，人类都扔掉了些什么呀!

It is astounding what people throw away!

这里遍地都是塑料瓶，你该怎么处理呢？

What can you do with bottles lying around here?

"我真搞不懂，人们怎么会被塑料比玻璃更好这样的说法给骗了。"住在同一地区的老鼠插话道。

"我敢说，你从没上过学。"第一只海鸥回答。"我上学时学过，塑料轻，玻璃重。所以你能在塑料瓶中装更多的东西。"

"但这里遍地都是塑料瓶，你该怎么处理呢？"老鼠对此很好奇。

"I don't understand how people could ever have been fooled into believing that plastic is better than glass," adds a rat who lives in the same area.

"You never went to school – I can tell," answers the first seagull. "I learned there that plastic is light and glass is heavy. So you can carry more in a plastic bottle."

"But what can you do with bottles lying around here?" wonders the rat.

"你可以用它们做新的瓶子。"年轻的海鸥吹了吹他的羽毛,对自己的答案很满意。

"你见过树叶在秋天从树上飘落吗?"老鼠问道。

"那当然!这又不是一件非得要上学才能理解的事情。"海鸥嘲笑道。

"You could make new bottles." The young seagull puffs out his feathers, happy with his answer.

"Have you ever seen leaves drop from a tree in autumn?" asks the rat.

"Of course, again something you do not need to go to school to understand," scoffs the seagull.

你能用它们做新的瓶子！

you could make new bottles!

旧树叶没法产生新树叶!

The tree does not make leaves out of leaves!

"那么你见过一棵树在春季重新长出秋天掉落的树叶吗?"

"别傻了。我们知道那行不通。即便行得通,也是耗费高昂、效率低下,而且绝不会好看。"

"既然旧树叶没法产生新树叶,为什么你希望用旧塑料瓶生产新塑料瓶呢?"

"But have you ever seen a tree in spring time, reattaching the leaves it dropped way back in autumn?"

"Now don't be ridiculous. We know that does not work, and if it ever were to work it would cost a lot, be inefficient and would never look good."

"So if the tree does not make leaves out of leaves, why do you want to make bottles out of bottles?"

"嗯,人家这样告诉我的。"海鸥坦白。

"你相信别人告诉你的一切?还是让我们一起讨论,把事情做得更好吧。"老鼠说。

"没问题,我会试一试。"海鸥答应了。

"玻璃熔化了会怎么样?塑料熔化了又会怎样?"

"Uhm, because I was told so," confesses the seagull.

"You believe everything they tell you? Let's put our heads together, let's do better," says the rat.

"Sure, I'll give it a try," promises the seagull.

"What happens if glass melts? And what happens if plastic melts?"

让我们一起讨论,把事情做得更好吧!

Let's put our heads together, let's do better!

燃烧塑料会产生黑烟

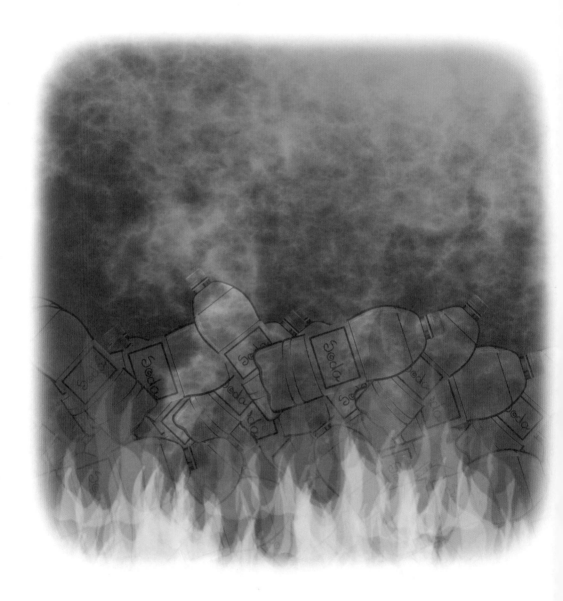

plastic burns creating black smoke

"嗯,燃烧塑料会产生黑烟,气味难闻。更糟糕的是,这对你我都不好。"

"那玻璃呢?"

"玻璃只是熔化。如果周围气体多(我们确实有很多气体),玻璃将变为泡沫。"

"Well, plastic burns, creating black smoke that smells terrible. Even worse, it is bad for you and me."

"And glass?"

"Glass just melts and if there is a lot of gas around – which we do have – then it turns into foam."

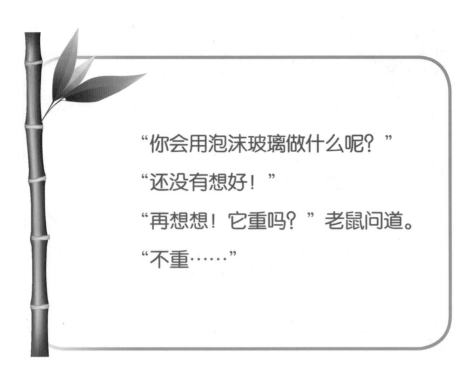

"你会用泡沫玻璃做什么呢?"

"还没有想好!"

"再想想!它重吗?"老鼠问道。

"不重……"

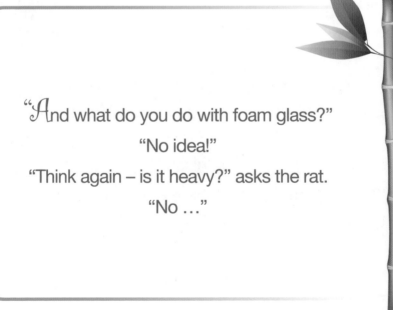

"And what do you do with foam glass?"

"No idea!"

"Think again – is it heavy?" asks the rat.

"No …"

你会用泡沫玻璃做什么呢?

What do you do with foam glass?

你能咬穿它吗?

Can you bite into it?

"你能咬穿它吗？"

"我不想这样干！你能吗？"

"不——如果我尝试这样做，会崩坏牙齿。"老鼠回答道，然后继续提问："水能通过这种泡沫吗？"

"不能。"

"你能坐在上面吗？"

"Can you bite into it?"

"I don't want to. Can you?"

"No – if I try I might lose my teeth," answers the rat before continuing with his questions by asking: "Does it allow water through?"

"No."

"Can you sit on it?"

"能，它很坚固！"海鸥点点头，很高兴知道答案。

"如果把它打成碎片，你能用它做什么呢？"

"嗯！玻璃不能被毁坏，只会变形，"海鸥回答道。

"看起来我们可以用泡沫玻璃造一座房子！"老鼠惊叹道。

"没错！现在我得回学校，去教会所有人用更多玻璃，这样我们就能拥有水晶宫殿啦！"

……这仅仅是开始！……

"Yes, it is strong!" The seagull nods, happy to know the answer.

"And if you crush it to pieces, what can you do with it?"

"Well, glass cannot be destroyed, only transformed," answers the seagull.

"It looks like we could build a house out of this!" exclaims the rat.

"Well, now I have to go back to school and teach everyone to use much more glass so we can have our crystal palace!"

... AND IT HAS ONLY JUST BEGUN!...

……这仅仅是开始!……

... AND IT HAS ONLY JUST BEGUN! ...

Did You Know?

你知道吗？

Seagulls are smart, work together, drop seashells on hard surfaces to crack them open, and even use bread as bait to catch fish.

海鸥很聪明，成群活动，将海贝从高空掷到坚硬地面来破开贝壳，甚至用面包当作诱饵来捕鱼。

Seagulls are found all over the planet, even in the Arctic and Antarctica. Rats can also be found everywhere on the planet.

海鸥分布在全球，甚至北极和南极也有。老鼠也是遍布全球。

Rats take care of the sick and wounded in their group. They spend hours cleaning themselves and are less likely to catch and transmit parasites or viruses than cats or dogs. They play a role in taking care of our health by serving as test animals in laboratories.

老鼠会照顾群体内患病或受伤的成员。它们肯花几小时清理自身，比猫和狗更少有机会接触和传播寄生虫。由于在实验室被作为实验动物，老鼠对人类健康也有所贡献。

Rats can be trained to detect land mines. They are small and light enough not to detonate land mines while searching for them.

老鼠能被驯化用来搜索地雷。它们的个头足够小而且轻，不会在搜寻过程中引爆地雷。

Rats have strong teeth which bite through bricks, cement and lead, but not through foam glass. The enamel of rat incisors is harder than platinum or iron.

老鼠牙齿坚固，能咬穿砖头、水泥和铅，但不能咬穿泡沫玻璃。老鼠门齿上的釉质比铂或铁更加坚硬。

Plastic is lighter than glass, and therefore considered more ecofriendly. However, when glass is recycled into glass foam and is used locally, substituting steel, insulation and chemicals, its resource efficiency is much higher.

塑料比玻璃轻，曾经被认为对生态环境更加友好。然而，当玻璃被再循环利用为玻璃泡沫，并能就地代替钢、绝缘材料和化学品，它的资源效率则会比塑料高得多。

Plastics do not degrade and are unsafe to burn at home or on a landfill. Ten percent of all plastics end up in the ocean where it accumulates. There are six kilograms of plastics for every kilogram of plankton in the sea.

塑料不容易降解，而且在家或垃圾填埋场燃烧时不安全。所有塑料垃圾的10%最终汇聚到海洋，平均每千克浮游生物对应6千克塑料垃圾。

Glass only has three ingredients, and glass foam can be made from a mix of brown, white and green bottles. Plastics have many unknown and secret additives.

玻璃只有3种成分，可以用棕色、白色和绿色的玻璃瓶生产泡沫玻璃。塑料有很多种不明确和保密的添加成分。

Think about It

想一想

Did you ever think that a seagull might use bait to catch fish?

你是否想到过海鸥有可能用诱饵捕鱼？

Rats are the animals most often used in laboratories to test the toxicity of chemicals and drugs. Are we grateful for their service?

老鼠是实验室中最常用于测试化学品和药物毒性的动物。我们要感谢它们的贡献吗？

Would you like to live in a house made of glass?

你愿意住在由玻璃建造的房子中吗？

Should we recycle bottles only into bottles, and paper only into paper, or can we do better?

我们应该只把旧瓶子循环为新瓶子，把废纸循环为再生纸，还是可以做得更好？

Do It Yourself!
自己动手！

How many bottles and containers do you and your family discard every day? How many are made from glass, and how many are made from plastics? How many different types of plastic do you recognise among the bottles? How many different plastics are combined in one bottle? Do you know the content of the plastic bottle? It is certainly more than the main polymer listed and will include UV-protectors, softeners, colour pigments, glue to stick the label on and caps made from yet another plastic. Now check how many years it takes for each of the plastics to degrade: polypropylene can take ten times longer than PET. You will be surprised to learn that before some of these plastics degrade, you will have already have grandchildren!

你和你的家人每天要丢弃多少瓶瓶罐罐？其中多少是玻璃制成，多少是塑料制成？你从这些瓶子中能识别出多少种不同的塑料？在同一个瓶子上，有多少种不同的塑料组合在一起？你知道塑料瓶含有哪些成分？显然不止罗列出的主要高分子聚合物，还会包括紫外保护剂、软化剂、色素、用于贴标签的胶水，以及用另一种塑料制作的瓶盖。现在检查这些塑料成分中每一种需要多少年才能降解，其中聚丙烯的降解时间是聚对苯二甲酸乙二醇酯（PET）的10倍。你可能会感到惊奇的是，在这些塑料中的某些成分降解前，你都可以有孙子了！

TEACHER AND PARENT GUIDE

学科知识
Academic Knowledge

生物学	海鸥属于肉食类动物，以环境中的腐肉为生；海鸥既可喝盐水，也可以喝淡水。
化 学	玻璃的口感是中性的，因为它不与口腔中任何成分发生化学反应；PET类塑料瓶会析出一种有毒的金属元素锑，后者被认为在室温下是安全的，但温度升高后有危害；塑料在低于900℃下燃烧时，会释放出剧毒的二噁英和呋喃。
物 理	塑料的孔隙比玻璃多，所以塑料瓶装的啤酒跑气更快；玻璃是无定形的非晶体；泡沫玻璃或蜂窝状玻璃有许多密封孔隙，可以阻断水和热量的扩散。
工程学	玻璃在1 100℃时变为液态，而PET塑料在260℃变为液态；泡沫玻璃由98%的玻璃和2%的无机盐混合后生成。
经济学	预制装配式住宅建设能够在工厂制造部件，然后现场快速组装；泡沫玻璃由于绝缘性能良好，替代传统绝缘材料后能节省30%—50%的能源；一个地区若能回收100万个容量一升的玻璃瓶，就可确保一家泡沫玻璃厂的生存。
伦理学	我们怎么可以丢弃能产生更多价值、无法毁坏而只能改变形态的产品呢？我们怎么可以继续制造和丢弃含有毒成分且只能一次性使用的塑料制品呢？我们怎么可以使用降解时间长达数百年，却只使用几个小时的塑料制品呢？
历 史	真正意义上的玻璃最早出现于公元前3000年的叙利亚、美索不达米亚地区和埃及；水晶宫是伦敦海德公园里一座由铸铁和平板玻璃构成的建筑，也是当时最大的玻璃建筑，1851年世界博览会在此举行；泡沫玻璃于20世纪30年代被发明出来，其目的是降低建筑成本和发展园艺学。
地 理	海鸥和老鼠生活在世界任何一个角落，甚至在北极和南极也有。
数 学	如何将线性的寿命周期评价（LCA）转换为系统性的寿命周期评价，即在一个产品寿命的末期，开发出能替代其他产品或服务的新功能，从而提供比最初所认为的最佳解决方案更多的效益和资金流？将生物可降解性作为时间和温度的函数来计算。
生活方式	在用后即弃型社会里，没人操心垃圾最终会带来什么后果。
社会学	邻避效应："不要在我家后院"式的事不关己态度；为了让城市适于居住，垃圾填埋场或焚化炉这类解决方案是需要的，但一旦靠近我们住处太近，就总会遭到反对。
心理学	无知的作用：我们往往不懂，因为我们不懂，也懒得问，我们就按过去的方式继续生活；你相信某项事物，是因为别人告诉了你，是因为你觉得就该那样，还是通过深思熟虑形成了自己的观点？
系统论	树叶被微小的生物、真菌和昆虫转化为含有树木所需营养成分的腐殖质。

教师与家长指南

情感智慧
Emotional Intelligence

海 鸥

海鸥处于绝望中,不理解怎么这么多垃圾的最终归宿是垃圾填埋场和大海。海鸥回答问题时,并未对老鼠表达过多的尊敬。他有固定的思维模式,只会复述他学过的东西。但仔细听了老鼠提出的逻辑后,海鸥被迫面对一个他过去就懂得的事实:树叶不能在春天重新长回树枝上。在这个阶段,他还是没有怎么尊重老鼠,只是宣称他们都知道将树叶长回树枝是不现实的。此后,海鸥不得不承认他的观点实际上不是自己形成的,是由其他人告诉他,而他过去也认为这是真的。现在傲慢让位于相互尊重和谦逊,说明现在是学习和分享海鸥获得的新见解的时候了。

老 鼠

老鼠重申海鸥的观察结果。当海鸥反对他的意见时,他并没表现出任何受冒犯的样子。老鼠更愿意挑战海鸥,让他跳出旧知识的窠臼。老鼠很有耐心,通过提问来温和地强调自己的观点。即便遭到嘲笑,老鼠还是坚持不懈地提出更多问题,把海鸥引入反思。老鼠没有愚弄或指责海鸥,也没有采用同样的行为报复海鸥。老鼠表现出自制力、尊重和合作的愿望,也演示了怎样才能共同寻找更好的解决方案。最后,老鼠成功地将海鸥由傲慢转变为愿意共同工作。老鼠通过提问把原本就存在却被割裂开的知识联系在一起。老鼠增强了海鸥的信心,并示范了一种关怀方式,能打动哪怕是原本行为非常傲慢的海鸥。

艺 术
The Arts

让我们去看电影吧!不止看一场,而是在尼尔·戴蒙德作曲和演奏的音乐声中看两场电影:阿尔弗雷德·希区柯克在《群鸟》中描绘的海鸥和理查德·巴赫在《海鸥乔纳森》中描绘的海鸥。看完电影后,思考一下电影艺术怎样塑造观点。

TEACHER AND PARENT GUIDE

思维拓展
Systems: Making the Connections

当我们使用瓶子时，我们很难意识到对地球资源的影响。每个塑料瓶都是用石油制造出来的，或许要过一百年才能降解，而它的可用寿命却只有短短几个小时或几天。更糟的是，只有不到5%的塑料瓶被回收，而且即便标签上写有"可回收"，也不能保证它会被回收。然而，这只是故事的一部分。某些热衷于降低成本、节约能源和水的商人，开始拿塑料瓶与玻璃容器对比，并得出结论，说塑料瓶对公司和环境更好。公司和学者采用一套寿命周期评价方法来看待塑料瓶的寿命。不幸的是，寿命周期评价忽略了如下事实，即玻璃只需用3种成分就可以制造出来，而且能一直被回收再利用。塑料却是含有许多有毒成分的"化学鸡尾酒"，虽然这些成分可能服务于某项用途（瓶子），但这些添加物质最好别用于另一项用途（与你的皮肤紧密接触的服装）。寿命周期评价并不认为重复使用有更高的价值。如果玻璃在本地收集，其中质量最好的玻璃被反复用于制造瓶子，质量差一些或破碎的玻璃转变为泡沫玻璃，那么我们就只需考虑资源效率、水的消耗和运输成本，而不用将诸如轻钢框架、矿物绝缘材料、阻燃剂等有毒化学物质之类的高性能产品组合在一起了。这个与泡沫玻璃集成的产品和服务组合，使得房屋更便宜、更舒适。虽然寿命周期评价是一项伟大的进步，它引入了纪律约束和透明度，但现在应该超越两种产品的比较，着手进行两种价值链的比较了。

动手能力
Capacity to Implement

要想建一个泡沫玻璃厂，每年必须要获得至少500万个玻璃瓶的供应量。我们来做一个计划，能确保原材料供应和最终产品的买家。那么，你将说服谁每天运送15 000只玻璃瓶到你的新工厂门口？然后，谁是产品买家？我们能保证所有好处都清楚地说明和计算出来吗？这样就不用把绝缘材料与绝缘材料或瓶子与瓶子作过于简单的比较了。我们需要做的，就是清楚地说出这项选择的所有好处，包括经济、社会和生态方面的好处。

教师与家长指南

故事灵感来自
This Fable Is Inspired by

奥克·莫德
Åke Mård

奥克·莫德出生于芬兰，1963年移居瑞典。1992年他创建了一家公司，为当时的施工难点寻找明智的解决方案。他的主要创新洞察力体现在 Koljern 项目上，该建筑的材料是含有 97% 空气的可回收玻璃。围绕这种泡沫玻璃，他耗费了10年时间来开发相关建筑技术，获得一项专利。原材料由比利时的匹兹堡康宁公司生产，但它首次被弗兰克·格里正式用在西班牙毕尔巴鄂的古根海姆博物馆。奥克意识到这种材料便于使用，性能优良。该博物馆古怪的形状需要把保温材料锯成特定大小，而对于泡沫玻璃而言，这不仅是可能的，而且很容易。奥克把泡沫玻璃改善到能用于短期施工，改变了绝缘材料的批量处理方式，使之成为一项预制建筑技术，能在3天内现场建造一座房子。匹兹堡康宁公司对这种实用方法印象深刻，决定出价购买以确保这项业务的连续性，并将这种创新技术集成运用到欧洲其他地区，甚至欧洲之外。由于在建筑技术上的创新，奥克·莫德于 2013 年获得瑞典环境奖。

图书在版编目（CIP）数据

冈特生态童书.第二辑修订版：全36册：汉英对照 /
（比）冈特·鲍利著；（哥伦）凯瑟琳娜·巴赫绘；
何家振等译.—上海：上海远东出版社，2021
书名原文：Gunter's Fables
ISBN 978-7-5476-1759-5

Ⅰ.①冈… Ⅱ.①冈…②凯…③何… Ⅲ.①生态
环境－环境保护－儿童读物—汉、英 Ⅳ.①X171.1-49

中国版本图书馆CIP数据核字（2021）第213075号

著作权合同登记号图字09-2021-0823

策　　划	张　蓉
责任编辑	祁东城
封面设计	魏　来　李　廉

冈特生态童书

水晶宫殿

[比]冈特·鲍利　著
[哥伦]凯瑟琳娜·巴赫　绘
唐继荣　译

记得要和身边的小朋友分享环保知识哦！
八喜冰淇淋祝你成为环保小使者！

Housing 50

遍地是金
Gold Everywhere

Gunter Pauli

[比]冈特·鲍利 著
[哥伦]凯瑟琳娜·巴赫 绘
唐继荣 译

上海远东出版社

丛书编委会

主 任：贾 峰

副主任：何家振 闫世东 林 玉

委 员：李原原 祝真旭 牛玲娟 梁雅丽 任泽林
　　　　王 岢 陈 卫 郑循如 吴建民 彭 勇
　　　　王梦雨 戴 虹 翟致信 靳增江 孟 蝶

特别感谢以下热心人士对童书工作的支持：

匡志强 宋小华 解 东 厉 云 李 婧 陈 果
刘 丹 熊彩虹 罗淑怡 旷 婉 杨 荣 刘学振
何圣霖 廖清州 谭燕宁 韦小宏 李 杰 欧 亮
陈强林 王 征 张林霞 寿颖慧 罗 佳 傅 俊
胡海朋 白永喆 冯家宝

目录

遍地是金	4
你知道吗?	22
想一想	26
自己动手!	27
学科知识	28
情感智慧	29
艺术	29
思维拓展	30
动手能力	30
故事灵感来自	31

Contents

Gold Everywhere	4
Did You Know?	22
Think about It	26
Do It Yourself!	27
Academic Knowledge	28
Emotional Intelligence	29
The Arts	29
Systems: Making the Connections	30
Capacity to Implement	30
This Fable Is Inspired by	31

两只海鸥正在垃圾堆里寻找残渣作为食物,并为这些破烂来自哪里感到好奇。

"我们过去在这里能找到好食物,"老海鸥哀叹道,"但现在人类用有机废物制造肥料。一些人甚至用剩下的咖啡渣种蘑菇!"

Two seagulls are scavenging a rubbish dump and wondering where all the junk comes from.

"We used to find good food here," laments the older seagull, "but now humans are making fertilizer from organic waste. Some are even farming mushrooms on their leftover coffee grounds!"

两只海鸥在寻找残渣

Two seagulls are scavenging

扔掉各种各样的东西

Throwing away all kinds of things

"对我们来说，那可不是个好消息。这些天他们扔掉各种各样的东西——在我的一生中，还从未看到过这么多的破手机。"另一只海鸥评论道。

"我甚至不得不把我的食物从旧电池上咬下来！"

"That's bad news for us. These days they're throwing away all kinds of things – I've never seen so many broken cellphones in my life." observes the other seagull.

"I've even been forced to pick my food off old batteries!"

"这些人类宣称要回收电池。那些又大又重的黑色电池已经没有了,但我过去从没见过这么多小电池。它们的味道真糟糕!"

"我听说人类打算烧掉所有废物,只把剩下的灰烬扔到垃圾场。"老海鸥难过地摇摇头。

"那就意味着我们将没有食物啦!"年轻的海鸥惊叫。"难道人类没有意识到,燃烧垃圾虽然节约空间,但会把所有的东西变得有毒?"

"These humans claim to recycle batteries. Those big, heavy black ones are gone, but I've never seen so many tiny ones. They taste awful!"

"I hear they're planning to burn all the waste, and only throw the leftover ash onto this dump." The old seagull shakes his head sadly.

"That means there'll be no more food for us!" the younger seagull exclaims. "Don't they realise that burning rubbish saves space, but turns everything toxic?"

烧掉所有废物

Burn all the waste

把金属，甚至是黄金，转变为气体

Turns solid metal – even gold – into gas

"他们不可能蠢到只担心空间够不够！但愿在用火烧掉这里之前，所有计算机、电话机、电视、打印机和电冰箱都先被拿走了。"

"我早就听说，他们将在一个巨大的炉子里烧掉所有的电子垃圾。"年轻的海鸥说。

"我也听说了！这个做法是先用强酸处理，然后把金属气化。这些人类懂得怎样把金属，甚至是黄金，转变为气体。"

"They can't be so dumb as to only be worrying about space! Let's hope that all the computers, phones, TVs, printers and refrigerators are taken out of here before they try burn it."

"There's a lot of talk that they're going to burn all electronic waste in a huge furnace," says the younger seagull.

"I've heard! The idea is to use strong acids and then to evaporate metals. These humans know how to turn solid metal – even gold – into gas."

"想一想那该多热呀!我曾经试着在妈妈的煎锅里煎一些电池……那真是个坏主意。"年轻的海鸥看上去有些负罪感。

"那太危险了!"老海鸥训斥道。"你不该那样干,那会让大家永远生病的!"

"Imagine how hot that must be! I once tried to fry some batteries in my mom's frying pan … That was a bad idea." The young seagull looks guilty.

"That is so dangerous!" scolds the older seagull. "You should never do that – it could make everyone sick forever!"

会让大家永远生病

Could make everyone sick forever

花几十年来收拾这个烂摊子

Derades to clean up this mess

"我肯定不会。但为什么他们一开始就要制造所有这些有毒的东西呢?"

海鸥叹了口气,说道:"即使他们停止制造所有的电子垃圾,即使他们终于开始重新利用它们,仍然要花几十年来收拾这个烂摊子。"

"你知道吗?一吨旧手机含有的金和铂比一吨开采出来的矿石中的含量还多。"

"Sure, but why do they make all this toxic stuff in the first place?"

The seagull sighs. "Even if they stop making all their electronic junk, it's still going to take decades to clean up this mess – even if they finally start reusing it."

"Did you know there is more gold and platinum in a ton of old cellphones than in a ton of ore excavated from a mine?"

"嗯,你知道吗?现在遍布全世界各个城镇的矿渣堆里还有大量贵金属,造成了许多健康问题。"

"嗯,你知道吗?旧的深矿井灌满了饮用水,但这些水现在只用来稀释被污染的水。"

"你是说饮用水中含有微小的金片?这是惊人的浪费。"

"Well, did you know that there's still lots of precious metals in the mine dumps that now litter the world's cities and towns, causing health problems?"

"Well, did you know that the old, deep mines are full of drinking water that is now only being used to dilute polluted water?"

"You mean drinking water with minute flakes of gold in it? Amazing that it's wasted."

深矿井灌满了饮用水

Deep mines are full of drinking water

在矿井周围被污染的土地上种庄稼，用来制造生物燃料……

Plant crops on the polluted land around the mines to make biofuels ...

"是呀!你知道吗?在老矿井有水从几千米高空落下,这可以用来发电。"

"嗯,你知道吗?人类可以在矿井周围被污染的土地上种庄稼,用来制造生物燃料。那才是真正的'金子'!"

"嗯,你还知道吗?这些废料堆将成为未来的'金矿'。没有人会再去挖洞了!与此相反,人类将对这些剩下的山体进行开发。"

"Yes, and did you know that water falling thousands of metres in the heat of an old mine can be used to generate power?"

"Well, did you know that humans could plant crops on the polluted land around the mines to make biofuels? Now that's real gold!"

"Well, did you know that these waste dumps will be the mines of the future? No one will dig holes any more – instead, humans will excavate these leftover hills."

"好吧！我的确懂得，我们身边遍地是金。"

"人类认为采矿业没有未来，许多人将失去工作，这是不是让人感到惊叹？我希望我来生不再是海鸥，而是一位现代矿工！"

……这仅仅是开始！……

"Well, what I do know is that there's gold all around us."

"Isn't it amazing that humans believe that there's no future in mining and that many will lose their jobs. I hope that in a next life I won't come back as a seagull, but as a modern-day miner!"

... AND IT HAS ONLY JUST BEGUN!...

……这仅仅是开始！……

... AND IT HAS ONLY JUST BEGUN! ...

The largest landfill of the United States is as tall as the skyscrapers of Los Angeles, with 1,500 trucks delivering 12,000 tons of waste per day.

美国最大的垃圾填埋场有洛杉矶的摩天大楼那么高,每天有1500辆卡车运来1.2万吨的废物。

The USA alone accumulates 250 million tons of garbage a year. The main problem is not the piles of waste but the leachate – a noxious brew of chemicals that leaks into the groundwater.

仅在美国,每年积攒的垃圾就有2.5亿吨。主要问题不在于废物的堆积,而是渗滤液,也就是由化学品混合后产生并泄漏到地表水中的有毒物质。

In Germany and Scandinavia, landfill is the last resort: today, only 5% of waste ends up in landfills as the majority is incinerated and recycled.

在德国和一些北欧国家，垃圾填埋是最后的手段。如今只有5%的废物最终到达垃圾填埋场，而绝大部分被焚化或回收了。

Americans discard 30 million computers each year, and Europeans dispose of 100 million phones each year. E-waste will increase by a factor of 5 in the next decade.

美国人每年丢弃3 000万台计算机，而欧洲人每年会处理掉1亿部电话机。在未来十年，电子垃圾将增至现在的5倍。

A ton of ore from a goldmine yields 5g of gold; a ton of discarded mobile phones could yield 150g, along with 100kg of copper and 3kg of silver.

一吨从金矿开采出来的矿石能产出5克黄金；一吨被丢弃的手机能产出150克黄金，以及100千克铜和3千克银。

Tailings (the residue of ore) still contains gold, and can be used to make bricks. The leftovers are good for making of stone paper, the production of which needs no cellulose and uses no water.

尾矿（矿石提炼后的残留物）仍旧含有黄金，也能用于制造砖头。矿石边角料也可以用来制造石头纸，后者的生产不需要纤维素，也不需要消耗水。

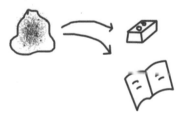

Deep mine shafts fill with warm water that can generate power: for example, in mines in South Africa, water flows downwards for 4,000 metres.

深矿井充满能用来发电的温水。例如在南非的矿山中，水的落差达4 000米。

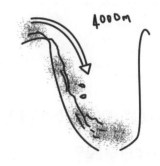

A particle accelerator boosts beams of carbon at the speed of light, using vast supplies of energy to convert lead into gold.

利用巨大的能量，粒子加速器能把碳离子束加速到（接近）光速，将铅元素转化为金元素。

Think about It
想一想

Would you rather mine gold from the Earth's crust or from your cellphone?

从地壳开采金矿或者从你的手机回收黄金，你选择哪一样？

可以从周围的废物堆和旧矿山赚钱，你同意吗？

Do you agree that there is money to be made around waste dumps and old mines?

If you were able to come back to the Earth in a distant future, what job would you like to have?

如果你能在遥远的未来回到人间，你希望从事什么工作？

把金属加热和气化像是一个危险的工作吗？

Does the heating and gasification of metals seem like a dangerous job?

Make an inventory of all the electronic equipment around your house. Everything needs to go on the list, including a small lantern that has a computer chip in it, your electric tooth brush and the quartz watch that has a button battery. Then ask how many of your appliances rely on batteries? Most importantly, what are you doing with the old batteries? What happens to the electronics at the end of their life? Who is in charge of recycling them? Check the weight of everything you have in the house and then estimate how much copper, silver and gold may be hidden in them. Urban mining may very well start at home.

列一份你房子周围的电子设备清单，涵盖所有的相关设备，包括有内置计算机芯片的小电灯、电动牙刷和有纽扣电池的石英手表。然后提问，这些家用电器中有多少依赖电池？更重要的是，你是怎样处理旧电池的？在那些电子产品过了使用寿命之后，该怎么处理它们？谁负责再循环利用它们？检查一下房子里所有电器的重量，估计它们中隐藏有多少铜、银和金。都市采矿业完全可以从家中开始！

TEACHER AND PARENT GUIDE

学科知识
Academic Knowledge

生物学	耐金属贪铜菌以纳米粒的形式在体内积聚金元素；某些细菌通过排出的代尔夫特肌动蛋白与金离子发生化学反应，把溶液中的金元素析出。
化 学	电子废物不仅包含重金属，还包括有毒的溴系阻燃剂和塑料添加剂；樱桃和苹果种子含有少量的天然氰化物。
物 理	金属在高度真空的条件下蒸发；胶质银和胶质金（纳米金）是存在于液体中的微观粒子；金元素具有弹性光散射特性，表现为红色、蓝色或金色；粒子加速器在获得巨大能量供应下，能把铅元素转变为金元素；金不会生锈或变色。
工程学	干电池或纽扣电池能分离出有价值的可回收金属；黄金可以先被蒸发，然后沉积为一层薄膜；尾矿坝是采矿废物被永久保存的场所；在公元前1200年发明出来、用于生产珠宝和艺术品的失蜡技术（即熔模技术），至今还被采用。
经济学	采用金本位制货币体系的国家，它们的货币与黄金重量挂钩；淘金热是一种早期投机形式：只有三分之一的投机者挖到黄金，且需要5年时间才能赚回到达淘金地的费用；供应的重要性：黄金在16世纪时比白银贵15倍，如今比白银贵50倍。
伦理学	当我们的电子垃圾就含有黄金时，我们怎么能继续开采金矿，把地球弄得满是伤疤，造成环境污染并制造健康危机呢？
历 史	埃及人在公元前3600年就融化了黄金；来自土耳其的吕底亚人在公元前564年首次铸造了金币，而中国在同一年有了首枚方形金币。
地 理	世界上最大的金矿在印度尼西亚、南非和巴布亚新几内亚；世界上最大的垃圾填埋场很可能是太平洋，而陆地上最大的垃圾填埋场位于韩国。
数 学	黄金价格和政府债务的变化趋势显然是有联系的。
生活方式	电子产品正在被快速消耗，却没有人关心它产生的废物，也没有从这些废物中回收那些贵金属。
社会学	在很多文化中，黄金已经被用作交换媒介、计价单位和价值符号；黄金象征着财富，而且在社会上是权力的保障。
心理学	黄金与男性能量、太阳的力量相联系，白银与女性能量、月亮的能量相联系；金色是获胜的颜色，被看作乐观和积极的颜色。
系统论	我们处置这么多废物，但在资源回收上所做甚少；我们太专注于我们的日常问题，错过了能带来收入和工作的机遇。

教师与家长指南

情感智慧
Emotional Intelligence

海鸥

海鸥很聪明，消息灵通。他们因现在的食物没有过去多而感到焦虑，并且他们的不适也与由电池释放的重金属有关。海鸥已经就谁将影响他们的未来进行了一次公开讨论：如果所有垃圾都被焚烧，他们就没有吃的了！海鸥展示了逻辑推理能力，不相信人类不愿意分享他们的见解和观点。海鸥精通垃圾管理技术，不带骄傲或争斗性地分享他们的知识。他们在沟通时毫无拘束，表达他们的意见时既无保留又不过分。海鸥对时间和空间很敏感，同时维持一种全球范围的视角。海鸥内在天性积极乐观，即便是处于压倒性的不利境地，也通过揭示没有被利用的机遇进行竞赛，鼓励彼此采取建设性的态度。这种积极的氛围激励其中一只海鸥成为新时代的矿工。

艺术
The Arts

你能找到一些旧的印制电路板（PCB）吗？把这种带有许多晶体管的绿色面板设想成拼图的一部分。你既可以把它看作废物，也可以把它视为艺术品。利用你的想象力，把印制电路板改造成一件艺术品。一定要把图片发给我们哦！我们很可能在下一版的《冈特生态童书》发表它。

TEACHER AND PARENT GUIDE

思维拓展
Systems: Making the Connections

　　我们不是消费者社会，而是用后即弃型社会。在进入我们经济体系的所有原材料中，平均只有10%是被实际消费掉的。我们真正消费的是既定企业成长战略的一部分，称为"计划性报废"，在这种战略中，功能性产品注定会由于软件更新、性能的边际改进或设计的改变而发生功能性失调。这些产品大多数属于电子产品，包括配有铜线和显示单元的印制电路板。产品的电子化已经降低了零件的数量，提高了装配速度，降低了生产和维护的成本；这也意味着电子元件现在已经渗透我们的家庭、办公室和工厂。当这些数字产品被抛弃时，同样意味着主要为聚合物形式的电子废物的增长。大部分电子废物没有得到循环利用。电子废物的循环利用自身是高度集中的，需要大规模运送，而这类金属和塑料的混合物的分离过程需要很多能量和化学品。新的机遇也在显现，能完善循环利用过程，消除酸浸污染。即便这些良性技术（例如螯合作用）已经众所周知并得到了坚实的证明，却很少有工业企业去推行，而是情愿坚持那些"别人都在干"的方式。现有企业已经感知到与采用其他可选技术相关的风险，却没有评估所有明显的机遇。在采矿业方面（来自垃圾填埋场、露天矿或绿地竖井矿的都市型开采），需要突破这种普遍的无知。对于像开采业这样的行业来说，有这么多的积极机遇可供利用，能很容易成为对社会有积极贡献的企业。

动手能力
Capacity to Implement

　　利用卫星地图软件近距离观察一些旧矿山，注意是竖井矿而不是露天矿，特别是那些运行至少达50年的老矿。列出至少10个由矿山造成的问题，包括矿产公司需要应付的负面议题。然后列出一份你在某处旧金矿现场观察到的机遇清单。系统地阐述你的观点：采矿业有未来吗？你认为采矿业可以激发年轻工人和工程师去寻求光明的职业生涯吗？

教师与家长指南

故事灵感来自
This Fable Is Inspired by

马克·库蒂法尼
Mark Cutifani

　　马克·库蒂法尼出生于澳大利亚悉尼市外的伍伦贡。他大学毕业时获得采矿工程专业的学位，并在澳大利亚的采矿业开始他的职业生涯。经过几年在加拿大的杰出工作，他先是成为世界上第二大黄金开采公司的首席执行官，后来在一家世界上最大的多元化矿业公司之一的公司担任首席执行官，后者几乎开采从铂、钻石到铜、铁、煤的所有矿产。他早就意识到采矿业面临的挑战，通过探索新的产业创造更大的价值，致力于给这一产业提供更可持续的发展基础。

图书在版编目（CIP）数据

冈特生态童书.第二辑修订版：全36册：汉英对照 /
（比）冈特·鲍利著；（哥伦）凯瑟琳娜·巴赫绘；
何家振等译.—上海：上海远东出版社，2021
书名原文：Gunter's Fables
ISBN 978-7-5476-1759-5

Ⅰ.①冈… Ⅱ.①冈… ②凯… ③何… Ⅲ.①生态
环境–环境保护–儿童读物—汉、英 Ⅳ.①X171.1-49

中国版本图书馆CIP数据核字（2021）第213075号

著作权合同登记号图字09-2021-0823

策　　划	张　蓉
责任编辑	祁东城
封面设计	魏　来　李　廉

冈特生态童书
遍地是金
［比］冈特·鲍利　著
［哥伦］凯瑟琳娜·巴赫　绘
唐继荣　译

记得要和身边的小朋友分享环保知识哦！
八喜冰淇淋祝你成为环保小使者！

Housing 51

快乐而健康
Happy and Healthy

Gunter Pauli

［比］冈特·鲍利 著
［哥伦］凯瑟琳娜·巴赫 绘
唐继荣 译

上海远东出版社

丛书编委会

主　任：贾　峰

副主任：何家振　闫世东　林　玉

委　员：李原原　祝真旭　牛玲娟　梁雅丽　任泽林
　　　　王　岢　陈　卫　郑循如　吴建民　彭　勇
　　　　王梦雨　戴　虹　翟致信　靳增江　孟　蝶

特别感谢以下热心人士对童书工作的支持：

匡志强　宋小华　解　东　厉　云　李　婧　陈　果
刘　丹　熊彩虹　罗淑怡　旷　婉　杨　荣　刘学振
何圣霖　廖清州　谭燕宁　韦小宏　李　杰　欧　亮
陈强林　王　征　张林霞　寿颖慧　罗　佳　傅　俊
胡海朋　白永喆　冯家宝

目录

快乐而健康	4
你知道吗?	22
想一想	26
自己动手!	27
学科知识	28
情感智慧	29
艺术	29
思维拓展	30
动手能力	30
故事灵感来自	31

Contents

Happy and Healthy	4
Did You Know?	22
Think about It	26
Do It Yourself!	27
Academic Knowledge	28
Emotional Intelligence	29
The Arts	29
Systems: Making the Connections	30
Capacity to Implement	30
This Fable Is Inspired by	31

城市里的水脏了,谁也不敢喝这种水。水上飘(水黾)拜访了青蛙一家,讨论发生的事情。

"一开始,人类只是把洗涤剂排入水中,但是现在情况更糟糕了,他们甚至把药品和激素也扔进我们珍贵的饮用水里。"水上飘叹息道。

The water in the city is dirty. No one dares to drink it. The water striders visit the frog family to discuss the situation.

"First, humans only threw soap in the water, but now it's worse – they're even throwing medicine and hormones into our precious drinking water!" laments the water strider.

城市里的水脏了

The water in the city is dirty

他们知道自己在干什么吗……

Do they know what they're doing ...

"他们知道自己在干什么吗？"青蛙很好奇。

"不！他们没有意识到即使吃下肚，药丸仍然有效。更糟的是，过期的药丸被冲进厕所下水道，以致我们也被迫吃下它们！"

"这么做不合理啊。对水进行清洁处理代价高昂，而且药丸本身也要花很多钱呢。"

"Do they know what they're doing?" wonders the frog.

"No, they don't realise that their pills continue to work even after they've taken them. Worse, when pills are out of date, people flush them down the toilet so that we have to take them!"

"It doesn't make sense. It's expensive to clean water, and pills cost a lot of money."

"对水进行清洁？没有哪种水处理设备和化学手段能强大到中和水里的人造化学物质。"水上飘难过地说。

"这真是坏消息！"

"那么，人们为什么不用更天然的药物？阿育吠陀医学已经流传了几千年，而且这种药能像食物那样溶解。"

"Clean water? No water treatment or chemical is strong enough to neutralise manmade chemicals in water," the water strider says sadly.

"That's bad news!"

"So why aren't people using more natural medicine? Ayurvedic medicine has been around for thousands of years, and it dissolves like food."

阿育吠陀医学

Ayurvedic medicine

……中国有一套独特的传统健康理念……

... Chinese have a unique health tradition ...

"阿育……什么？就是那种来自印度的植物药吗？"青蛙问道。

"阿——育——吠——陀。没错，就是它。"

"我希望这种伟大的药物有个我能记住的名字。但我的确知道中国有一套独特的关于强身健体的传统健康理念，让人少生病。"

"Ayu– what? Is that the plant medicine from India?" asks the frog.

"A-yu-r-ve-dic. Yes, it is."

"I wish this great medicine had a name I could remember. But I do know that the Chinese have a unique health tradition of strengthening the body so that it doesn't get sick."

"中国人有伟大的智慧,非洲人、阿拉伯人以及许多其他著名或不著名的文化也是。"水上飘边沉思边说道。

"问题在于西方对天然药物持怀疑态度!"

"嗯,西方医学也有许多改善生活质量的发明。"

"The Chinese have a great wisdom, so do Africans, Arabs and many known and unknown cultures," the water strider muses.

"The problem is that the West is suspicious of natural medicine!"

"Well, Western medicine also has many innovations that improve quality of life."

对天然药物持怀疑态度

Suspicious of natural medicine

多少物种灭绝了

How many species go extinct

"这些发明没有改善我的生活质量。"青蛙说道。"你有没有意识到由于他们这种生活方式,有多少物种灭绝了?"

"你可以关注坏的方面,但你也可以看到好的方面。要知道,没有哪件事情是完美的,任何事物都可以被改善。"

"They haven't improved my quality of life," says the frog. "Do you realise how many species go extinct because of their way of life?"

"You can focus on the bad, but you can also look for the good, knowing that nothing is perfect, and everything can be improved."

"那是不是意味着没有哪种药物能对所有人在所有时间都有效？"

"首先，我们要记住，许多人生病是因为他们有压力又生气。"

"你的意思是，生病并不总是因为细菌或病毒？"

"Does that mean there's no medicine that works for everyone, all the time?"

"Firstly, we need to remember that many people get sick because they are stressed and angry."

"You mean it's not always because of bacteria or viruses?"

乐观和快乐

Be positive and happy

"没错！其次，某些人觉得自己病了，只是他们认为自己生病了而已。"

"天啊！你是说这些人想象某件事情要发生，然后这件事情就真的发生了？"

"第三，当人生病时，如果相信自己会好起来，他们就会更快恢复健康！"

"这么说，最好的药物就是乐观和快乐，即便身边有坏消息也是如此？"青蛙问道。

"So! And secondly, some people feel sick because they think they're sick."

"Oh dear! You mean they imagine something that then comes true?"

"Thirdly, when people are sick, they get healthy faster if they think they can get better!"

"So the best medicine is to be positive and be happy, even when there's bad news around you?" asks the frog.

"噢，如果听新闻，我保证你听到的一切都是坏消息。请记住，在中国，危机意味着机遇！"

"有能让人快乐的药吗，水上飘？"

"可能有！你可以从讲笑话开始每一天，经常笑，尤其是自己犯错时更要笑。还要一直想着，今天能为其他人做些什么！"

……这仅仅是开始！……

"Oh, I guarantee that if you listen to the news, bad news is all you hear. Remember, in China crisis means opportunity."

"Is there a medicine to be happy, Water Strider?"

"Maybe. You can start the day by telling a joke, laugh regularly – especially about your own mistakes – and always wonder what good you can do for others today!"

... AND IT HAS ONLY JUST BEGUN!...

……这仅仅是开始！……

… AND IT HAS ONLY JUST BEGUN! …

Did You Know?

你知道吗？

When people take pills, part of the medication passes through their bodies and is flushed down the toilet. Water-treatment plants cannot remove all traces of this medication.

当人们吃药时，一些药物成分通过身体后，被冲入厕所。水处理设备不能去除所有残留的药物成分。

Pharmaceuticals in water cause damage to people and wildlife. Drinking water can contain an accumulated cocktail of drugs.

水中的残留药物对人类和野生生物造成伤害，而饮用水可能含有各种各样积累的药物，成了"药品鸡尾酒"。

Drugs are active beyond their expiration date.

过了有效期之后,药品仍具有活性。

Soap in water reduces water tension. This means that water becomes "wetter". And very wet water eliminates the natural protection we all have, while it permits water to penetrate fibers and take away the dirt.

水里的肥皂成分降低了水的张力。这就意味着水变得"更湿"了。非常湿的水打破了我们原有的天然保护,但也让水渗透纤维,把污垢带走。

Ayurvedic medicine originated in India 3,000 years ago. The term means "The science of life".

3 000 年前，阿育吠陀医学起源于印度，这个名词的原意是"生命的科学"。

People who are happy and have a positive outlook on life are less likely to get sick.

那些快乐和有积极人生观的人不容易生病。

In China, a crisis is also considered an opportunity. The key is to have the capacity to see the opportunity in a crisis.

在中国,危机也被看作机遇,关键是要具备从危机中看到机遇的能力。

机遇

If all else fails, we should at least have the capacity to learn a lesson from that failure.

如果一切都失败了,我们至少应该有能力从失败中汲取教训。

Think about It
想一想

When you are sick in bed, do you feel happy?

当你卧病在床的时候，你感到快乐吗？

当你面临无法解决的麻烦时，你会有压力吗？

Do you get stressed when you have a problem you cannot resolve?

When everything goes wrong, do you spare a moment to think about the lessons you could learn, or do you prefer to forget the situation as soon as possible?

当事事不顺心时，你是会抽空思考能够汲取的教训，还是宁可尽快忘记这一困境？

一个国家能通过计算它的国民有多么幸福来衡量它的进步吗？

Can a nation measure progress by calculating how happy its citizens are?

Check the medicines you have at home. Does your family only take prescription drugs, or do some members of your family use alternative medicines? Check that your family's medicine is stored in a safe place, and then separate the bottles into two groups: those prescribed by a medical doctor, and those bought over the counter. Look at the expiration date of the drugs. Are some too old to be taken? The medicine that is too old needs to be disposed of. How would you do that?

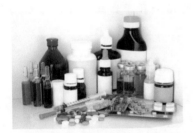

检查家里的药品。你的家人只服用处方药,还是有些人会服用非传统药品?检查并确保你家的药物放在一个安全的地方,然后将药瓶分为两组,一组是医生开的药,另一组是在药店买的药。注意药物保质期,有些是不是已经过期了?过期的药需要正确处置。你会怎样处置它们?

TEACHER AND PARENT GUIDE

学科知识
Academic Knowledge

生物学	由腺体分泌的激素调节生理和行为；只要水的张力维持正常，水黾（俗名水上飘）就能在水面上行走；青蛙的皮肤有一层天然的保护膜，只要水的表面张力正常，就能在水中防止重金属的侵入；芦荟可以治愈烧伤；山金车是消炎药；小檗消解肾结石；大蒜素有很强的杀菌作用。
化 学	合成分子被设计成具有稳定的性能，即生物降解过程缓慢，能够长时间维持活性，以延长其有效期；植物合成化合物来发挥生物功能；植物化学品是植物体内自然生成的化合物；番木瓜能够用于处理伤口，茶树可以抵抗真菌侵袭，姜黄能改善肝功能。
物 理	孤立现象的因果关系；安慰剂效应。
工程学	废水通过沉淀、曝气和需氧消化几个过程进行处理；只有反渗透方法才能去除饮用水中几乎全部（90%）的溶解药物和来自香烟的尼古丁。
经济学	城市公共服务消耗的能源中，20%用来净化污水；医疗保健（尤其是药物）的成本正在上升，促使药房为患者提供与品牌药效果完全一样的通用药；能有效治疗人类疾病的12 000种植物化合物创造了相关产业和工作岗位。
伦理学	制药业生产出昂贵的药物，但这些药物只有部分被患者消耗，却在水中保持很长时间的活性。人们明明知道药物含量随着消费增长而积累，更大剂量的化学药品没有经过任何方式处理就被释放到水体中，怎么能宣称在水中被稀释的药物没有危险？
历 史	阿育吠陀医学起源于3 000年前，它包括早在中世纪就闻名于世的制剂和手术步骤；中医起源于2 000多年前。
地 理	不丹通过幸福感来衡量进步。
数 学	不丹政府进行了详细的调查，来计算国民幸福指数。
生活方式	中医包括药物、针灸、推拿、气功和食疗；把将要过期的药物等废物通过厕所冲走的习惯，是一个主要但相对不引人注意的健康危机；法国人不怎么注重个人卫生，但很关注肝脏健康；美国人的保健成本是世界上最高的；持续的压力和恐惧可以改变生物系统，甚至导致各种疾病，如心脏病、中风和糖尿病；长期的愤怒和焦虑可以加速动脉粥样硬化和炎症；积极乐观的生活态度会让患冠心病的危险降低一半；持续暴露在抗生素环境下会降低免疫系统应对自然变化的能力。
社会学	医生所穿的白大褂表示效率和卫生，而医院里哗哗作响的医学机器显示了高科技的超凡能力；如果患病是因为抽烟所致，那么患者获得同情的可能性会低一些。
心理学	积极乐观的态度有助于身体恢复；个人态度创造积极或消极的身心联系；不能把某种疾病归咎于任何人，人们需要下决心挺过它。
系统论	身心一体；如果心灵不幸福，身体不会健康。

教师与家长指南

情感智慧
Emotional Intelligence

青 蛙

青蛙控制着他的情绪，即便得到坏消息也不会过于压抑。他确认了水上飘所说的现象属实。青蛙对自身的局限性诚实而坦率。然而，当讨论转向医学的时候，青蛙显示出理解力，并分享他的意见，强烈表达他对人类和生物多样性遭受破坏的负面看法。青蛙向水上飘学习，渴望去了解更多，他不满足于大致的了解，提出了与个人请求相关的问题：是否有一种让人快乐的药物？

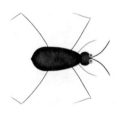

水上飘

水上飘心烦意乱，焦躁不安。他抱怨人类所犯的错误，以及在解决问题方面的技术不足。水上飘脑海中有清晰的解决方案，也就是从合成药物转向天然药物，并歌颂许多文化的智慧。但水上飘也平衡了自己的意见，指出西医改善了生活质量。他的关注点从水转向更冷静达观的生活态度。他的逻辑结构有条理，不是在药物方面阐述他的观点，而是转而讨论态度和生活方式。水上飘概述了怎样管理危机，并提供在他自身行为基础上的建议，表明他有一种快乐的生活方式。

艺术
The Arts

想象一下你很生气，你会选用哪些颜色来绘画？现在，设想你感到快乐和放松，你会选什么颜色？想象你生病了，需要很长时间才能恢复，那么哪些颜色能表达你的感觉？再设想你有严重的感冒，但相信明天就会恢复，那么你现在想选用哪些颜色？颜色反映我们的心理状态、态度和期待，要用让你感到快乐和健康的颜色来绘画。哪些颜色能让你有这种感觉呢？与你的朋友们讨论你偏爱的颜色，并发现他们对颜色的选择。

TEACHER AND PARENT GUIDE

思维拓展
Systems: Making the Connections

　　水资源有限，被污染的水会危及生命。合成药物不易降解，被人体消化且过了有效期之后，仍然能在水中发挥药效。这看似没有意义，但它证明了现代社会的线性思维，即只考虑单一、可测变量的简单因果逻辑关系，在本例中就是用药物杀死病毒或传染性细菌。虽然某种药品在水体中可能是被稀释得浓度太低，很难被检测到，但成千上万种合成药品都被释放到同一水体中时，问题就出现了。随着时间推移，药物分子积累超过一个危险的阈值；即便这种积累效果还不足以直接影响人类健康，成百万的水生物种有可能受到冲击。海洋中的塑料制品已经积累到可以形成塑料岛屿，而鱼类可能因为错把这些塑料当作浮游生物食用，导致肠道堵塞。水中的药品还包括能从根本上改变身体和活细胞的功能与强度的激素和抗生素，这涉及从免疫系统到繁殖能力等各个方面。虽然我们这个社会总体上没有意识到由现代医学导致的挑战，然而它也积极接受由非传统医学所提倡的健康福祉。现代医学关注因果关系，而其他医学系统关心心灵、精神、生活方式和态度。正如我们不能再在生活上忽视心理的重要性那样，我们也不能忽略由合成化学品造成的破坏。越来越多的科学证据表明我们的心理状态和态度确实影响我们的健康。我们是该学会变得快乐，并且在面对一连串负面消息时也能努力去寻找机遇。我们必须保持头脑清醒、积极和专注，同时定期找时间笑一笑，有时甚至可以自嘲。

动手能力
Capacity to Implement

　　努力把快乐带给不快乐的人。这样做的难度有多大？想想你比较亲近的家庭和朋友圈中那些不快乐的人。你不需要知道他们为什么不快乐，而是需要制订计划去振奋他们的精神。试着讲一个笑话，但需要注意，仅仅打扮成一个小丑是不够的，你需要做更多。

教师与家长指南

故事灵感来自
This Fable Is Inspired by

皮尔丽特·弗朗德兰
Pierrette Flandrin

皮尔丽特·弗朗德兰出生于摩洛哥的卡萨布兰卡，在这座城市长大，她父亲是宫廷的著名摄影师。她在学习秘书服务专业之后，决定去巴黎生活。当她结束在巴黎银行长期从事的银行业工作并退休后，决定将她的余生投入促进和平和理解的事业中，特别是（包括摩洛哥、阿尔及利亚和突尼斯在内的）马格里布地区和法国之间的和平与理解。她受了挫折也不气馁，特别是不被健康因素打败，总是乐观积极、平易近人、努力工作。在寻找机遇时，她保持冷静，控制好自己的情绪。她确保了《冈特生态童书》第一辑法文版的翻译工作顺利完成，对翻译细节精益求精，动员志愿者并依靠他们完成了严谨的出版目标。她在辞世之前享受到了阅读《冈特生态童书》法文版和英文版的乐趣，给我们留下了无限的回忆。

图书在版编目(CIP)数据

冈特生态童书.第二辑修订版:全36册:汉英对照/
(比)冈特·鲍利著;(哥伦)凯瑟琳娜·巴赫绘;
何家振等译.—上海:上海远东出版社,2021
书名原文:Gunter's Fables
ISBN 978-7-5476-1759-5

Ⅰ.①冈… Ⅱ.①冈…②凯…③何… Ⅲ.①生态
环境-环境保护-儿童读物—汉、英 Ⅳ.①X171.1-49

中国版本图书馆CIP数据核字(2021)第213075号

著作权合同登记号图字09-2021-0823

策　　划　张　蓉
责任编辑　祁东城
封面设计　魏　来　李　廉

冈特生态童书
快乐而健康
[比]冈特·鲍利　著
[哥伦]凯瑟琳娜·巴赫　绘
唐继荣　译

记得要和身边的小朋友分享环保知识哦！
八喜冰淇淋祝你成为环保小使者！

Health 52

可以喝，还可以穿

Drink It, Wear It

Gunter Pauli

［比］冈特·鲍利 著
［哥伦］凯瑟琳娜·巴赫 绘
何家振 李欢欢 译

上海远东出版社

丛书编委会

主 任：贾 峰

副主任：何家振 闫世东 林 玉

委 员：李原原 祝真旭 牛玲娟 梁雅丽 任泽林
　　　　王 崂 陈 卫 郑循如 吴建民 彭 勇
　　　　王梦雨 戴 虹 翟致信 靳增江 孟 蝶

特别感谢以下热心人士对童书工作的支持：

匡志强 宋小华 解 东 厉 云 李 婧 陈 果
刘 丹 熊彩虹 罗淑怡 旷 婉 杨 荣 刘学振
何圣霖 廖清州 谭燕宁 韦小宏 李 杰 欧 亮
陈强林 王 征 张林霞 寿颖慧 罗 佳 傅 俊
胡海朋 白永喆 冯家宝

目录

可以喝，还可以穿	4
你知道吗？	22
想一想	26
自己动手！	27
学科知识	28
情感智慧	29
艺术	29
思维拓展	30
动手能力	30
故事灵感来自	31

Contents

Drink It, Wear It	4
Did You Know?	22
Think about It	26
Do It Yourself!	27
Academic Knowledge	28
Emotional Intelligence	29
The Arts	29
Systems: Making the Connections	30
Capacity to Implement	30
This Fable Is Inspired by	31

兔子正在为那场著名的赛跑做准备。乌龟观察着这位赛跑明星，发现他正在往腋下喷东西。

"你是在给自己喷兴奋剂吗？"乌龟好奇地问道，"没有必要啊，你比我们跑得都快。"

"喷兴奋剂？当然不是，我只是为了确保在比赛后身体不会有异味。"

A hare is preparing himself for the famous race. A tortoise observes the star of the track and notices that he is spraying something under his arms.

"Are you doping yourself?" wonders the tortoise. "There is no need to. You run faster than any of us."

"Doping? Of course I am not doping, I'm just ensuring that I will not smell bad after the race."

兔子正在为那场著名的赛跑做准备

A hare is preparing himself for the famous race

很多草本植物都可以控制异味

Many herbs control odour

"真的吗？你要用化学品来控制异味？有很多草本植物也可以达到同样的效果，而且不含会堵塞毛孔的金属成分，你知道吗？"

"我不了解你的情况，乌龟先生。以你的跑步速度，你可能从来都没有在运动中出过汗。但是我必须保证自己在冲过终点线时全身清爽。"

"Really? Do you need chemicals to control odour? Do you know there are many herbs that can do exactly the same – without plugging your pores with metals?

"I do not know about you, Tortoise. You have probably never, ever worked up a sweat at the pace you run at. But I have to be sure I am fresh when I cross the finishing line!"

"为什么你不用咖啡来达到同样的效果呢?"乌龟问道。

"咖啡含有咖啡因!它也许会使我跑得更快,但不会让我的体味闻起来更好!"

"不,我说的不是可以喝的咖啡,而是可以穿的咖啡。"

"Why don't you do that with coffee?" asks the tortoise.

"Coffee – with caffeine! That may make me run faster, but not smell better!"

"No, I'm not talking about drinking coffee. I'm talking about wearing coffee."

咖啡使我跑得更快

Coffee makes me run faster

用咖啡来种蘑菇

Coffee used to farm mushrooms

"对不起,你一定是搞错了。我知道喝完咖啡后,你可以用咖啡渣来种有益健康的蘑菇。所以当你喝完咖啡后,你可以用余下的残渣来种植食物,但这并不能帮我控制异味!"

"Excuse me, you must be mistaken. I know that after you drink coffee, you can use the waste to grow healthy mushrooms. So after you drink it you use it to grow food to eat – but I still smell!"

"你说的完全正确。但仔细想想,难道咖啡的味道不好闻吗?"

"我喜欢新鲜咖啡的气味。"

"你以前知道咖啡能吸收异味吗?"

"That is absolutely correct. But think about it – doesn't coffee smell great?"

"I love the smell of fresh coffee."

"Did you know that coffee absorbs foul odours?"

我喜欢新鲜咖啡的气味

I love the smell of fresh coffee

咖啡在面料里

Coffee is in the fabric

"我并不会感到惊讶,但是如果有全是咖啡渣的喷雾除臭剂,我会感到吃惊。"

"啊,可是咖啡不在喷雾式或滚珠式除臭剂里,而是在面料里。"

"我才不信呢!那样可能一两次还管用,可当你洗衣服的时候,汗水和咖啡渣就会一起被冲洗掉。"

"I wouldn't be surprised, but I would be amazed if there is a deodorant spray full of coffee grounds."

"Ah, but the coffee is not in the sprays or roll-ons, it is in the fabric."

"I'm not convinced! That may work once or twice, but the moment you wash it, my sweat and the coffee grounds will be flushed away."

"那我问你：你跑步时会喝水吗？"

"会喝啊。你知道我们需要补充水分。"

"那你会怎样处理装水的塑料瓶呢？"

"Let me ask you this: Do you drink water while running?"

"Yes, I do – you know we have to hydrate ourselves."

"And what do you do with the plastic bottle?"

那你会怎样处理装水的瓶子呢?

What do you do with the bottle?

咖啡可以消除所有异味

Coffee eliminates all odour

"放到废物回收桶里，但是好像大部分塑料瓶都被烧掉了。"

"听着，信不信由你，咖啡可以掺到回收的塑料瓶中，加到做成服装的面料里。咖啡不仅能消除衣服的所有异味，还包括你鞋子的异味。"

"Off to the recycling bin – but it seems they burn most of them."

"Listen, believe it or not, coffee can be mixed with recycled plastic bottles and added to fabrics for clothing. It not only eliminates all odours of your clothes, but also your shoes!"

"哦，这下可好了，我的跑鞋真是太需要它了！我太爱咖啡了，可以供我喝，供我吃，供我穿，穿上它跑步还能助我取胜！而这一切都不会产生任何异味！"

……这仅仅是开始！……

"Well, what a relief, my running shoes are really in need of it! I love the fact that I can drink it, eat it, wear it, run with it and win! All without any smell!"

... AND IT HAS ONLY JUST BEGUN! ...

……这仅仅是开始！……

...AND IT HAS ONLY JUST BEGUN!...

Did You Know?
你知道吗？

We only ingest 0.2% of the bean when we drink a cup of coffee: 99.8% is wasted.

当我们喝一杯咖啡时，只能吸收咖啡豆的 0.2%，99.8% 都被浪费了。

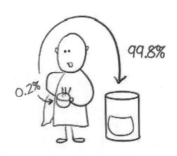

If we were to use 100% of the bean, then we could create 500 times more value.

如果我们能够 100% 利用咖啡豆，那么我们将会创造出 500 倍的价值。

Coffee originally comes from Ethiopia where, as early as the 9th century, farmers noted that goats were more energetic after eating the red coffee cherries.

咖啡起源于埃塞俄比亚，早在公元9世纪，那里的农民就注意到，山羊在吃完红色的咖啡果后会变得更加兴奋、有活力。

Worldwide people drink 1.4 billion cups of coffee every day that amounts to 500 billion cups per year.

全世界人们每天喝掉14亿杯咖啡，一年加起来就是5 000亿杯。

Coffee is sterilized after being brewed. The fibre-rich substrate is then ready for farming mushrooms.

咖啡经过冲泡后就消毒了。然后其中富含纤维的物质可用于种植蘑菇。

The cultivation and processing of beans for one cup of coffee requires 130 litres of water.

种植和加工一杯咖啡所需的咖啡豆，需要消耗130升水。

People drink coffee for the stimulating effect the caffeine molecule has on the brain.

人们喝咖啡是因为咖啡因分子对大脑产生刺激作用。

The farmers pick the coffee cherry from the coffee plant, and we brew our coffee from with the pips of the coffee cherry.

农民从咖啡树上采摘咖啡果,而我们就是用咖啡果的果仁来冲泡咖啡。

Think About It

Why would we use chemicals when natural products could accomplish the same job?

如果天然产品可以实现同样的效果，那么我们为什么还要用化学产品呢？

What do you do when your sportswear and shoes smell after exercise?

运动后，如果你的运动服和运动鞋有异味，你会怎样做？

What does the hare have to say when he realises that a blend of waste plastic and waste coffee can improve his quality of life?

当兔子意识到废塑料和咖啡渣的混合物能够改善他的生活质量时，兔子会怎么说呢？

Is the tortoise worried about winning - or is he more concerned about something else?

乌龟是担心比赛的胜负，还是更关心其他事情呢？

Do It Yourself! 自己动手!

Collect coffee grounds from a coffee machine at home, or stop by your local restaurant, café or coffee shop. Collect at least one large jar. Then add some fruit and vegetables and allow it to decompose. You can also use fish heads. When it begins to decompose, put the jar in the refrigerator. What do you notice after a day? Now put the coffee grounds in the refrigerator again. What do you notice the next day?

从家里的咖啡机，或者到当地的餐厅、咖啡厅或咖啡店中收集咖啡渣。至少要收集一大罐。然后，加入一些水果和蔬菜，放置使其腐烂。你也可以用鱼头替代。当它开始腐烂时，把罐子放进冰箱。一天后，你会有什么发现？现在把咖啡渣再放进冰箱里，第二天你会有什么发现？

TEACHER AND PARENT GUIDE

学科知识
Academic Knowledge

生物学	运动员使用人工兴奋剂；汗液对身体的清洗作用；用天然气味来控制身体的异味；咖啡因对神经系统的作用；用纤维种植蘑菇；运动时摄入水分对于身体的重要性。
化 学	咖啡因分子的特殊性质；金属氧化物在止汗方面的应用；咖啡中多糖转化为淀粉的过程。
物 理	如何制造出不用气体的喷雾；与咖啡混合后，合成纤维由疏水性变为亲水性；通过加入咖啡纤维增强织物抵抗紫外线的能力；临界二氧化碳的提取；咖啡渣的孔隙多，具有高吸水性。
工程学	如何从冲泡后的咖啡中提取油；如何把咖啡渣碾碎成3—4微米的小粒子。
经济学	当资源的100%而不是0.2%被充分利用时，资源利用率对经济的影响；更多更快的资金循环对当地经济的影响；天然产物对合成物的替代效果；产生多重效益和大量现金流的商业模式；气候变化对于区域经济尤其是农业的影响，是什么导致了种植户卖出农作物的价格和消费者付费之间的价格差；努力提高土地生产率与让最终产品产生更多价值的对比。
伦理学	当一种原材料可以提供健康、具有价格竞争力的食物和化学物质时，却将原材料的99.8%用于焚烧发电，企业怎么证明这种做法的合理性？少做一件坏事仍是坏事，拒绝做好事也是坏事，做更多的好事才是最好的！
历 史	咖啡是如何风靡全球的；谁是第一个咖啡种植户。
地 理	荒漠化：第一棵咖啡树生长的地方已经变成了一片沙漠，荒漠化破坏了当地的经济。
数 学	如果资源利用率提升500倍，巴西、越南和哥伦比亚的经济会发生什么变化？
生活方式	对待个人卫生的态度转变：过去我们习惯一周洗一次澡，但是由于空气污染，现在的城市居民每天要洗两次澡；我们更多地使用化学合成物来控制身体的异味。
社会学	咖啡聚会的文化；法国咖啡文化的产生、美国咖啡店的出现以及世界各地的新兴咖啡文化。
心理学	相对在一个简单的环境中进行咖啡的创新和纯度制造，咖啡品牌的营销指通过营销促使消费者花更多的钱来支付咖啡公司的品牌宣传。
系统论	作为饮品的咖啡在冲调后的残渣可以变成一种原材料，提供食物和化学能量，与单纯地增加产量相比，会给种植户带来更多的收入。

教师与家长指南

情感智慧
Emotional Intelligence

兔子

兔子有自我意识，并且知道自己是赛场明星，但是仍有时间和乌龟对话，乌龟是比赛中的一个对手，但谈不上竞争关系。兔子与乌龟开玩笑，说乌龟从没有在跑步后出过汗。兔子没有因乌龟的问题而生气，而是解释了自己的需求。他仔细听取了乌龟提出的建议，并直接给出答案，展示了自己的知识。与此同时，兔子始终关注着对话内容，充满好奇，但不做任何评判。当找到机会学习时，他一边学习一边发笑。

乌龟

乌龟很耐心，尊重兔子，同时保持距离，从表面的问题下手循循善诱，为兔子寻求解决方案，以免太过直接。乌龟认真听取每个回答，以有逻辑条理的方式来进行信息提炼。乌龟意识到兔子的难以置信，所以将话题转移到咖啡的主题上，从喝水的话题开始寻求解决方案。乌龟用这个话题吸引了兔子，并直接暗示兔子的回答是错误的（瓶子回收后大部分都被焚烧掉了），这使得兔子抛弃成见，接受了乌龟的解决方案。作为回报，乌龟得到了兔子的尊重和共鸣。

艺术
The Arts

咖啡不仅仅是一种饮品，还是纺织品的原材料以及一种重要的着色剂。因此，收集一些咖啡渣（咖啡冲泡之后留下的废弃物）吧。用手挤干其中残留的水或者油脂。现在，拿起画笔在T恤上画出你喜欢的图案，用咖啡作为你的油墨。确保你使用的是旧T恤，这样你妈妈就不会因为T恤的永久染色而介意哦。

TEACHER AND PARENT GUIDE

思维拓展
Systems: Making the Connections

　　咖啡种植户追求更高的产值。他们清理竹林和小菜园，以便有效地利用每一寸土地来收获更多的咖啡，这增加了生产力，但是随着市场上咖啡数量的增多，咖啡价格下跌。此外，由于山顶上的竹子越来越少，土地变得贫瘠，需要灌溉、施肥和投药，这些都增加了生产成本。高成本降低了利润，导致种植户虽然产量上升，但收入却变少了。情况糟糕时，很多种植户甚至买不起自己的食物，以前他们会在自己的土地上种植这些食物，而现在却不得不去市场上购买。结果，在世界上咖啡的高产地区出现了更多的营养不良和贫穷。第一个战略是划分出多个不同特色品牌，使高品质产品能够获得更高的价格。我们的挑战是种植户没有一个有力的品牌。烘焙店、酿酒厂和咖啡店将不同类型的咖啡按比例混合，它们控制、影响着品牌。种植户只有一种选择：在原有的基础上做更多的事情。除了市场上销售的咖啡豆，咖啡果的核是种植蘑菇的理想材料，利用当地种植户手中的废弃物就地培育食物是解决问题的第一步。蘑菇收获之后剩下的残留物是很好的鸡饲料。种植户还可以同时做其他事情，比如生产速溶咖啡。这种生产仅仅提取了咖啡中的可溶物质，剩下了15%的油脂和85%的固体成分。这些固体成分可以与聚合物混合，加工成制作衣服和鞋子的纤维；油脂部分可以加工成泡沫和涂料。这样的话，种植户应对咖啡危机便有了一个明确的答案：利用你所拥有的东西，创造出比想象中更多的价值。

动手能力
Capacity to Implement

　　有的纺织品公司会这样推销自己的产品："可以让你保持凉爽、温暖和干爽，保护你避免日晒……"找到这类纺织品公司并针对其服装产品做一个调查，看看这些服装产品是不是对生态友好，因为它提供了额外的功能。同时，查看一下有没有提及环境效益。然后与掺有咖啡纤维的纺织品在产品和性能上比较一下。拟定一份简单的销售方案：给出至少10个理由来说服你的潜在客户购买掺入了咖啡纤维的运动衣和运动鞋。

教师与家长指南

故事灵感来自
This Fable Is Inspired by

陈国钦
Jason Chen

陈国钦获得纺织工程学位之后，成为一名纺织品制造商。1989年，他决定创建自己的公司，并从那时起开始专注于纺织品的功能性。与世界各地那些寻求低成本的生产商不同，他想为人类制造出舒适、健康的衣服。2005年，陈国钦与他在台湾兴采实业（Singtex）的团队一起，开始致力于研究由400纳米咖啡颗粒和塑料纤维混合而成的融合物。今天，100多个国际户外服装品牌采用了该材料。最初他们在当地商店收集咖啡渣，但是随着需求的增加，他们与哥伦比亚国家咖啡种植户联合会（世界上最大的种植户合作联合会）建立了合作。现在他们合作用咖啡制造纱线和纺织品供应当地市场。这个想法只是陈国钦众多创意之一。他认为，茶叶、稻壳和菠萝纤维也将会为创建可持续发展的纺织行业带来更多突破性的新组合。

图书在版编目(CIP)数据

冈特生态童书.第二辑修订版:全36册:汉英对照/
(比)冈特·鲍利著;(哥伦)凯瑟琳娜·巴赫绘;
何家振等译.—上海:上海远东出版社,2021
书名原文:Gunter's Fables
ISBN 978-7-5476-1759-5

Ⅰ.①冈… Ⅱ.①冈…②凯…③何… Ⅲ.①生态
环境-环境保护-儿童读物—汉、英 Ⅳ.①X171.1-49

中国版本图书馆CIP数据核字(2021)第213075号
著作权合同登记号图字09-2021-0823

策　　划　　张　蓉
责任编辑　　程云琦
封面设计　　魏　来　李　廉

冈特生态童书
可以喝,还可以穿
[比]冈特·鲍利　著
[哥伦]凯瑟琳娜·巴赫　绘
何家振　李欢欢　译

记得要和身边的小朋友分享环保知识哦!
八喜冰淇淋祝你成为环保小使者!

Health 53

不用画的色彩

Colour without Paint

Gunter Pauli

[比]冈特·鲍利 著
[哥伦]凯瑟琳娜·巴赫 绘
何家振 李欢欢 译

上海远东出版社

丛书编委会

主　任：贾　峰
副主任：何家振　闫世东　林　玉
委　员：李原原　祝真旭　牛玲娟　梁雅丽　任泽林
　　　　王　岢　陈　卫　郑循如　吴建民　彭　勇
　　　　王梦雨　戴　虹　翟致信　靳增江　孟　蝶

特别感谢以下热心人士对童书工作的支持：

匡志强　宋小华　解　东　厉　云　李　婧　陈　果
刘　丹　熊彩虹　罗淑怡　旷　婉　杨　荣　刘学振
何圣霖　廖清州　谭燕宁　韦小宏　李　杰　欧　亮
陈强林　王　征　张林霞　寿颖慧　罗　佳　傅　俊
胡海朋　白永喆　冯家宝

目录

不用画的色彩	4
你知道吗?	22
想一想	26
自己动手!	27
学科知识	28
情感智慧	29
艺术	29
思维拓展	30
动手能力	30
故事灵感来自	31

Contents

Colour without Paint	4
Did You Know?	22
Think about It	26
Do It Yourself!	27
Academic Knowledge	28
Emotional Intelligence	29
The Arts	29
Systems: Making the Connections	30
Capacity to Implement	30
This Fable Is Inspired by	31

炽热的沙滩上,一只蜣螂(屎壳郎)正沿着上坡把大象的粪便滚成粪球。这时一只蜂鸟飞了过来,注意到了蜣螂的辛勤工作。

蜂鸟钦佩地说:"看到你是如何把一堆粪便变成一个个粪球的,真是太神奇了。"

"我别无选择。"蜣螂回答道。

A dung beetle is rolling a ball of elephant dung uphill, across the hot sand when a hummingbird comes along and notices his hard work.

"It's amazing to see how you turn this pile of dung into round balls," says the hummingbird with admiration.

"I have no choice," responds the beetle.

沿着上坡把大象的粪便滚成粪球

Rolling a ball of elephant dung uphill

为刚孵出来的孩子们提供好的食物

Good food when they hatch,

"嗯,我明白,因为你要为那些刚从卵里孵出来的孩子们提供好的食物。"
"谁告诉你这些的?"

"Well, I do understand that you need to provide your little ones with good food when they hatch from their eggs."
"Who told you that?"

"这是关注你独特生活方式的人都知道的事情啊。这种生活方式甚至使你在埃及法老和后妃中很受欢迎呢！"

"你知道这沙子有多热吗？"蜣螂问道。

"不知道，我的朋友。我通常飞在花丛中啜饮花蜜，几乎从不飞到地面上。"

"It's something everyone who pays attention to your strange lifestyle knows. It even made you a favourite among the Egyptian pharaohs and their consorts!"

"Do you have any idea how hot this sand is?" asks the beetle.

"No, my friend. I fly from flower to flower to sip nectar and hardly ever come down to sit on the ground."

在埃及法老中很受欢迎

A favourite among the Egyptian pharaohs

我用这堆粪便!

I use this pile of dung!

"你不到地面来,太好了。沙子实在是太热了。"

"那么你怎样才能让你的脚保持凉爽呢?你会舔它们吗?"

"当然不会。我可没有那么多唾液来保持凉爽,我用这堆粪便!"

"It's better you don't. It's burning hot."
"So what do you do to keep your feet cool? Do you lick them?"
"No ways. There would never be enough saliva around for that. I use this pile of dung!"

"什么？我从没听说过有谁坐在一堆粪便上来保持凉爽！"

"这是一堆新鲜的粪便，所以它是湿润的。"

"真的吗？"

"Excuse me? I've never heard of anyone sitting on dung to keep cool."

"This is fresh dung. So it's moist."

"Really?"

坐在糞便上保持清爽

Sitting on dung to keep cool

……水分蒸发……

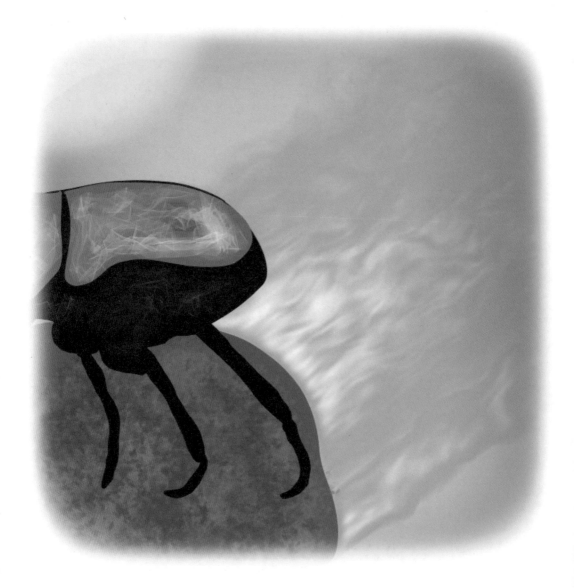

... water evaporates ...

"那么告诉我,蜂鸟,潮湿的东西暴露在太阳下会发生什么?"

"呃,它会变干,尤其是在这么热的环境中。"蜂鸟回答道。

"那么,表面的水分蒸发后会发生什么呢?"

"So tell me, Hummingbird, what happens to something that is wet and exposed to the sun?"

"Well, it dries out, especially in this hot environment," answers the little bird.

"So, what happens to a surface where water evaporates?"

"当然是温度降低。"

"你说对了。我这样做不仅仅是为孩子们储备食物,同时也可以让我的脚保持凉爽。记住,在大自然中我们做事情永远不是只有一个理由。"

"你说得太对了,"蜂鸟感叹道,"可不幸的是,人们认为我的羽毛只是看起来漂亮,却忽略了它能调节我的体温,让我保持干燥。"

"It cools down, of course."

"You've got it. I'm not just getting food ready for my little ones, I'm keeping my feet cool. Remember, in Nature we never have only one reason for doing something."

"You are so right," sighs the hummingbird, "but unfortunately people think I have feathers just to be beautiful. They ignore the fact that feathers regulate my temperature and keep me dry."

羽毛调节我的体温!

Feathers regulate my temperature!

你的色彩十分艳丽

Your colours are simply gorgeous

"呃，我不得不承认你的色彩十分艳丽，你的美丽几乎把我的眼睛都弄花了。"蜣螂坦承道。

"你呢，蜣螂先生，你的那些令人难以置信的圣甲虫兄弟们同样有着漂亮的色彩，而且天然、无毒。"

"的确，圣甲虫的亮绿色在古埃及很受欢迎，著名的奈费尔提蒂王后还将一只甲虫放进了她的项链中，其色彩在她去世后几千年间始终保持亮丽。"

"Well, I have to admit that your colours are simply gorgeous. I am almost blinded by your beauty," admits the dung beetle.

"And you, Dung Beetle, you have some incredible scarab brothers who have spectacular colours as well, natural and non-toxic."

"Yes, that's true. My brothers' bright green colouring was so popular in Ancient Egypt that even the famous Queen Nefertiti put one of us in her necklace and it stayed bright for millennia, until long after she passed away."

"在一些文化中,我们的羽毛会世代相传,这些羽毛也永远不会失去光泽。"

"这是因为我们的亲戚,甲虫和蜂鸟们,已经学会了仅仅利用光泽度,而不用任何颜料来制造色彩。"

"要是再有一位奈费尔提蒂王后来彰显我们的美丽就好了,我的蜣螂朋友!"

……这仅仅是开始!……

"And in some cultures our feathers are kept for generations, and these never lose their sparkle either."

"Why yes, that's because our kin, the beetles and the birds, have learned how to make colour without any pigments, by merely making use of the play of light."

"If only we had another Nefertiti to celebrate our beauty, my beetle friend!"

... AND IT HAS ONLY JUST BEGUN!...

……这仅仅是开始!……

... AND IT HAS ONLY JUST BEGUN! ...

Did You Know?
你知道吗?

Dung beetles eat the droppings of other animals, preferably herbivores, which means that they are eating mainly undigested plants.

蜣螂以动物（尤其是食草动物）的粪便为食物，这说明它们主要吃未消化的植物。

蜣螂是称职的父母，有些甚至一辈子在一起。它们共同筑巢，共同照看它们的后代。母蜣螂把每个卵放在一个形似香肠的小卵房里，这样当幼虫孵出来时，便有充足的食物了。

Dung beetles are good parents, and some even stay together for life. They construct a nest together, and share in the care of their offspring. The female deposits each egg in its own tiny sausage-shaped casing so when a larva emerges, it has ample food.

In Australia, dung beetles raise on the droppings of local animals, such as kangaroos, refuse to feed on the droppings from horses, sheep or cattle.

在澳大利亚，蜣螂食用本地动物（如袋鼠）的粪便，而不吃外来的马、羊或牛的粪便。

Dung beetles want fresh droppings as dried ones are less palatable. A fresh pile of elephant dung will attract 4,000 dung beetles within 15 minutes, and as many as 16,000 in half an hour.

蜣螂喜欢新鲜的粪便，因为干燥的粪便不够"可口"。一堆新鲜的大象粪便在15分钟内会吸引4 000只蜣螂，在半小时内会吸引多达16 000只蜣螂。

Dung beetles are guided by the stars to find their way home.

蜣螂依靠星星的指引寻找回家的路。

Dung beetles can pull or push a load equivalent to 1,141 times their own body weight. In human terms that is like a person of 70 kg pulling or pushing 80 tons.

蜣螂可以推拉相当于自己体重1141倍的重物。对人类而言，相当于一个70千克的人推拉80吨重的东西。

ℬeetles that eat dung do not need to eat or drink anything else as the dung provides enough water and all the necessary nutrients.

蜣螂除了粪便外不需要吃喝其他东西，因为粪便已经给它们提供了充足的水和所需的营养。

一些蜣螂拥有超常的嗅觉，能够帮助它们找到新鲜的粪便。另一些蜣螂则把自己吸附在粪便制造者身上，当有粪便排出时再跳下来。

𝒮ome dung beetles have an exceptional sense of smell, which enables them to find fresh dung. Others attach themselves to a dung-provider and jump off when their reward is delivered.

Think about It
想一想

Would you be able to survive on the droppings of others?

你能靠其他生物的粪便生存吗?

你同意我们应该为蜣螂坚持回收粪便长达 3 000 万年的行为给予称赞吗?

Do you agree that we should celebrate the dung beetle for relentlessly recycling waste for 30 million years?

Who was the stronger: Hercules in Ancient Greece or the dung beetle today?

谁更强大,是古希腊的大力神还是今天的蜣螂?

Have you ever thought of insects as good parents?

你想到过昆虫也可以做称职的父母吗?

Do It Yourself!
自己动手!

Time for a little mathematics. The speed of light is 300,000 kilometres per second. The sun is 149 million kilometres from the earth. If you ride your bicycle at 25 kilometres per hour, how many years would it take to get there? The speed of light is the fastest speed that we know of. That is why a change of angle immediately gives a different colour to a bird's feathers. So play with feathers in the sunlight and see how fast you can make the colours switch.

做一个简单的计算。光速是 3.0×10^5 千米/秒，太阳距离地球 1.49×10^8 千米。如果你以每小时 25 千米的速度骑自行车，需要多少年才能从地球骑到太阳？光速是人类已知的最快速度，这就是角度变换会立刻使鸟类的羽毛显现出不同色彩的原因。所以，拿着羽毛在阳光下玩吧，看看能让色彩变换得多快。

TEACHER AND PARENT GUIDE

学科知识
Academic Knowledge

生物学	蜣螂没有骨头，所以很少在化石中出现，但粪便化石表明在3 000万年前蜣螂就已经存在了；蜣螂通过把粪便转化成土壤，传播种子，促进营养循环，抑制蝇类病菌的滋生，提供了重要的生态系统服务；蜣螂以粪便的液体成分为食，如果无法找到粪便，则以真菌和腐烂的水果为食；蜣螂一生要经历卵、幼虫、蛹和成虫四个阶段。
化 学	蜣螂的外壳由几丁质组成，是一种含氮的多糖类物质（同样存在于虾壳中）；外壳中还含有弹性蛋白，一种名为节肢弹性蛋白的橡胶状物质。
物 理	水分蒸发使粪球的温度略低于环境温度；蜣螂利用太阳、月亮和银河系里光的不同强弱的位置来导航；蜣螂用特殊的触角来捕捉牛粪味；色彩因为光的折射而形成；不需要颜料而由光的干涉形成的颜色叫作结构色，光的干涉是指两波叠加，形成高低不一致的光波幅度。彩虹的颜色是随着光线角度的改变而引起的色彩改变，这是两个或更多的薄膜表面结合后，光线进出膜层发生折射、反射现象的干涉结果。
工程学	在照相机、扫描仪、望远镜、DVD播放器、影印机、夜视镜甚至人造卫星中都利用到光的折射和颜色控制原理。
经济学	古代地中海人和中东人从埃及进口圣甲虫。
伦理学	对于埃及人来说，如果你通过了审判判决，你就能继续活下去，否则你将不复存在；爱的对立面不是恨，而是无视对方的存在。
历 史	埃及人认为蜣螂使地球像一个巨大的球一样不停运转；奈费尔提蒂是埃及著名法老阿肯那顿的妻子，因其半身像而闻名，这是古埃及复制最多的艺术品之一；阿兹特克人在身上戴蜂鸟标本作为护身符，秘鲁纳斯卡线遗址上还有蜂鸟的绘画。
地 理	蜂鸟生活在除南极洲外的各大洲；古埃及人依靠尼罗河的潮起潮落生存，尼罗河是他们唯一的水和食物来源。
数 学	古埃及历法将一年分为12个月，每月30天，年末另加5天来配合季节，共365天；公元前238年，托勒密王朝下令每四年增加一天，但是在公元前22年才生效。
生活方式	据说护身符可以保护自己远离危险和伤害，因而被认为是福物。
社会学	在逻辑学和科学实验盛行之前，人们创造了神话，来解释周围的世界；埃及人有三个季节：冬季、夏季和洪水期；羽毛作为装饰物被应用在不同文化的时装和传统服饰中。
心理学	人们排斥改变，价值观基本不变；我们的免疫系统、睡眠质量以及对抗疼痛的忍耐力会受到自身信念的影响，所以护身符在古代医学里具有重要寓意。
系统论	蜣螂提供了广泛的生态系统服务。

教师与家长指南

情感智慧
Emotional Intelligence

蜣螂

蜣螂专心地、反复地将粪球推上山,他的回答简短扼要,没有任何寒暄。蜣螂的计划是给出建议,尽管他的解释起先听起来有些神秘。随着蜂鸟对可行的解决方案表示出了兴趣,蜣螂进一步扩充答案,回答了蜂鸟的疑问。起初蜣螂本能地从个人情况的角度出发回答问题,后来态度发生了转变,与蜂鸟产生了共鸣,并赞美了蜂鸟。同时随着蜂鸟的进一步恭维,蜣螂展开了更深层次和宽度的对话,展示了自己的知识,显示了自信。

蜂鸟

蜂鸟钦佩蜣螂的辛勤工作和娴熟技巧。蜂鸟没有因为蜣螂的回答简单而受挫,相反地,蜂鸟继续自信地与蜣螂对话,谈到蜣螂家族在古代的荣耀。蜂鸟在交谈中感到惊讶,并从提问者变为了回答者,进行了简短、切中要点的回答。蜂鸟诉说了他的伤心之处,大家只是因为他的美丽而欣赏他。蜂鸟承认了自己的美丽,同时投桃报李称赞了蜣螂,表示希望能和蜣螂一起因为本色得到奈费尔提蒂王后的赞美,引起了蜣螂的共鸣。

艺术
The Arts

在我们的家园中,我们不会因为甲虫或蜂鸟的美丽而杀掉它们。我们能做的是利用它们的瑰丽颜色来创作绘画。为了展现生物的金属光泽,需要采用特殊涂料。要确保亮度是从金属而非化学品中提取而来。你可以尝试在新的材料上绘画,比如石头纸。这种纸的表面不吸水,需要使用油性涂料创作。通过这种方式,你可以了解到油性涂料和水性涂料的区别。

TEACHER AND PARENT GUIDE

思维拓展
Systems: Making the Connections

蜣螂是生态系统中的一个缩影，几乎所有动物的新陈代谢都会包含摄食、消化和排泄。当过多的物种聚集在一个有限空间时，如一个公寓或一个城市，我们设计的解决方案是在污水处理厂以高成本的运输（管道）和处理（污泥曝气）方式集中处置排泄物。现在的方案是在空气充足的条件下对废弃物进行集中处置，促进细菌将生物残留转化成矿化沉淀。不同的是，在自然界中，粪便没有被看作废物，而是一种食物。一种生物的排泄物可能是其他生物的食物来源。把粪便有效转化为食物的关键是把垃圾碎片化，提供给需要的人。大象粪便会吸引数以万计的蜣螂，它们将粪便滚成球，然后在形似香肠的壳里产卵，以保障幼虫的营养。为了维持生计，它们可以在几分钟内将一堆粪便分解成若干小块，这是生态系统服务的一部分。快速处理掉粪便能够控制蝇类传染病的传播，粪便还可以作为肥料，肥化土地，补充表层土壤，确保生物的多样性。事实上这能够保证在接下来的季节里，可以生长出喂养动物们的草。城市里我们面临的最大挑战是，我们从土壤中获取营养物质，但从没有利用自然界中的废物来给土壤补充营养。这导致了土地养分的耗尽和农业产量的减少，从而要求更多的合成肥料和更高强度的耕种。这是一个恶性循环，而蜣螂却向我们展示了一个良性循环。

动手能力
Capacity to Implement

找出一块玻璃，但是小心不要划伤自己！确保它不是凸透镜，因为那样在太阳下会有着火的危险。如果你能找到几块厚度不同的玻璃就更好。现在，开始寻找合适的角度让阳光透过玻璃照在一张白纸上，观察纸上的颜色。你会发现，在同样的角度下，照射出的颜色是一样的。随着太阳的移动，你会看到不同的颜色。如果是下雨天不能到室外，尝试着在室内做。你将看不到任何颜色，这是因为没有阳光的照射，玻璃不能发生折射现象。等太阳出来了，试着再做一遍。你可以尝试利用不同的玻璃，直到在纸上显现出独特的颜色来。这需要耐心，但是你会得到回报的！

教师与家长指南

故事灵感来自
This Fable Is Inspired by

安德鲁·帕克
Andrew Parker

安德鲁·帕克生于英国，23岁时为了学习海洋生物学和物理学移居澳大利亚。10年后，他回国成为牛津大学的研究人员，并很快有了"新世纪科学家"的名声。他认为，大约5.2亿年前的生物多样性大爆发是由眼睛的进化造成的，因为这使捕食者和猎物可以看到彼此。安德鲁的见解为模拟自然界中的色彩提供了很多机遇。他用天然纳米技术模仿蝴蝶金属色的翅膀；利用鸟类的彩虹色生产安全设备来取代全息图；发明了无需反光板的太阳能电池板。他通过研究沙漠中收集水的甲虫，发明了从空调中回收水的装置。他的作品有《创世之谜》《眼睛一眨》和《七种致命的色彩》。他是伦敦自然历史博物馆的研究负责人，那里收藏有上百万种不同的甲虫。

图书在版编目(CIP)数据

冈特生态童书.第二辑修订版:全36册:汉英对照/
(比)冈特·鲍利著;(哥伦)凯瑟琳娜·巴赫绘;
何家振等译.—上海:上海远东出版社,2021
书名原文:Gunter's Fables
ISBN 978-7-5476-1759-5

Ⅰ.①冈… Ⅱ.①冈… ②凯… ③何… Ⅲ.①生态
环境-环境保护-儿童读物—汉、英 Ⅳ.①X171.1-49

中国版本图书馆CIP数据核字(2021)第213075号

著作权合同登记号图字09-2021-0823

策　　划	张　蓉
责任编辑	程云琦
封面设计	魏　来　李　廉

冈特生态童书
不用画的色彩
[比]冈特·鲍利　著
[哥伦]凯瑟琳娜·巴赫　绘
何家振　李欢欢　译

记得要和身边的小朋友分享环保知识哦!
八喜冰淇淋祝你成为环保小使者!

Health 54

无需开采的金属

Metals without Mining

Gunter Pauli

[比]冈特·鲍利 著
[哥伦]凯瑟琳娜·巴赫 绘
何家振 李欢欢 译

上海远东出版社

丛书编委会

主　任：贾　峰

副主任：何家振　闫世东　林　玉

委　员：李原原　祝真旭　牛玲娟　梁雅丽　任泽林
　　　　王　岢　陈　卫　郑循如　吴建民　彭　勇
　　　　王梦雨　戴　虹　翟致信　靳增江　孟　蝶

特别感谢以下热心人士对童书工作的支持：

匡志强　宋小华　解　东　厉　云　李　婧　陈　果
刘　丹　熊彩虹　罗淑怡　旷　婉　杨　荣　刘学振
何圣霖　廖清州　谭燕宁　韦小宏　李　杰　欧　亮
陈强林　王　征　张林霞　寿颖慧　罗　佳　傅　俊
胡海朋　白永喆　冯家宝

目录

无需开采的金属	4
你知道吗？	22
想一想	26
自己动手！	27
学科知识	28
情感智慧	29
艺术	29
思维拓展	30
动手能力	30
故事灵感来自	31

Contents

Metals without Mining	4
Did You Know?	22
Think about It	26
Do It Yourself!	27
Academic Knowledge	28
Emotional Intelligence	29
The Arts	29
Systems: Making the Connections	30
Capacity to Implement	30
This Fable Is Inspired by	31

细菌们聚集在一起参加每周的聚会。和往常一样,他们嘲笑着人类。

"你们知道吗?他们用炸药来炸开地壳。"一个细菌说道。

The bacteria are gathering for their weekly get-together. As usual, they are poking fun at humans.

"Do you know they are using dynamite to break open the Earth's crust?" says one bacterium.

细菌们聚集在一起参加每周的聚会

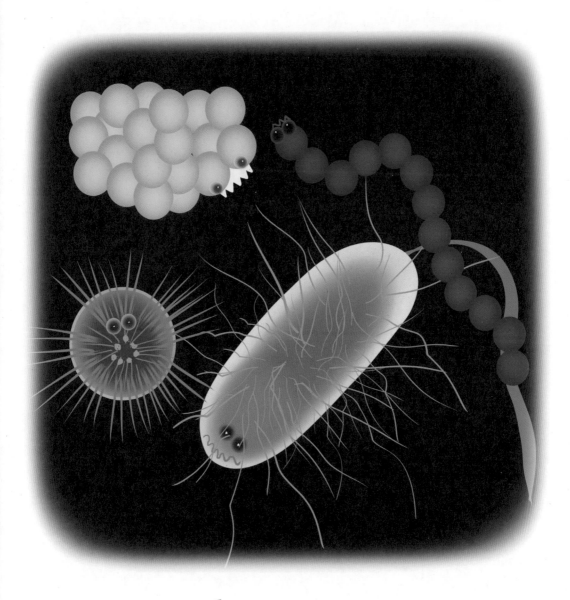

The bacteria are gathering for their weekly get-together

在地面上炸出大洞

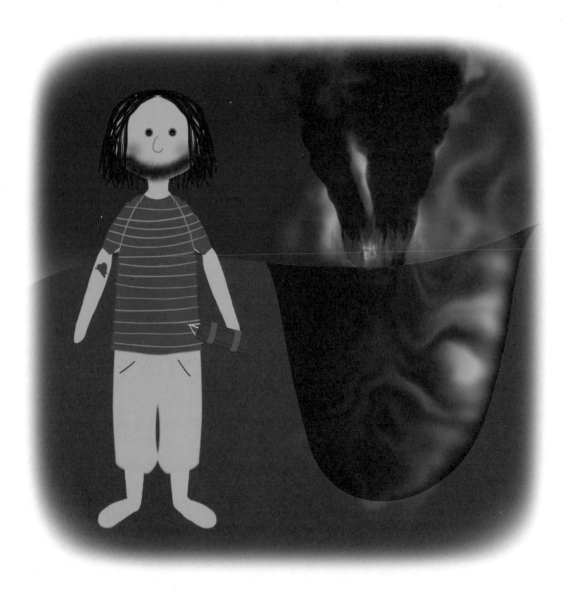

Big holes in the ground

"他们有什么权利这么做？他们发明了取火，就开始焚烧一切他们碰得到的东西，现在他们想毁灭一切吗？"

"有些人用炸药捕鱼，毁坏珊瑚，但大部分人用炸药在地面上炸出大洞。"

"What gives them the right to do that? They invented fire and started burning everything they could get their hands on, and now they want to blow everything up?"

"Some humans use dynamite to catch fish and destroy corals, but most use it to blow big holes in the ground."

"我想知道，"另一个细菌说道，"他们从来没有和地衣——我们的专业矿工聊过吗？"

"想也别想！大多数人根本不知道地衣是什么！他们误以为它是植物，事实上，它是菌类和藻类的共生联合体，给双方都带来了巨大利益。"

"I wonder," says another bacterium. "Have they never talked to the lichens, our professional miners?"

"No ways! Most humans don't even know what lichen is! They mistake it for a plant, when really it's the friendship between mushroom and algae that brings great benefits to both."

和地衣聊过

Talked to the lichens

……最沉默的开采方式

... most silent type of mining

"人类如此无知,太丢人了!地衣能在岩石上挖出小沟,一天能挖几百米,然后把矿物质输送给植物。"

"他们从没认识到地衣只有两个细胞的厚度。这是最温和最沉默的开采方式。想想吧,如果用挖掘机而不是手术刀做胃部手术——光是这个想法就足以吓跑他们了!"一个细菌咯咯地笑起来。

"These humans are so ignorant, it's embarrassing! Lichens can dig tiny tunnels in rocks, hundreds of metres a day, and deliver minerals to plants."

"They've never figured out that lichens are only two cells thick. It's the most benign and silent type of mining. Imagine proposing surgery of the stomach with an excavator instead of a scalpel – the idea alone will scare them off!" giggles a bacterium.

"听着,人类本来也可以成为温和的开采者。问题是他们根本不知道自己擅长什么。"

"你的意思是人类也有擅长的东西?别忘了,我们称他们为'无智慧的人'!"细菌们大笑起来。

"Look, humans could also be benign miners. The problem is they have never realised what they are good at."

"You mean humans are good at something? Remember, we call them the homo non sapiens!" The bacteria burst out laughing.

无智慧的人

Homo non sapiens

富含铁质的食物而不是垃圾食品

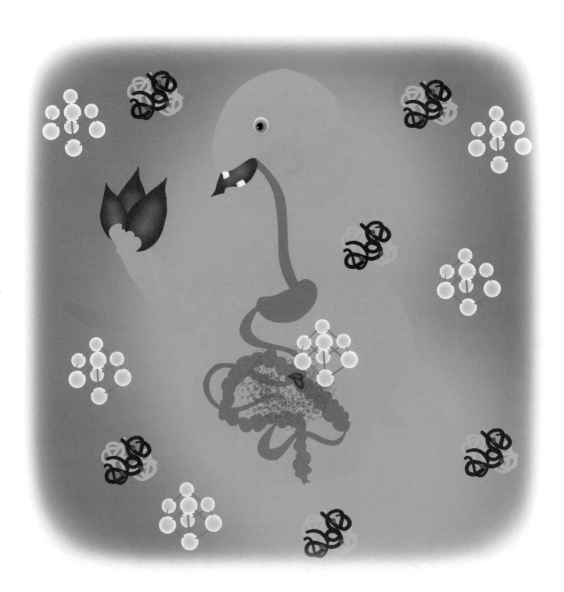

Iron-rich food instead of junk food

"听着,他们的祖先可没这么蠢,"一个年长的细菌说道。"他们吃富含铁质的食物,而不是现在的垃圾食物。他们的身体可以吸收少量铁元素,使其和氢、碳、氮和氧相结合,形成血红蛋白。"

"血红什么?哦,别讲了。这是晚会,不是化学课!"

"不,说真的。就像我们细菌可以从一堆垃圾中辨别和吸收铜、金、银或锌,人类可以利用铁元素做同样的事情。"

"Look, their ancestors weren't so dumb," says an older bacterium. "They ate iron-rich food instead of the junk food of today. And their bodies can take a minute amount of iron and mix it with hydrogen, carbon, nitrogen and oxygen to make haemoglobin."

"Haemo– what? Oh, stop it. This is an evening out, not a chemistry lesson!"

"No, really. Just as our bacteria friends can identify and absorb copper, gold, silver or zinc from a pile of dust, people can do the same with iron."

一些细菌对此提起了兴趣。"那为什么人类的血液需要铁元素呢？"

"嗯，铁元素帮助血液接收和输送更多的氧气，这样会使身体充满活力，也能让大脑更好地运转。"

Some of the bacteria are becoming interested. "So why do humans want iron in their blood?"

"Well, iron helps blood to pick up and transport more oxygen, which in turn energises the whole body, and makes the brain work better."

让大脑更好地运转

Makes the brain work better

如果他们真的如此聪明，干吗要……

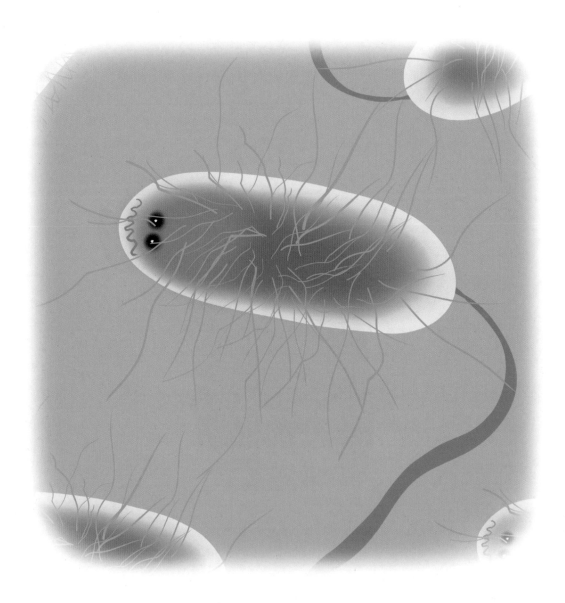

If humans are that intelligent, how come …

"对于人类来说是个不错的设计！"年长的细菌喃喃地说。

可是一个细菌急于发问："如果他们真的如此聪明，干吗要用炸药把地球炸出巨大的坑？他们为什么要破坏海洋？又是谁提议把金属转化成气体？"

"Not a bad design for a human being!" mumbles the old one.

But one of the bacteria is eager to ask a question. "If they are that intelligent, how come they use dynamite to blow huge holes in the Earth? And why do they destroy the sea? And whoever proposed turning metals into gas?"

"我想知道,"又一个细菌开口道,"人类是愚蠢吗,或者他们只是无知?他们丢弃的电话和电脑里所含的金和其他金属,比他们能从地球开采的还多,他们没有意识到吗?"

"嗯,"另一个细菌若有所思地说,"他们确实行事鲁莽,就像闯进了瓷器店的牛。"

……这仅仅是开始!……

"I wonder," says another. "Are people stupid, or are they just ignorant? Don't they realise that they already have more gold and other metals in the phones and computers that they dump, than could ever be extracted from the Earth?"

"Hmm," another bacterium muses. "They certainly do behave like a bull in a china shop."

... AND IT HAS ONLY JUST BEGUN!...

……这仅仅是开始！……

…AND IT HAS ONLY JUST BEGUN!…

Did You Know? 你知道吗?

Dynamite is destructive as an explosive, but its main ingredient, nitroglycerine, is also used as a medicine to treat heart conditions.

甘油炸药是一种极具破坏性的炸药，不过甘油炸药的主要原料硝酸甘油也是一种治疗心脏病的药物。

Lichens are a combination of algae and fungi: one makes food for the other.

地衣由藻类和真菌共同组成：两者相互供给养料。

Lichens inhibit the growth of ringworms and tuberculosis. Roccella lichen produces litmus, which is the standard method to measure pH.

地衣能抑制皮癣和结核病。从染料衣属地衣中可以提取石蕊，它是测量溶液酸碱度（pH值）的标准方法。

Lichens are so strong, they can pulverise rock solely with the water pressure in their cells.

地衣非常强大，通过其细胞内的水压可以压碎岩石。

Iron, – a mineral which is abundant in meat, tofu and spinach – is a component of haemoglobin, the substance in red blood cells that carries oxygen through the body.

铁是肉类、豆腐和菠菜中富含的一种矿物质，是血红蛋白的组成成分。血红蛋白是红血球内的一种物质，负责将氧气输送到全身。

叶绿素分子和血红蛋白分子几乎是完全一样的，但有一个原子不同：镁代替了铁。

Chlorophyll is exactly the same molecule as haemoglobin, but with one atom difference: the iron is replaced by magnesium.

Bacteria and their synthetic counterparts can remove heavy metals from our bodies and can separate platinum group metals from ore.

细菌和细菌合成物能清除人体内的重金属，还能从矿石中分离出铂族金属。

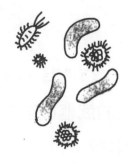

Junk food has little nutritional value and offers high fat, sugar, salt and calories.

垃圾食品几乎没有营养价值，而且脂肪、糖、盐和热量含量都很高。

Think About It 想一想

Can you manufacture weapons while also honouring peace?

你能一边崇尚和平，一边制造武器吗？

How can a tiny alga supported by a small fungus be strong enough to crush a rock to pieces?

有没有一种威力特别强大的武器以至于人们害怕使用它？

Could a weapon ever be so powerful that people would be afraid to use it?

在微小的真菌帮助下，微小的藻类如何强到足以研碎岩石？

How is it possible that something bad can also be good?

有些坏事也可能又是好事，这怎么可能呢？

Do It Yourself !   自己动手！

Have a good look around your bedroom. How many products do you own that have been made using dynamite or other explosives? Make a list of everything you suspect is a result of the use of explosives. Present your list to your friends and family and, if you are ready, state that you do not want the face of the earth to be destroyed. Will anyone listen to you?

仔细查看你的卧室。你有多少件物品在生产过程中用了甘油炸药或其他炸药？将所有你怀疑使用过炸药的物品列成一个清单。让你的朋友和家人看看你的清单，如果你准备好了，请告诉他们你不希望地球表面遭到破坏。会有人听你的吗？

TEACHER AND PARENT GUIDE

学科知识
Academic Knowledge

生物学	地衣是混合型微生物，以藻类为基础，生存于真菌的藻丝中，这种共生关系出现于6亿年前，两种微生物都从中受益；组成地衣的藻类可以独立生长，或和真菌共同生长，但组成地衣的真菌只能和藻类共同存在；地衣覆盖了8%的地球表面，大部分分布于北极冻原；富含铁质的食物促进红血球生成。
化　学	安全炸药的基础是硝化甘油；医学上，硝酸甘油用于治疗心脏病和前列腺癌；石蕊是一种从地衣提取的染料混合物，渗入纸张制成试纸来指示pH值：酸性环境下，石蕊纸由蓝色转为红色，碱性环境下，红色石蕊纸又转为蓝色；如果缺铁，人体不能制造血红蛋白，因而患上贫血；螯合作用是细菌对特定的正离子有亲和力，借此有选择地离析金属物质。
物　理	硝化甘油不耐冲击，因此需要土壤或木屑吸收；安全炸药捕鱼，即用炸药炸晕或炸死鱼群，以便捞捕，这样毁坏了环境，尤其是珊瑚礁；在潮湿的情况下，地衣能逐渐分解岩石。
工程学	在德国、意大利和瑞士，地衣用作空气污染的生物监测器；医学手术已经从能让手伸进人体做手术的切口，发展到显微镜和精密仪器的微创切口；通过高温蒸发，合金能重新利用。
经济学	本地天然的原料越来越少，取而代之的是来自世界各地的合成原料。
伦理学	你怎么能一边呼吁和平一边支持武器工业呢？在有其他方法可供选择时，采矿和矿石加工怎能继续依赖破坏性的做法呢？
历　史	诺贝尔从克里米亚半岛油田赚到的钱比他的安全炸药专利赚的要多。
地　理	地衣到处都有，从雨林到沙漠，只有深海没有；在一些不适宜居住的环境下，地衣可能是唯一出现的植被。
数　学	含有75%硝化甘油的0.19千克柱状安全炸药包大概含有1兆焦耳能量，相当于0.8千克TNT炸药的威力。
生活方式	石耳属地衣是美味佳肴，英语中叫岩石内脏，日本叫岩茸，韩国和中国叫石耳。
社会学	诺贝尔曾希望能生产一种极具毁灭性的材料，通过其威慑作用阻止战争发生。
心理学	幽默在批判中能否发挥作用？习语能更清楚巧妙地表述某事吗？
系统论	地衣和细菌等自然系统能开采岩石，而没有炸药毁坏性的副作用。

教师与家长指南

情感智慧
Emotional Intelligence

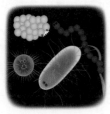

细菌

细菌热爱生活，乐于参加晚会，爱开玩笑。他们有批判性的眼光，展示了逻辑性来支持自己的观点。细菌们联系之前人类滥用火石、毁坏珊瑚的经历，没有给人类狡辩的权利。细菌们自娱自乐时，展现了学问、严谨和对技术细节的精通。他们有时间和位置感，利用对比来澄清争议。人类自称为智者，对此细菌们进行了批判，他们笑称人类为"无智慧的人"。当他们意识到人类的优点和弱点后，话题引向了解决方案，细菌们提出了一些深远的问题，但没有立刻得到回复。细菌们用一个习语进行总结，使得其语言听起来意思明确、说法聪明而且减轻了攻击性。

艺术
The Arts

怎么模仿爆炸声？让我们试着用嗓子。首先，自己想象一下爆炸听起来应该是怎么样的，然后大家一起喊出来。之后，每个人再次喊出自己认为的爆炸声。你注意到听到别人的声音后，自己的声音是怎么改变的吗？大家能发出一样的声音吗？当然，不可能每个人都发出一样的声音，尽管我们试着发出一样的声音，因为每个人的表达和艺术创造力不同。

TEACHER AND PARENT GUIDE

思维拓展
Systems: Making the Connections

> 采矿是一个具有破坏性的行业。几千年来实践证明，一旦矿产挖空了，矿区就成了鬼城，只有极少数例外。更糟的是，只重视利润的公司往往不负责任：它们可能破产倒闭，留下一片惨遭破坏的环境。对高产量的追求导致使用炸药。用于提取矿物和稀有金属的采矿技术极其依赖资源。不断地追求产量和经济规模，采矿过程用的人力越来越少，导致经济崩溃。环境恶化导致生态系统崩溃，环境污染影响着附近居民的健康和生活。如今，最大的金属来源是废弃的电子和医疗设备。但回收利用旧电子产品需要和采矿业相同的技术，同时极其依赖能源，却并没有创造更多的工作，因此不是非走那条路不可——还有其他可选的技术。工业过程自始至终都需要大量能源和原料，所以不会考虑小规模加工的商业模式。但如果我们能大大减少全球矿石和废弃物的运输，那就能证明其他技术也可以是有效的。无论如何，利用细菌对特定金属和矿物质的吸附力已成功开辟了一个市场。只有办公复印机大小的螯合机器每天能处理100到500部手机，以细微粉末形式回收金属，再循环利用。这为都市采矿开创了一条大道：有效利用资源，把危险废弃物转换成经济增长点。

动手能力
Capacity to Implement

> 你家附近有潜在的矿山吗？不要想着大矿山，想一想大量金属在哪里——也许就在你家里！金属无处不在：电线、冰箱电路板、电脑、打印机、电视、手机和玩具……到处都有！想象这是你的潜在矿山，开始策划一个当地的回收公司吧。想象有一台办公复印机大小的螯合机器，想一想你需要多少电子废弃物才能让公司运转起来。这不是写商业计划，而是设计一个商业模式。你会惊奇地发现原来身边有那么多金属，又有那么多金属闲置在家，却没能给你或社区带来收益。

教师与家长指南

故事灵感来自
This Fable Is Inspired by

亨利·科勒辛斯基
Henry Kolesinski

亨利·科勒辛斯基被朋友们称为汉克,他获得了位于美国马萨诸塞州波士顿的东北大学的化学专业学位。汉克有多年合成化学方面的背景。在拍立得公司,他开发了大量用于摄影行业的特殊化学品,并获得专利。担任密理博沃特公司的咨询师期间,他获得了生命科学的经历。他创办了团结科技公司,提供色谱仪,该企业后被热电公司并购。他有多项获批的专利和尚待批准的专利,发表了多篇技术论文。汉克和罗伯特·科里一起创立了 Prime Separations 公司,生产分子分离和分子净化机器,既可以是微小的有机化合物、稀有金属的分子,也可以是蛋白质和污染物质的分子。

图书在版编目（CIP）数据

冈特生态童书.第二辑修订版：全36册：汉英对照 /
（比）冈特·鲍利著；（哥伦）凯瑟琳娜·巴赫绘；
何家振等译.—上海：上海远东出版社,2021
书名原文：Gunter's Fables
ISBN 978-7-5476-1759-5

Ⅰ.①冈… Ⅱ.①冈…②凯…③何… Ⅲ.①生态
环境-环境保护—儿童读物—汉、英 Ⅳ.①X171.1-49

中国版本图书馆CIP数据核字（2021）第213075号
著作权合同登记号图字09-2021-0823

策　　划　　张　蓉
责任编辑　　程云琦
封面设计　　魏　来　李　廉

冈特生态童书
无需开采的金属
[比]冈特·鲍利　著
[哥伦]凯瑟琳娜·巴赫　绘
何家振　李欢欢　译

记得要和身边的小朋友分享环保知识哦！
八喜冰淇淋祝你成为环保小使者！

Health 55

秸秆的妙用

Jobs and Beauty from Straw

Gunter Pauli

[比]冈特·鲍利 著
[哥伦]凯瑟琳娜·巴赫 绘
何家振 李欢欢 译

上海远东出版社

丛书编委会

主　任：贾　峰

副主任：何家振　闫世东　林　玉

委　员：李原原　祝真旭　牛玲娟　梁雅丽　任泽林
　　　　王　岢　陈　卫　郑循如　吴建民　彭　勇
　　　　王梦雨　戴　虹　翟致信　靳增江　孟　蝶

特别感谢以下热心人士对童书工作的支持：

匡志强　宋小华　解　东　厉　云　李　婧　陈　果
刘　丹　熊彩虹　罗淑怡　旷　婉　杨　荣　刘学振
何圣霖　廖清州　谭燕宁　韦小宏　李　杰　欧　亮
陈强林　王　征　张林霞　寿颖慧　罗　佳　傅　俊
胡海朋　白永喆　冯家宝

目录

秸秆的妙用	4
你知道吗？	22
想一想	26
自己动手！	27
学科知识	28
情感智慧	29
艺术	29
思维拓展	30
动手能力	30
故事灵感来自	31

Contents

Jobs and Beauty from Straw	4
Did You Know?	22
Think about It	26
Do It Yourself!	27
Academic Knowledge	28
Emotional Intelligence	29
The Arts	29
Systems: Making the Connections	30
Capacity to Implement	30
This Fable Is Inspired by	31

水稻种子拼命地在找一个地方来处理秸秆，那些秸秆是由废弃的水稻茎秆组成的。胡萝卜一直陪伴着她。

水稻种子叹了口气，说道："你知道的，以前有很多房子是用秸秆盖的，但近来，人们更喜欢用钢筋、水泥和红砖盖房子了。"

A rice seed is desperately looking for a place to get rid of the straw made of leftover rice stalks. A carrot is accompanying her on her search.

The rice sighs. "You know, there used to be so many houses built with straw, but lately people prefer steel, concrete and red bricks."

……找一个地方来处理秸秆……

... looking for a place to get rid of the straw ...

稻草屋

Straw-bale houses

"是呀,"胡萝卜说,"不过有很多人对稻草屋感兴趣。"

"当然了,那很好,但是每年秸秆产量有几百万吨,却没有造上百万座稻草屋。连日本的榻榻米都越来越不受欢迎了。"

"Yes," comments the carrot, "but there are lots of people interested in straw-bale houses."

"Of course, and that's great, but we produce millions of tons of straw every year, and there are not millions of straw-bale houses. Even Japanese tatami mats are losing favour."

"你说得对。但起码中国人在利用秸秆种植蘑菇呀。"

"他们确实有变废为宝的传统。要不然他们怎么能养活十几亿人口呢？"

"You're right. But at least the Chinese use straw to farm mushrooms."

"They do have a culture of converting waste into food. How else would they feed more than a billion citizens?"

用秸秆种植蘑菇

Straw to farm mushrooms

……短杆水稻？

"... rice with shorter stalks?"

"水稻,你听说了吗?在埃及首都开罗,人类正试图改变水稻的基因,生产短杆水稻。"
"为什么要这么做呢?"

"Rice, have you heard that in Cairo, the capital of Egypt, the humans are fiddling with your genes to make rice with shorter stalks?"
"Why would you ever want to do that?"

"呃，人类曾经用秸秆造房子，但现在这不流行了。因此他们决定焚烧堆积如山的秸秆。"

"焚烧？！人类什么时候才会彻底认识到焚烧并不是解决之道呀？"

"Well, they used straw for construction, but it wasn't popular. They had so much piled up that they decided to burn it."

"Burn it?! When will humans finally learn that burning is not a solution?"

决定焚烧秸秆

Decided to burn it

影响孩子们的健康

Risks the health of children

胡萝卜明智地点点头。"这么大量的焚烧，空气里弥漫着浓烟，没人能看清尼罗河的对岸。"

"别说了。焚烧秸秆会影响孩子们的健康。"

The carrot nods sagely. "The burning was so intense, the air got thick with smoke and no one could see across the Nile."

"Don't tell me. Burning straw risks the health of children."

"绝对不利,越来越多的孩子得病,呼吸困难。人类可以想出的唯一解决办法就是改变水稻品种,只种植短杆水稻。"

"所以,人类又想干涉我,就像以前他们给我加入胡萝卜基因一样吗?我有一个更好的主意。埃及人和意大利人聊过吗?"

"为什么他们要这么做?"

"Absolutely, more and more children got sick and couldn't breathe. The only solution they could come up with was to modify rice and only grow the type with short stalks."

"So, these humans want to interfere with me like they did when they added some carrot genes to me? I have a better idea. Have the Egyptians talked to the Italians?"

"Why would they?"

埃及人和意大利人聊过吗？

Have the Egyptians talked to the Italians?

利用秸秆来制造塑料

Make straw into plastics

"意大利人种植水稻有2 000多年的历史了。他们生产粮食的能力巩固了罗马帝国的战斗力。"

"埃及人并不想打仗——他们为什么要和意大利人聊聊呢?"

"噢,意大利人已经学会了如何利用秸秆来制造塑料。"

"The Italians have been farming rice for more than 2,000 years. Their capacity to produce food powered the war machine of the Roman Empire."

"The Egyptians aren't thinking about war - so why talk to Italians?"

"Well, the Italians learnt how to make straw into plastics."

"抱歉，水稻，不过塑料是由石油制成的。"

"过去确实如此，但现在人类可以把秸秆分裂成微粒，就像提炼油厂提炼石油一样。"

"如果那是真的，那就让大家多吃大米，制造更多的塑料吧！我想起来了，我知道有人甚至可以用水稻做出珠宝……"

……这仅仅是开始！……

"Sorry, Rice, but plastics are made from petroleum."

"They used to be, but now humans can crack rice straw into dozens of molecules, just like a refinery does with oil!"

"If that's possible, let's all eat more rice and make plastic! Now that I think about it, I know someone who can even make jewellery out of rice …"

... AND IT HAS ONLY JUST BEGUN!...

……这仅仅是开始！……

... AND IT HAS ONLY JUST BEGUN! ...

Did You Know?
你知道吗？

$\mathcal{S}$traw-bale houses are built to last. There are straw-bale houses that are 140 years old, and those constructed today could last for at least 100 years.

稻草屋经久耐用。有些稻草屋已有140年的历史。今天建造的稻草屋至少能使用100年。

$\mathcal{A}$ single rice straw is not strong, but baled rice straw can last for decades, provided that it is protected from rain and humidity.

一根秸秆不结实，但成捆的秸秆只要不受雨淋，不受潮，就可以用上几十年。

Tatami flooring was originally reserved as seating area for the most important aristocrats in Japan. It was popularised in the 17th century, and only lost favour after the Second World War, though most Japanese homes still have one tatami room.

最初，榻榻米是为日本最负名望的贵族预留的专属座区。榻榻米盛行于17世纪，第二次世界大战之后才渐渐不受欢迎，不过大多数日本家庭依旧保留着一间榻榻米屋。

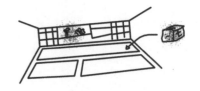

Straw mushrooms are also called paddy straw mushrooms and have been farmed in China for hundreds years. These mushrooms could secure more food than could be supplied by rice.

草菇也叫杆菇，已在中国种植了几百年。这些菇类所能提供的食物比大米提供的还多。

The first book describing 11 edible mushrooms was written in 1245, and described mushrooms in Chekiang Province (now Zhejiang Province), China. The first time straw mushrooms were documented was in 1822.

第一本记载了11种可食用菇的书写于1245年，描述的是中国浙江省的菌菇。1822年，第一次出现草菇的文字记载。

Every year there is as much straw produced by plants as there is petroleum extracted from the Earth.

各种作物每年产出大量的秸秆，几乎相当于每年开采的石油量。

Straw is rich in cellulose, which can be converted into plastics. We could potentially make all the plastic we need from the waste that is burnt today.

秸秆富含的纤维素可以转变成塑料。我们今天焚烧的废弃物很有可能生产出我们需要的所有塑料。

The company Novamont in Italy pioneered the methods of converting agricultural waste into the chemical products that they and society need.

意大利诺瓦曼特公司将农业废弃物制成人类和社会所需的化工产品，他们在这一领域处于领先地位。

Think About It

Would you burn straw if you knew that you could make houses, plastics, food and fuel from it?

如果你知道可以利用秸秆建造房子、生产塑料、种植食物、提供燃料，你还会焚烧秸秆吗？

你更喜欢科学家做哪件事：是改变水稻基因，发明短杆水稻，还是充分利用秸秆的所有可能用途？

What would you prefer that scientists did: genetically modify rice to have short stalks, or to capitalise on all the possible uses for naturally occurring straw?

Would you like to live in a straw-bale house?

你愿意住在稻草屋里吗？

Is burning waste a good solution? Or should we try to create products and jobs with what we consider to be waste?

焚烧废弃物是个好办法吗？还是应该利用我们所认为的废弃物，努力创造产品和工作机会？

Do It Yourself!

自己动手!

You may not have rice straw lying around at home, but you'll certainly have drinking straws handy. After plastic bags, drinking straws present one of the greatest problems for aquatic life. We need to find something to do with them! There are thousands of ideas, but try to think of your own idea for straws that you can do at home. My favourite is making a straw aeroplane. Figure out how to do it yourself – it flies better than a paper plane!

你家附近可能没有秸秆,但吸管肯定随手可得。水生生物面临的最大问题之一就是吸管,仅次于塑料袋。我们必须找到处理它们的方法!虽然已经有成千上万种方法,但请努力想出自己的方法,你在家就可以实行的方法。我最喜欢的方法是做成吸管飞机。想想看,你该怎么做——它飞得比纸飞机棒多了!

TEACHER AND PARENT GUIDE

学科知识
Academic Knowledge

生物学	草菇菌盖被顶破之前的卵形期是最佳采收期，此时的草菇含有最丰富的氨基酸；空气湿度超过20%时秸秆分解；秸秆适合用于建造牛舍，不适合用于喂养；除了甘蔗和玉米，大米是最广泛食用的主食。
化学	成熟的水稻作物40%的氮、30%的磷、80%的钾和40%的硫存在于秸秆中；秸秆富含锌元素，为水稻的硅元素平衡发挥重要作用；燃烧秸秆导致所有的氮和50%的硫流失；水稻外壳的硅和碳可以被转化成金刚砂。
物理	一根秸秆的抗压强度很弱，但成捆的秸秆可以和任何建筑材料媲美；成捆秸秆是天然的阻燃剂；谷粒通过水稻外壳的小孔进行呼吸，这意味着外壳的硅元素可以通过气孔渗透出来，应用于工业。
工程学	草砖房利用小麦、水稻、黑麦和燕麦的茎秆作为结构要素或隔热材料，或既是结构要素又是隔热材料；传统的埃及建筑使用尼罗河的泥和秸秆混合建造而成。
经济学	草菇种植量全球排名第三，仅次于蘑菇和香菇；秸秆可以转化成生物碳；水稻秸秆水解后可转化成生物柴油；木糖喂养的真菌可生产乙醇；埃及水稻农场平均每公顷产量9.5吨，而世界水稻平均产量仅为4.5吨。
伦理学	秸秆能够用来种植蘑菇、生产塑料、生产清洁燃料和补充土壤，这种情况下，如何能够证明焚烧秸秆或者改变基因生产短杆作物是合理的？
历史	非洲自旧石器时代起就建造草房；马里共和国的杰里大清真寺是世界上最庞大的泥砖和秸秆结构建筑；公元前7000年至公元前3300年期间，南非已使用泥砖；8 200至13 500年前，中国珠江流域已经移植水稻；非洲水稻品种于3 500年前移植于尼罗河流域；水稻经由西亚传入欧洲和美洲。
地理	草菇种植于湿热的东南亚地区；泥砖在西班牙叫作风干土坯，广泛用于西班牙和拉丁美洲。
数学	25亿吨秸秆可以转化成：（1）多少蘑菇？（2）多少吨塑料？（3）多少升乙醇？这些行业可以创造多少个就业机会？
生活方式	在电影《幻想曲》里沃特·迪斯尼给草菇戴上棕色草帽，配上了舞动的双脚，成了快乐蘑菇；蘑菇已经逐渐成为世界各地主食的一部分。
社会学	现代科学对转基因感兴趣，然而关于废弃物管理的研究同样应被认为科学；在亚洲，水稻被视为好运和繁荣的象征。
心理学	如果总是抽到下签，你会觉得自己运气一直不好。
系统论	秸秆是农业废弃物，与其烧掉，还不如把它转化成有用的产品，正如几百年来一直在做的。

教师与家长指南

情感智慧
Emotional Intelligence

水稻种子　水稻种子分享了她的担忧。她充分认识到除了大米，水稻产出了更多的东西：大量的秸秆和谷壳。她察觉到她的副产品越来越不受欢迎。水稻种子认为种植蘑菇是一个必不可少的选择。她非常诧异人类为什么要发明短杆水稻，当得知通过焚烧来处理过量的秸秆时，她感到紧张。她对焚烧的后果很敏感，尤其是对孩子们健康的影响。她觉得有责任积极寻找解决方案。她知道一个现成的解决方法，并坚持用这种方法把秸秆转化成一系列天然化学品。水稻种子的建设性提议激起了胡萝卜更多的积极思考。

胡萝卜　胡萝卜有着积极乐观的态度，想要鼓励水稻种子。当水稻种子对他的积极言语不做回应时，他提供了更多的积极消息。他俩试图寻找实用的可持续性解决方法。首先他们否定了人类的解决方法，因为这些解决方法引起了更多的新问题：焚烧不可取，事实说明焚烧与人类健康风险相关；转基因不适合，会导致资源永久流失。胡萝卜没有跟上水稻种子的思路，一时无法理解从石油到生物塑料的跳跃。然而，一旦他明白了，就立刻恢复了积极的心态，提醒说水稻甚至可以制成珠宝首饰。

艺术
The Arts

现在我们要加大难度了。准备好了吗？我们要开始在大米上作画了。大米很小，因此你得有一枝很细的铅笔。试着在大米上画一条线，手指千万不要抹掉画好的线。感到受挫之前，再试一次吧。大米作画艺术非常盛行。如果一直画不好，你可以玩另一个画画游戏！取100粒大米，看看它们的色泽有什么不同。把它们放在太阳底下，你会看到细微的差别。把米粒分类，设计一个由明到暗的圆圈。

TEACHER AND PARENT GUIDE

思维拓展
Systems: Making the Connections

全球每年产生的秸秆总量相当于全球石油开采量。秸秆有很多用途，首先是种植蘑菇，这已有200年左右的历史，促进了中国的食品保障。如果所有秸秆都用于种植平菇或草菇，我们只需花一半的力气，就能多生产12.5亿吨食物。这意味着我们可以把2020年的粮食预测产量从21亿吨提高至33.5亿吨，让我们能够消除饥饿。秸秆不易消化，却可以在酶的作用下分解成富含氨基酸的物质，成为极好的动物饲料。不仅如此，秸秆还可以用于塑料制造。如今，秸秆经过生物炼制——类似于把石油分子转换成酯，可以制成具备各种价值和功能的商品。秸秆还可以转换成燃料、乙醇或者生物柴油；由此产生的废弃物又可以被视为另一种资源。水稻秸秆用于建造房屋已有上千年历史，秸秆捆在耐用性、防火性和能源效率方面受到认可。谷壳富含二氧化硅，由此提取的硅元素能转化成各种产品，包括电池的硅电极和珠宝首饰。目前受到推崇并广泛应用的秸秆处理方法不仅是不可持续的，还剥夺了人类创造性地利用原材料的可能性，秸秆利用是满足食物、住房、能源等工业和社会需要的重要途径。这就把改变水稻基因，生产短杆水稻的决策放在质疑的聚光灯下。社会和决策者是否能容忍，仅因某家公司的利益就丢弃淘汰如此常见的资源——一种能创造产品、价值和无数就业机会的资源？

动手能力
Capacity to Implement

把所有能用秸秆制成的产品列成清单。这则寓言已提供了一些可行的应用，可一旦你想到了塑料，你还会大有收获。现在，罗列出哪些不可持续性产品可用水稻产品代替，或者应该创造哪些新产品，以至于地球资源不被耗尽。我们建议，企业要更善于利用地球现有的资源。

教师与家长指南

故事灵感来自
This Fable Is Inspired by

伊江玲美
Remi Ie

伊江玲美认为自己是一名可持续发展型的企业家。她热衷于改善农民生活水平，提高农产品行业的竞争力。她与冲绳科技研究所合作，尝试了多项创新，包括创立美人鱼咖啡馆，那儿的有机食品由水稻、小麦和糖蜜制成。伊江对谷壳很好奇，从商人那儿学会了怎么把二氧化硅转变净化成珠宝首饰。这些首饰只是用水稻废弃物制成的，却透着珍珠的光泽。

图书在版编目(CIP)数据

冈特生态童书.第二辑修订版：全36册：汉英对照 /
(比)冈特·鲍利著；(哥伦)凯瑟琳娜·巴赫绘；
何家振等译. —上海：上海远东出版社，2021
书名原文：Gunter's Fables
ISBN 978-7-5476-1759-5

Ⅰ.①冈… Ⅱ.①冈…②凯…③何… Ⅲ.①生态
环境-环境保护-儿童读物—汉、英 Ⅳ.①X171.1-49

中国版本图书馆CIP数据核字(2021)第213075号
著作权合同登记号图字09-2021-0823

策　　划	张　蓉
责任编辑	程云琦
封面设计	魏　来　李　廉

冈特生态童书
秸秆的妙用
［比］冈特·鲍利　著
［哥伦］凯瑟琳娜·巴赫　绘
何家振　李欢欢　译

记得要和身边的小朋友分享环保知识哦！
八喜冰淇淋祝你成为环保小使者！

Health 56

细菌汤

A Soup of Bacteria

Gunter Pauli

［比］冈特·鲍利 著
［哥伦］凯瑟琳娜·巴赫 绘
何家振 李欢欢 译

上海远東出版社

丛书编委会

主　任：贾　峰

副主任：何家振　闫世东　林　玉

委　员：李原原　祝真旭　牛玲娟　梁雅丽　任泽林
　　　　王　岢　陈　卫　郑循如　吴建民　彭　勇
　　　　王梦雨　戴　虹　翟致信　靳增江　孟　蝶

特别感谢以下热心人士对童书工作的支持：

匡志强　宋小华　解　东　厉　云　李　婧　陈　果
刘　丹　熊彩虹　罗淑怡　旷　婉　杨　荣　刘学振
何圣霖　廖清州　谭燕宁　韦小宏　李　杰　欧　亮
陈强林　王　征　张林霞　寿颖慧　罗　佳　傅　俊
胡海朋　白永喆　冯家宝

目录

细菌汤	4
你知道吗？	22
想一想	26
自己动手！	27
学科知识	28
情感智慧	29
艺术	29
思维拓展	30
动手能力	30
故事灵感来自	31

Contents

A Soup of Bacteria	4
Did You Know?	22
Think about It	26
Do It Yourself!	27
Academic Knowledge	28
Emotional Intelligence	29
The Arts	29
Systems: Making the Connections	30
Capacity to Implement	30
This Fable Is Inspired by	31

一条金枪鱼在塔斯曼海游来游去,他发现了一小片红藻林。

"你有和我一样的颜色。"金枪鱼说出了自己的看法。

A tuna fish swirls around the Tasman Sea and finds a small forest of red seaweed.

"You are the same colour as me," observes the tuna.

……一小片红藻林……

... a small forest of red seaweed ...

我的名字叫美味

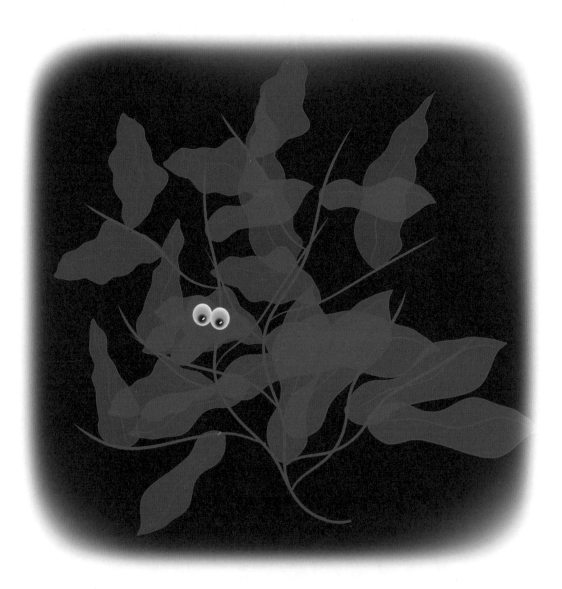

My name is delicious

"是呀,那你知道我的名字叫美味吗?"红藻回答道。

"哦,那咱俩都很美味,"金枪鱼边说边钻进了海藻丛,"不过你好粗糙呀。你一点都不柔软。"

"Yes, and did you know that my name is delicious?" responds the seaweed.

"Oh, so we both are delicious," remarks the tuna as he swims through the weeds. "But you feel so coarse. You're not soft at all."

"啊，那是因为我是无菌的呀。"

"世界上根本就不存在无菌的东西！地球上的生命起源于海洋，而最早出现的物种就是细菌。细菌王国是自然界里最大的！"

"Ah, that is because I am bacteria-free."
"There's no such thing in the world! Life on Earth emerged from the sea, and the first living species were bacteria. It's the biggest kingdom in nature!"

我是无菌的

I am bacteria-free

一碗细菌汤……

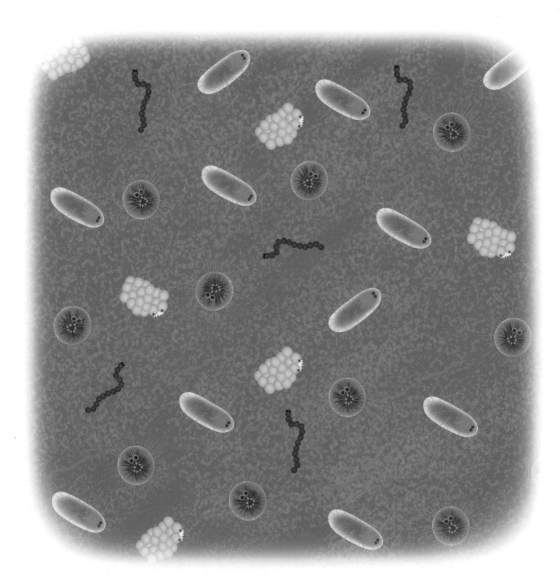

A soup of bacteria...

"那还用说嘛——海洋简直就像一碗细菌汤！"

"既然你知道你就生活在细菌汤里,你怎么能说自己是无菌的呢?"

"因为我把细菌变成了聋子。"

"Tell me about it – the sea is like a soup of bacteria!"

"So if you know you're living in a bacteria soup, how can you claim to be free of them?"

"Because I make them deaf."

"哦，别胡扯了！我听说过鲸会交流，植物会唱歌，可你居然告诉我你能叫得如此尖锐，以至于细菌们都无法听见彼此的声音了？"

"不，我不是尖叫——我只是塞住了他们的耳朵。"红藻说。

"细菌没有耳朵，因此你不可能塞住他们的耳朵。"金枪鱼坚持道。

"Oh, come on! I've heard about whales communicating and plants singing, but you're telling me you can scream so loud that the bacteria can't hear each other?"

"No, I'm not screaming – I'm plugging their ears," says the seaweed.

"Bacteria don't have ears, so you can't plug them," insists the tuna.

……我听说过……

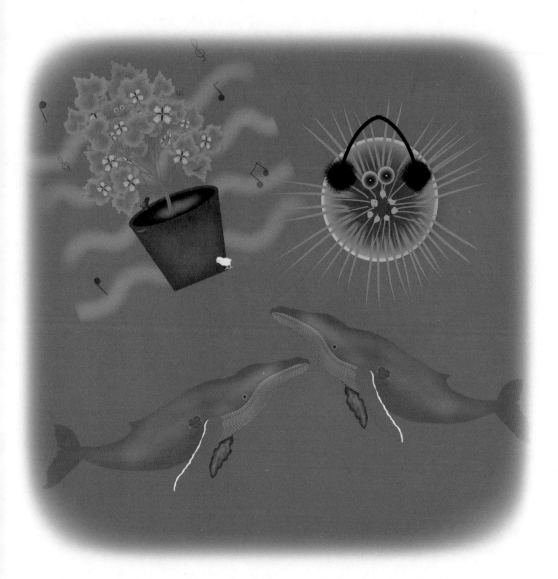

...I've heard about...

它们叫作受体,不叫耳朵

They're called receptors, not ears

"呃，它们叫作受体，不叫耳朵，不过这是一回事。我释放这些微小的化学物质，塞住了他们的耳朵。"

"那你为什么要这么做呢？"

"Well, they're called receptors, not ears, but it's the same thing. I release these small chemicals that block their ears."

"And why do you do that?"

"嗯，"红藻解释道，"细菌无处不在，侵入一切生物，包括我。它有时让我们难以生存。"

"你有没有尝试过杀死他们？"金枪鱼好奇地问。

"杀死细菌？这种事还是留给人类吧——他们一直在尝试消灭细菌，可结果却是在慢慢杀死自己。他们永远不会成功的。"

"Well," says the seaweed, "bacteria are everywhere and invade everything, including me. It sometimes makes life impossible."

"Have you ever tried to kill them?" wonders the tuna.

"Killing bacteria? Leave that to humans – they try all the time but are slowly killing themselves. They'll never succeed."

……杀死细菌……

... killing bacteria ...

细菌喜欢群居

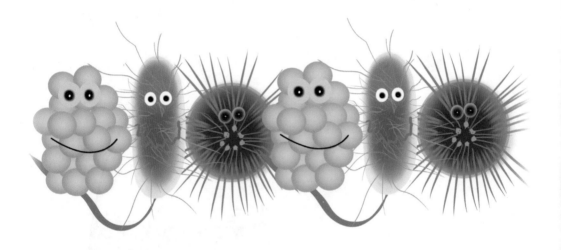

Bacteria love to cuddle together

"因此你决定塞住他们的耳朵。接下来怎么办呢？"

"细菌喜欢群居。细菌群会形成一层薄薄的软膜，但只有细菌家族壮大到足够的数量，才能控制他们所寄生的寄主。"

"So you decided to plug their ears. Then what?"

"Bacteria love to cuddle together. They make a thin, soft film, but they need a quorum in the family so that they can control their host."

"因此当你塞住了他们的耳朵,他们就不知道家在哪儿,只能到别的地方定居,这样你就安全了。"

"我们是不是很聪明?你也应该试试——这些细菌最喜欢享用鱼肉了,他们会破坏你的肉质。这样也许你闻着都不像鱼了!"

……这仅仅是开始!……

"So when you plug their ears, they have no clue where their family is and settle somewhere else, leaving you at peace."

"Aren't we clever? You should try it too – these bacteria love feasting on fish, and they spoil your meat. And then maybe you wouldn't smell like fish!"

… AND IT ONLY HAS JUST BEGUN!…

……这仅仅是开始!……

... AND IT HAS ONLY JUST BEGUN! ...

Did You Know?
你知道吗?

A single bacterium is not a problem. It's when bacteria work together to create a biofilm that sometimes try to control their host.

单个细菌不是问题。当一群细菌共同作用形成菌膜时,才有可能控制它们所寄生的寄主。

*Y*ou need an antibiotic perhaps 1,000 times stronger to kill a colony of bacteria formed into a biofilm than you would for a single bacterium.

比起消灭单个细菌,你需要使用功效可能要高 1000 倍的抗生素才能消灭一个已形成菌膜的细菌群落。

Bacteria communicate with each other using a chemical language. Bacteria not only talk, they also coordinate and undertake joint initiatives.

细菌使用化学语言进行沟通。细菌不仅相互交流，它们还相互协作，采取联合行动。

Bacteria need to have a minimum number of members to start coordinating. This is called "quorum sensing". They also sense how many "other" bacteria are present.

细菌成员必须达到一个最低数值，才能开始协作，这叫群体感应。它们还能感知到有多少其他种类的细菌存在。

The type and quantity of microorganisms in your gut can prevent or encourage disease, including mental health, making your gut function like a second brain.

人体肠道内的微生物种类和数量不同，有些能预防疾病，有些却会诱发疾病，甚至影响心理健康，这就使得肠道起到了人体第二大脑的作用。

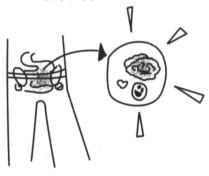

A swimmer swallowing a mouthful of seawater may be consuming more than 1,000 types of bacteria.

一名游泳者吞下一口海水，也许就喝下了1000种细菌。

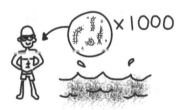

Pickled food, is fermented food, like miso, natto, tempeh, kimchee, kombucha, sauerkraut, pickles, olives and raw milk cheeses.

腌制食品是一种发酵食品，如味噌、纳豆、豆豉、韩国泡菜、康普茶、德国酸菜、各式腌菜、腌橄榄和生奶酪。

Tears are the best natural antibacterial substance produced by a human being.

眼泪是人体产生的最佳天然抗菌物质。

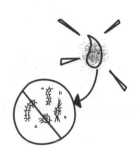

Think about It
想一想

Plants, animals and now bacteria seem to communicate with each other. So which language will you learn next?

植物、动物和细菌似乎都能相互交流。那么，你下一步打算学哪种语言呢？

你认为，你能单靠自己就获得成功，还是需要一个团体（至少有一定数量的朋友）才能让你立于不败之地？

Do you think you can win on your own, or do you need a quorum (a minimum number of friends) to make you invincible?

Will people ever win the war against bacteria and kill them all?

人类是否可以在对抗细菌的战争中获胜，并且消灭所有细菌？

如果你知道自己很聪明，那么炫耀自己的聪明是个好主意吗？

When you know when you are clever, is it a good idea to brag about it?

Try to get hold of a strong microscope. Take some water from the tap, and look for life in the sample. Then collect some saliva from your mouth, and look for life in there. Next, take some water that has been boiled, and look for life in there. Now leave the boiled water standing for 2 days. What do you see under the microscope? It seems impossible to get rid of bacteria, they are everywhere, so it's best to learn to live with them.

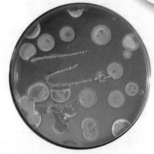

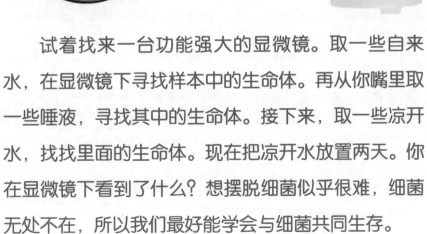

试着找来一台功能强大的显微镜。取一些自来水，在显微镜下寻找样本中的生命体。再从你嘴里取一些唾液，寻找其中的生命体。接下来，取一些凉开水，找找里面的生命体。现在把凉开水放置两天。你在显微镜下看到了什么？想摆脱细菌似乎很难，细菌无处不在，所以我们最好能学会与细菌共同生存。

TEACHER AND PARENT GUIDE

学科知识
Academic Knowledge

生物学	海洋红藻是一种红色的海草；细菌是地球上最原始的生命形式；细菌喜欢成群结队，以菌膜的形式，而不是自由浮动的细胞个体存在；超级病菌是对多种抗生素有抗性的细菌菌株；一升海水含有2万种不同的微生物；海洋中，90%的生命体是微生物；细菌会在压力之下变异。
化　学	呋喃酮由海洋红藻制成，可以使细菌"失去听力"，这样不用杀死细菌就能避免细菌感染，还不会引起细菌的抗药性；肠道菌群对抗生素、加氯水、农药和抗菌皂非常敏感；眼泪中的溶解酶是人体最佳的抗菌抗病毒药剂，10分钟内能杀死90%—95%的细菌；饮食中，发酵食品的种类越丰富越好，发酵食品可以增加肠道内微生物种类，既提供微量营养素，还有排毒作用；天然的杀菌剂包括氧气、碘酒、酒精和臭氧。
物　理	呋喃酮与细菌受体部位结合，会阻止细菌间的交流，但不会引起其他任何副作用。
工程学	维护石油和天然气管道包括去除管道内的菌膜；受到菌膜威胁的食品行业利用刺激性化学物质来控制细菌，尤其是奶制品、鱼类、家禽和即食食品行业。
经济学	呋喃酮可以应用于疾病治疗和医疗设施、受菌膜侵蚀的油气管道、空调、去污产品和污水处理方面。
伦理学	如果有其他选择不会引起细菌抗药性，我们怎么能用会引起细菌变异和产生抗体的化学物质呢？让生物变聋比杀死它要更好吗？改变比拒绝现状要好吗？
历　史	法国科学家于1844年最早记载了海洋红藻。
地　理	海洋红藻在塔斯曼海域大量生长，也见于太平洋沿岸地区；生物地理学研究显示没有相同的细菌汤，但在海洋的部分地区，有些细菌很常见，有些却很罕见。
数　学	数学模型可以预测传染病如何发展，如何预防传染病，何种公共卫生干预措施可以达到防治目的。
生活方式	世界上多个国家把发酵食物作为饮食的一部分：日本的味噌、豆豉、纳豆、康普茶，韩国的泡菜，德国/法国的酸菜，比利时的酸乳酒。
社会学	整个社会一味地想要消灭细菌，有着一套严谨的消灭细菌的策略，这样最终可能会导致人的免疫系统失效。
心理学	洁癖或细菌恐惧症是对污染物和细菌的病态恐惧，属于强迫性精神障碍。
系统论	滥用抗生素导致细菌变异，形成超级病菌；其实有很多天然的方法可供选择，如紫外线、臭氧和呋喃酮，这些都不会引起细菌变异。

教师与家长指南

情感智慧
Emotional Intelligence

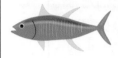

金枪鱼　　　金枪鱼善于社交，积极地开启话题，但他对海藻的回答有些自负。他对环境很敏感，并乐于分享他的观点和经历。金枪鱼不相信海藻说的话，并通过列举事实，来展示他的博学多才。金枪鱼有些自高自大，并开始质疑海藻，当他获得新信息时，还讥讽海藻。他否定海藻的解释，认为细菌没有耳朵，也没法堵上耳朵。他质疑海藻说的每一点，但最后终于明白了海藻的逻辑，确定细菌防治是可行的，而且他的话讲得好像早就知道这点似的。

海　藻　　　海藻欢迎金枪鱼来闲聊，她很了解自己，毫无保留地分享自己的秘密。海藻并没有进行长篇大论的解释，而是简洁清晰地表述重点。当金枪鱼表示质疑时，她阐明了科学上的正确用词，指出让细菌耳聋而非杀死他们的原因。最后，明知听起来很自负，海藻仍表示她对自己的方法很满意。

艺术
The Arts

找几张海洋红藻的图片：这些红色的分支相当漂亮。如果你住在太平洋沿岸，尤其是澳大利亚，你可以试着摘一片海洋红藻，不过要小心哦，这种海藻生长缓慢，你取下的任何一点红藻，都需要几年时间长回去。把红藻图片投影到银幕上，试着把红藻画下来。如何来显示这是无菌叶片呢？

TEACHER AND PARENT GUIDE

思维拓展
Systems: Making the Connections

　　人类已向细菌宣战：我们的唯一策略就是消灭细菌。然而，细菌分很多种，其中大部分是有益菌：没有它们无穷无尽的有效支持，我们可能没法存活。可惜，我们很少想到细菌独特的贡献或我们对其的依赖性。我们嘴里帮助消化的细菌数量比地球上的人口还多。人和细菌是共生关系，一个人身上约有100万亿个细菌细胞。人体内的细菌细胞数量是身体细胞的10倍。但是我们仍然把所有的细菌归类为病菌，并从很早开始，执着地想要消灭细菌。问题在于消灭所有的细菌，不可能不影响到我们自己的生活质量。我们已经介绍了那么多细菌防治措施，过量消耗抗生素或加氯水使我们变得虚弱：体内的有益菌受到化学物质的攻击，变异菌种和产生抗药性的菌种生存下来。近几年超级病菌的出现就是几十年来大量使用抗生素的后果。因此人类应该借鉴自然界的方法：盲目地消灭一切细菌，还不如阻止细菌彼此交流，从而使其不可能联合进攻其寄主。

动手能力
Capacity to Implement

　　我们需要控制有害的细菌，但我们不想杀死有益的细菌。列出所有不杀害微生物就能防治微生物的自然方式。我们想让大家理解：尽管抗生素是一种选择，但并非首选，它也许是治疗疾病的最后一种手段。勤洗手也许是最佳的预防措施。列出你的清单，确定不同的情况下，哪个是最好的选择。只有我们清楚地认识到如何展开无毒防治，我们才能维护地球上的生命免受细菌侵害。

教师与家长指南

故事灵感来自
This Fable Is Inspired by

皮特·斯坦伯格
Peter Steinberg

皮特·斯坦伯格曾在美国马里兰大学学习动物学,并获得加州大学博士学位。他在澳大利亚工作生活了20年,在跳水时注意到了海洋红藻的纹理。他的研究巅峰是一系列关于细菌群体效应的研究论文,还有几项专利。之后,他和新南威尔士大学的同事一起开创了生物信号公司。尽管公司没能把产品推向市场,但让人们对细菌防治有了新的看法。现在,皮特是悉尼海洋科学研究所的主任以及澳大利亚新南威尔士大学海洋港口项目负责人。

图书在版编目(CIP)数据

冈特生态童书.第二辑修订版:全36册:汉英对照/
(比)冈特·鲍利著;(哥伦)凯瑟琳娜·巴赫绘;
何家振等译.—上海:上海远东出版社,2021
书名原文:Gunter's Fables
ISBN 978-7-5476-1759-5

Ⅰ.①冈… Ⅱ.①冈…②凯…③何… Ⅲ.①生态
环境-环境保护-儿童读物—汉、英 Ⅳ.①X171.1-49

中国版本图书馆CIP数据核字(2021)第213075号

著作权合同登记号图字09-2021-0823

策　　划　　张　蓉
责任编辑　　程云琦
封面设计　　魏　来　李　廉

冈特生态童书
细菌汤
[比]冈特·鲍利　著
[哥伦]凯瑟琳娜·巴赫　绘
何家振　李欢欢　译

记得要和身边的小朋友分享环保知识哦!
八喜冰淇淋祝你成为环保小使者!

Energy 60

风吹来的蛋糕
Cakes from Wind

Gunter Pauli

[比]冈特·鲍利 著
[哥伦]凯瑟琳娜·巴赫 绘
王菁菁 译

上海远东出版社

丛书编委会

主　任：贾　峰
副主任：何家振　闫世东　林　玉
委　员：李原原　祝真旭　牛玲娟　梁雅丽　任泽林
　　　　王　岢　陈　卫　郑循如　吴建民　彭　勇
　　　　王梦雨　戴　虹　翟致信　靳增江　孟　蝶

特别感谢以下热心人士对童书工作的支持：

匡志强　宋小华　解　东　厉　云　李　婧　陈　果
刘　丹　熊彩虹　罗淑怡　旷　婉　杨　荣　刘学振
何圣霖　廖清州　谭燕宁　韦小宏　李　杰　欧　亮
陈强林　王　征　张林霞　寿颖慧　罗　佳　傅　俊
胡海朋　白永喆　冯家宝

目录

风吹来的蛋糕	4
你知道吗？	22
想一想	26
自己动手！	27
学科知识	28
情感智慧	29
艺术	29
思维拓展	30
动手能力	30
故事灵感来自	31

Contents

Cakes from Wind	4
Did You Know?	22
Think about It	26
Do It Yourself!	27
Academic Knowledge	28
Emotional Intelligence	29
The Arts	29
Systems: Making the Connections	30
Capacity to Implement	30
This Fable Is Inspired by	31

一天，在岛上的农场里，一只小猪一边嚼着胡萝卜，一边向他的邻居抱怨着。
"为什么人类要给我们吃这么多胡萝卜啊？"他叹了一口气。"我已经厌倦了整天嚼胡萝卜！"

One day, on a farm on an island, a pig is chewing and chewing on a carrot, and complaining to his neighbour.

"Why do they give us so many carrots to eat?" he sighs. "I'm tired of chewing on them!"

一只小猪嚼着胡萝卜

A pig is chewing on a carrot

我们长着很多牙齿

We have lots of teeth

"别抱怨了,好好享用吧!"他的朋友回答道。"这些胡萝卜已经被切成块了。"
"唉,人类只知道我们长着很多牙齿,但是我想他们不知道我们没有耐心把这些都嚼完。"

"Don't complain, enjoy!" says his friend. "These carrots have already been shredded to pieces."

"Well, the humans know we have lots of teeth, but I suppose they don't know we lack the patience to chew through all of them."

"那么，他们为什么要给我们啃那么多胡萝卜啊？"

第一只小猪哼了一声："因为超市不想要这些胡萝卜呗！"

"为什么？这些胡萝卜怎么了？"第二只小猪问。

"So then why do they overload us with carrots?"

The first pig snorts. "Because the supermarkets don't want them!"

"Why? What's wrong with them?" the second pig asks.

超市不想要这些胡萝卜

The supermarkets don't want them

只想要长得笔直的胡萝卜

Only want carrots that are straight

"就因为这些胡萝卜长弯了!"

"可还是很好吃啊!"

"那当然,但是超市的顾客们看了太多电影,他们只想要长得笔直、长成完美圆锥形的胡萝卜。"

"Only that they're crooked!"

"But still tasty!"

"Of course, but buyers at supermarkets have watched too many movies and they only want carrots that are straight and perfectly conical."

"太可笑了。"第二只小猪摇摇头,"也许人们应该把它们拿去做胡萝卜蛋糕。"

"是呀,他们应该收割所有胡萝卜,把它们储存起来,这样一年到头就有的吃啦!"

"Ridiculous." The second pig shakes his head. "Perhaps these people should make carrot cake instead."

"Yeah, they should harvest all carrots, store them somewhere, and have a supply all year round."

人们应该把它们拿去做胡萝卜蛋糕

People should make carrot cake instead

鲜榨胡萝卜汁

Freshly squeezed carrot juice

"没错，很有道理。我就很喜欢胡萝卜。吃胡萝卜让我看起来漂亮又粉嫩。"

"人类应该把那些块头大的胡萝卜榨成胡萝卜汁，每天都喝鲜榨的，然后把那些被榨完的碎末喂给我们吃。"

"Yes, that makes a lot of sense. I like my carrots. And they make me look nice and pink."

"They could turn the biggest carrots into juice, freshly squeezed daily, and leave the nicely shredded leftovers for us."

"你的想法很高明,可是,在这座岛上储存胡萝卜需要很多能源——夏天要保持凉爽,冬天要防止冰冻。"

"我们是住在岛上的,对吧?"

"是呀,没错。那又怎样呢?"

"是这样的。在白天,地面变热,可海水能够保持低温;而在夜间,地面变凉,可海水又是热的了。"

"You are smart, but it'll require a lot of energy to store carrots on this island – to keep them cool all summer and stop them from freezing in winter."

"We live on an island, right?"

"Yes, we do. So?"

"Well, during the day the land gets hot and the water stays cool, and during the night the land gets cool and the water is warm."

我们是住在岛上的，对吧？

We live on an island, right?

冷藏，然后用船把胡萝卜蛋糕运到世界各地

Freeze and ship carrot cake around the world

"这和胡萝卜蛋糕有什么关系吗?"

"这意味着我们全年都有风啊。这样我们就有能量来切胡萝卜,利用免费的能源烘焙、冷藏,然后用船把胡萝卜蛋糕运到世界各地。"

"别忘了,在地球的另一端,胡萝卜蛋糕还得解冻呢。"

"噢,那样他们就可以卖新鲜出炉的美味蛋糕了。"小猪微笑着说。

"What does that have to do with carrot cake?"

"It means that we have wind all year. We have the power to shred carrots, bake, freeze and ship carrot cake around the world with free energy."

"Don't forget that on the other side of the world, the carrot cake will have to be unfrozen."

"Ah, but then they can sell great cake oven-fresh," smiles the pig.

"这些销售员总能找到好听的词。你怎么能把已经冷冻了几周甚至几个月的东西叫成'新鲜出炉'的呢?"

"不要担心这些用词。我们的胡萝卜蛋糕能让我们的风出口。"小猪说。

"我希望你是指空气中的风,而不是我们肚子里的风!"他的朋友大笑着说。

……这仅仅是开始!……

"These sales people find nice words for everything. How can you call something that's been frozen for weeks and months 'oven-fresh'?"

"Don't worry about words. Our carrot cake will allow us to export our wind," says the pig.

"I hope you mean wind from the air, not wind from our tummies!" laughs his friend.

... AND IT HAS ONLY JUST BEGUN!...

……这仅仅是开始！……

... AND IT HAS ONLY JUST BEGUN! ...

Did You Know?
你知道吗?

Supermarkets don't buy fruits and vegetables unless they are a prescribed shape, even if they taste great and the price is right.

超市进货时不会选择那些外形不规则的水果和蔬菜,就算味道鲜美、价格公道也不行。

Newborn piglets will leave the nest to go to the ablution area within hours of birth. Pigs can by potty-trained like dogs – as a matter of fact, pigs are easier to train than dogs.

小猪在出生后几小时内就会离开猪窝,被送去冲洗干净。猪可以像狗那样被训练如厕,事实上,猪比狗更容易训练。

Pigs have 4 downward-pointing toes on each foot, so a pig walks on the tips of its toes rather than on its whole foot.

猪的每只蹄子上都有4个趾尖向下的脚趾头，因此，猪是靠脚尖而不是整个脚掌走路。

Pigs need to chew their food, otherwise it cannot be digested. Pigs have a tremendous sense of smell.

猪必须咀嚼食物，否则不能消化。猪的嗅觉非常灵敏。

Pigs constantly communicate with each other. They have a range of different oinks, grunts and squeals, each with its own meaning.

猪时常会和同伴交流。它们会发出哼哼声、咕噜声和尖叫声等不同的声音,每一种声音都代表着不同的含义。

The pig is the last of the 12 animals in the Chinese zodiac. It represents fortune, honesty, happiness and virility.

在中国的十二生肖中,猪排在最后一个。其寓意是财富、诚实、幸福和刚强有力。

Coastal zones always have wind: it is created by a difference in temperature and pressure between the land, which absorbs more heat during the day, and the cooler sea. The result is a permanent flow of wind.

海岸地带总是有风。这个风是由陆地与海洋之间的温差和气压差产生的，白天，陆地吸收热量较多，海洋则相对凉爽，结果产生了持久的气流。

The term "oven-fresh" actually means recently unfrozen and thus does not at all refer to fresh food.

"新鲜出炉"这个词实际指的是刚刚解冻，根本不是指新鲜食物。

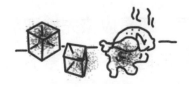

Think about It
想一想

Do you like eating fruits and vegetables because they look good, or because they taste good?

你喜欢吃水果和蔬菜，是因为它们外形好看，还是因为它们味道鲜美？

如果一个地方总是刮风，那有什么理由不将风力利用起来呢？

If it's always windy somewhere, what reason is there for not harvesting wind power?

Do you think that bread called "oven-fresh" is really fresh?

你认为号称"新鲜出炉"的面包真的新鲜吗？

我们应该直接把不要的胡萝卜给猪吃，还是应该先榨取胡萝卜汁然后再把剩下的胡萝卜渣喂给猪吃呢？

Should we feed left-over carrots to pigs, or should we first get the juice out and then feed the pigs the waste?

Do It Yourself!
自己动手!

Go to the supermarket and look for carrots. Are they all the same shape? Then go to a farmers' market and look for carrots. Do they look the same as the ones at the supermarket? Then go back to the supermarket and ask some customers if they would mind tasty but odd-shaped carrots. Interview at least 10 people, starting with your mom and dad. Record their reaction when you ask them if they like "odd-shaped" or "ugly-looking" carrots. The interview will be more interesting if you have a few unusually shaped carrots at hand.

去超市里观察一下胡萝卜。它们都是相同形状的吗？再到农贸市场去观察一下，这里的胡萝卜和超市里的一样吗？再回到超市找几位顾客，问一下他们是否介意购买那些味道好但形状古怪的胡萝卜。至少要采访10个人，可以先从自己的爸爸妈妈开始。当你问他们是否喜欢"奇形怪状"或"长相丑"的胡萝卜时，记录下他们的反应。如果你手里拿着几个奇形怪状的胡萝卜，采访会变得更加有趣。

TEACHER AND PARENT GUIDE

学科知识
Academic Knowledge

生物学	一个拥有5 000头猪的养殖场产生的粪便量与一座拥有5万人口的城市产生的相当；在欧洲，法律禁止用肉或骨粉饲料来喂猪；在猪作为禽流感病毒的中间宿主时，该病毒可以感染人类。
化 学	吃3根胡萝卜就能提供足够走完5千米的能量；胡萝卜中的β-胡萝卜素让猪肉呈现出更饱满、更深的颜色。
物 理	胡萝卜呈橙色是因为它吸收了特定波长的光线，尤其是蓝色和靛蓝色；猪为了凉快而在泥里洗澡；科里奥利效应是指在旋转的坐标系统（旋转的地球）中运动，物体的路径会发生偏移。
工程学	胡萝卜收割机被设计成连刚长出的嫩胡萝卜都可以收割。
经济学	中国和俄罗斯的胡萝卜产量分别名列世界第一和第二；在鸡尾酒派对上提供小胡萝卜作为点心是一种营销策略，通常超市不卖这种小胡萝卜，这种做法盈利丰厚，以至于一些公司开始购买大胡萝卜然后切成小块出售；经商的秘诀，在于要有库存（仓储）和物流（逐日供应）。
伦理学	我们怎么能拒绝出售那些完全能吃，只是长得不好看的水果和蔬菜呢？
历 史	猪在11 000年前被驯化家养；胡萝卜的种植起源于中亚和中东地区，最初的颜色有红色、紫色和黄色；胡萝卜蛋糕从中世纪的胡萝卜布丁发展而来，因其天然甜美的味道广受称赞。
地 理	作为向王室进贡的贡品，橙色的胡萝卜在荷兰非常流行，还有一个原因是橙色胡萝卜比其他颜色的胡萝卜更甜，肉质更饱满。
数 学	圆锥形的几何结构是什么样子？
生活方式	全世界大约有20亿头猪，而且猪越来越被人们接受作为宠物；猪的每个部位都能吃；德国泡菜（酸卷心菜）是由包括明串珠菌和乳酸杆菌的乳酸菌发酵而成的，通常配着猪肉或猪肉香肠吃。
社会学	胡萝卜最早是当作药物来种植，而不是食物；关于猪的习语有很多，但是在西方世界，这些习语基本上没有正面含义，而且也经常是错误的："吃得像猪似的"——其实猪吃东西很慢而且咀嚼地非常仔细，"像猪一样全身冒汗"——其实猪没有汗腺，根本不会出汗。
心理学	画猪人格测试，是指你在纸上画一只猪，这是一种让人们发现自己隐性人格的有趣方式，但是在人格评估中是不准确的；习语"胡萝卜加大棒"是指提供物质奖励（胡萝卜）辅以惩罚（大棒）的威胁来激励某人去做事，来源于马车夫在骡子面前悬挂一根胡萝卜来驱使其往前走。
系统论	那些因长得"丑"而被扔掉的胡萝卜除了为制作胡萝卜蛋糕提供丰富的原料外，还为猪提供了高质量的饲料。

教师与家长指南

情感智慧
Emotional Intelligence

猪

第一只小猪向同伴抱怨时非常有自信，提醒后者咀嚼完切碎的胡萝卜也是非常耗费时间的。他很好奇为什么有这么多美味的胡萝卜可以吃，而且也想知道为什么超市不喜欢这些胡萝卜。小猪意识到吃胡萝卜的好处：他的肉会变得粉嫩，很讨消费者喜欢。小猪们发起了一段积极的对话，而且好奇心很强。他们选择寻求让各方都受益的解决方案：省下胡萝卜汁，然后将胡萝卜残渣作为饲料。在探寻过程中，小猪们非常现实，考虑到了方案的缺点，比如为保证一年的供给而储存胡萝卜和冷藏胡萝卜所需的能源成本。下定决心的小猪们观察到岛上常年刮风的现象，于是想出了利用风力来满足加工胡萝卜所需的额外能源的解决办法。小猪们每想出一个方案，自信就会增加一些，他们接受了有创意的营销策略，并开始设想独到的销售计划——用风力出口胡萝卜。小猪们很知足，十分自在地开着玩笑。

艺术
The Arts

猪呼噜噜的叫声在不同的语言里有不同的表达。你想知道都有哪些表达吗？实际上，猪能发出20种声音。在以下语言中，猪叫声听起来像什么呢？试着来些和声吧。

中文：呼噜呼噜
丹麦语：øf
英语：oink oink
芬兰语：röh röh
法语：groin groin
匈牙利语：röf-röf-röf
日语：buubuu

韩语：kkool-kkool
挪威语：nøff-nøf
波兰语：chrum chrum
俄语：khryu-khryu
西班牙语：oink-oink
瑞典语：nöff

TEACHER AND PARENT GUIDE

思维拓展
Systems: Making the Connections

　　对于水果和蔬菜的形状和外表，消费者心中已经有了固定的想法。可能有多达25%的水果和蔬菜会被视为长相稀奇古怪，因而不被超市购买。这对农民来说是非常大的损失，评级高的胡萝卜售价可高达每吨1 000欧元，而作为动物饲料的胡萝卜售价只有每吨15欧元。如果超市和消费者在喜欢味道和质量的同时，也能够接受那些小疙瘩和瑕疵，那么同样的收成，农民可以多赚不少钱。但这不意味着养殖场的农民要去买高价饲料，而是说我们要在获取食物和营养方面变得更明智些，在能源利用方面变得更高效些。果汁产品是价值链中的重要组成部分，可以创造更多的就业机会：胡萝卜汁是价值较高的饮品，而剩下的胡萝卜渣通常却被扔掉了，虽然它们对于动物（尤其是猪）来说很容易消化。把奇形怪状的胡萝卜做成胡萝卜蛋糕和腌胡萝卜，为消费者提供了价值更高的产品。尽管加工过程需要更多能源，但是我们却常常忽略对本地可再生能源的充分利用，也就失去了特有的竞争优势。瑞典哥特兰岛生产的本地胡萝卜蛋糕就受益于这个策略，即将价值较低的原材料转化成备受欢迎的产品——胡萝卜蛋糕，烘烤后出口到全世界。收割、储存、加工、烘焙、冷冻、包装、运输、解冻等工序所需的能源都是由岛上的风车提供的，因此创造了远超过想象的收入。风电在该国的售价和电网电价一致，使得风力发电的投资在不到5年的时间内便可收回。这种以当地资源为主导的生产和消费一体化设计，可以在农民、公司和社区之间创造更好的灵活性。

动手能力
Capacity to Implement

　　胡萝卜蛋糕很容易做，腌胡萝卜就更简单了！将干净的胡萝卜放进罐子里，将罐子填满到再也塞不进胡萝卜为止。将浓盐水倒在胡萝卜上，直到全部没过，然后盖上松紧合适的盖子。常温放置24小时，时不时检查一下是否有气泡，这些气泡闻着有新鲜的酸味。当出现气泡后，把罐子放进冰箱，再过一两个星期就可以吃了。如果一次吃不完，就把剩下的放进冰箱里最凉快的地方。为什么要吃发酵了的胡萝卜呢？提出论点，然后给出一个逻辑性强、有说服力的答案。同时，你还可以开启自己的腌萝卜事业，将那些奇形怪状的胡萝卜做成美味健康的食物。

教师与家长指南

故事灵感来自
This Fable Is Inspired by

哈肯·埃尔斯腾
Håkan Ahlsten

哈肯·埃尔斯腾生来就注定会成为一名银行家。他出生于瑞典的哥特兰岛,一直承诺要为下一代将这座岛发展到最好。他受自己看到的机遇的启发,决定离开银行,成为一名企业家。他的创举之一便是接管了当地的一家小面包房,并将其发展成瑞典胡萝卜蛋糕供应商中的龙头企业。他让农民送来长相奇怪的胡萝卜,动员另一位企业家创建了胡萝卜储藏企业 Ryftes,以确保每日供应。出口海外的胡萝卜的价格里包含了收集风力所需的成本。现在,哈肯和他的妻子玛利亚正在哥特兰岛上经营一家出色的精品酒店。

图书在版编目(CIP)数据

冈特生态童书.第二辑修订版:全36册:汉英对照/
(比)冈特·鲍利著;(哥伦)凯瑟琳娜·巴赫绘;
何家振等译.—上海:上海远东出版社,2021
书名原文:Gunter's Fables
ISBN 978-7-5476-1759-5

Ⅰ.①冈… Ⅱ.①冈… ②凯… ③何… Ⅲ.①生态
环境–环境保护–儿童读物—汉、英 Ⅳ.①X171.1-49

中国版本图书馆CIP数据核字(2021)第213075号
著作权合同登记号图字09-2021-0823

策　　划	张　蓉
责任编辑	祁东城
封面设计	魏　来　李　廉

冈特生态童书
风吹来的蛋糕
[比]冈特·鲍利　著
[哥伦]凯瑟琳娜·巴赫　绘
王菁菁　译

记得要和身边的小朋友分享环保知识哦!
八喜冰淇淋祝你成为环保小使者!

Energy 61

兔子的燃料
Rabbit Fuel

Gunter Pauli

［比］冈特·鲍利　著
［哥伦］凯瑟琳娜·巴赫　绘
王菁菁　译

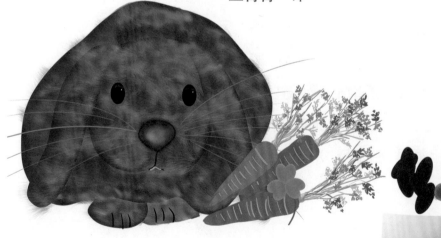

上海远东出版社

丛书编委会

主　任：贾　峰

副主任：何家振　闫世东　林　玉

委　员：李原原　祝真旭　牛玲娟　梁雅丽　任泽林
　　　　王　岢　陈　卫　郑循如　吴建民　彭　勇
　　　　王梦雨　戴　虹　翟致信　靳增江　孟　蝶

特别感谢以下热心人士对童书工作的支持：

匡志强　宋小华　解　东　厉　云　李　婧　陈　果
刘　丹　熊彩虹　罗淑怡　旷　婉　杨　荣　刘学振
何圣霖　廖清州　谭燕宁　韦小宏　李　杰　欧　亮
陈强林　王　征　张林霞　寿颖慧　罗　佳　傅　俊
胡海朋　白永喆　冯家宝

目录

兔子的燃料　　　　　　4

你知道吗?　　　　　　22

想一想　　　　　　　　26

自己动手!　　　　　　27

学科知识　　　　　　　28

情感智慧　　　　　　　29

艺术　　　　　　　　　29

思维拓展　　　　　　　30

动手能力　　　　　　　30

故事灵感来自　　　　　31

Contents

Rabbit Fuel　　　　　　　　　　4

Did You Know?　　　　　　　　22

Think about It　　　　　　　　26

Do It Yourself!　　　　　　　　27

Academic Knowledge　　　　　28

Emotional Intelligence　　　　　29

The Arts　　　　　　　　　　　29

Systems:
Making the Connections　　　　30

Capacity to Implement　　　　　30

This Fable Is Inspired by　　　　31

在新西兰的海岸边，一只兔宝宝正在享用美味的三叶草和胡萝卜，甚至还有荨麻。兔妈妈则和她的家人一起分享着她肚子里保存的食物。家里最小的兔宝宝热切地想跟妈妈学习。

On the coast of New Zealand, a baby rabbit is eating its wonderful clovers and carrots, and even nettles. The rabbit mother is sharing food that she keeps in her tummy with the family. The youngest rabbit in the family is keen to learn from Mom.

美味的三叶草和胡萝卜

Wonderful clovers and carrots

我们需要反复咀嚼

We need to chew a lot

"这么说，我们整天都要吃东西吗？"兔宝宝问道。

"确实如此，而且我们可以吃地里生长的任何作物。但是，在将食物吞到胃里之前，我们需要反复咀嚼。不用担心会磨坏牙齿，因为我们的一生中牙齿一直在不断生长。"

"So, we eat all day?" asks the baby.

"Indeed, and we can eat just about anything that grows in the field. But, before we send the food down to our stomach, we need to chew a lot. Don't worry about grinding up your teeth because they keep growing all our lives."

"我们最爱啃蔬菜纤维了！"兔宝宝说。

"可是有一些特别不容易消化，哪怕你费劲嚼了很久。所以我们需要让这些食物发酵。"

"发酵是不是意味着我们会放很多屁啊？"兔宝宝调皮地问。

"We love to munch on fibres!" the baby says.

"Some are tough to break down, even if you chew them for a long time. So we need to ferment our food."

"Does fermenting mean that we fart a lot?" the baby asks cheekily.

需要让这些食物发酵……

Need to ferment our food ...

我们要小心挑选我们的食物

We have to be careful with our food

"拜托!我们把它叫做'胃穿风'!"妈妈笑道。

"我们是不是要小心挑选我们的食物,这样我们就会少点'风'?"

"I beg your pardon! We call it 'wind in the tummy'!" laughs Mom.

"Do we have to be careful with our food so we have less wind?"

"要时刻小心，认真检查你吃的食物。但是我们兔子的胃里有个非常特别的袋子，里面装满了细菌、微生物和酵母菌。我们的胃不能处理的食物就交给它们来处理。"

但是，兔宝宝似乎害羞起来。"妈妈，我看见你在吃……你自己的便便！"

"Always be cautious and check what you're eating. But we rabbits have a very special bag in our tummy that's full of bacteria, microbes and yeast. Whatever our stomach can't handle gets taken care of there."

But the baby seems ashamed. "Mom, I saw you eating … your own poo!"

……我看见你在吃……你自己的便便!

...I saw you eating ... your own poo!

是很美味的哦!

It's a delicacy!

"小宝贝,那可不是便便。那是从我们的特殊胃袋里掉出来的食物。是很美味的哦!"

"你怎么会让美味的东西从你的后面出来呢?"

"听着,你要学会区分不同的圆形排出物,一种是便便,一种是这些湿湿的小球球。一旦这些小球球掉出来,你应该马上吃掉它们。"

"Little one, that's not poo. That's food from our special tummy bag. It's a delicacy!"

"How can you make a delicacy out of something that comes out of your other end?"

"Look, you have to learn the difference between round droppings, which is poo, and these moist pellets. The minute these pop out, you should eat them again."

"太奇怪了。"

兔宝宝仔细回想着。"我可不知道我能不能分清楚……"

"酵母菌和细菌是我们的朋友。它们生存在这个特殊的袋子里，而且是独一无二的。我们的这些好朋友让我们能够吃下任何食物。人类甚至已经研究出如何用它们来制造能源。"

"That's weird," reflects the baby. "I don't know if I can handle that …"

"Yeasts and bacteria are our friends. They live in this special bag and are unique. These buddies of ours make it possible to eat anything we have. Humans have even figured out how to make energy with thcm."

酵母菌和细菌是我们的朋友

Yeasts and bacteria are our friends

将有害气体转化成能源

Convert bad gasses into energy

"用细菌和酵母菌制造能源?人类会偷走我们的健康排出物吗?"

"不,不会。但是,受我们制造特殊食物小球的启发,他们能将工厂排放的有害气体转化成能源。"

"Energy from bacteria and yeast? Are people stealing our healthy droppings?"

"No, they aren't. But, inspired by the way we make our special food pellets, they can convert bad gases from factories into energy."

"您的意思是利用我们的特殊器官里的物质将工业生产排放的有害气体和黑烟转化成燃料?"

"将黑烟转化成洁净能源!"妈妈肯定地说。

"嗯,如果可以那样,"兔宝宝说道,"我会努力把便便和有用的小球球区分开,这样我就能够展示给人类看,如何用我原以为闻起来臭臭的东西来制造燃料!"

……这仅仅是开始!……

"You mean what's in our special gut is now converting dirty gas and black smoke from industry into fuel?"

"Pure energy from black smoke!" confirms Mom.

"Well, if that works," says the baby, "I'll make an effort to tell my droppings from my pellets so I can show people how to make fuel with something that I thought just smelt bad!"

...AND IT HAS ONLY JUST BEGUN!...

……这仅仅是开始！……

...AND IT HAS ONLY JUST BEGUN!...

Did You Know?
你知道吗？

Rabbits have a nearly 360° vision, allowing them to see from behind. Rabbits can jump 1 metre high and 3 metres across.

兔子的视野范围几乎接近360度，这使得它们能够看到自己的后面。兔子可以跳1米高、3米远。

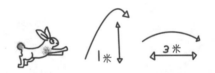

Rabbits communicate with each other. When there is danger, they thump on the ground. Rabbits show when they are happy by running, jumping, twisting and even flicking their feet.

兔子可以相互交流。当遇到危险时，它们会咚咚跺地。兔子通过奔跑、跳跃、扭动甚至是抖脚来表示开心。

Rabbits are born without any fur, and with their eyes closed. Rabbit mothers nurse their babies for only 5 minutes per day. Wild bunnies may be left alone for the rest of the day.

兔子刚出生时没有毛发，眼睛紧闭。兔妈妈每天只花5分钟照顾兔宝宝。剩下的时间，没有人照料的兔宝宝可能都只能独自待着。

The rabbit is one of the 12 animals in the Chinese Zodiac. It represents graciousness, kindness, sensitivity, compassion, tenderness, and elegance.

兔是中国的十二生肖之一。它代表亲切、善良、敏感、同情心、温柔和优雅。

Rabbits are used in experiments to test cosmetics – this is because rabbits sleep with their eyes open and cannot produce tears.

兔子被用于测试化妆品的实验，这是因为兔子睡觉时也睁着眼睛，而且不会流泪。

Rabbit meat is sold because of its high-quality protein. Some rabbits are sheared and their fur is harvested like sheep wool.

兔肉因为含有优质蛋白质而被出售。有些兔子的兔毛可以剪下来使用，就像羊毛一样。

蛋白质

Flatulence, the release of gas from the gut, occurs in all animals – including insects, fish, lizards, snakes, elephants and rabbits. It happens whether you like it or not.

肠胃胀气（放屁），即从胃肠道释放出气体，会发生在几乎所有动物身上，包括昆虫、鱼、蜥蜴、蛇、大象和兔子。不管你是否喜欢，它都会发生。

The air emissions from steel mills are similar to hydrothermal vents, and include carbon monoxide, carbon dioxide, hydrogen sulphide and methane. These emissions are converted to fuel by bacteria, some of which you also find in the guts of rabbits.

从炼钢厂排出的废气与海底温泉喷出的气体相似，都含有一氧化碳、二氧化碳、硫化氢和甲烷。这些排放物可以通过细菌转化成燃料，你在兔子的肠道里也能发现一些这样的细菌。

Think about It

想一想

Do you chew your food enough, or do you expect your stomach to do that job?

你会仔细咀嚼食物吗？还是你指望你的胃来完成这项工作？

你认为细菌对你有好处吗？

Do you consider bacteria to be good for you?

How easy would it be to convert emissions in the air to a liquid fuel for your car?

将空气中的废气转化成你车里的液体燃料会很容易吗？

你认为什么是美味佳肴？你喜欢的美味跟兔子喜欢的种类有什么不同吗？

What do you consider to be a delicacy? Are your delicacies different to the kind enjoyed by rabbits?

Do It Yourself!
自己动手!

Time for some mathematics. A pair of rabbits – 1 male and 1 female – are put in a field. Rabbits can mate at the age of 1 month, and a female can give birth to a new pair of rabbits at the end of the next month. Suppose that these rabbits never die and that the female always produces one new pair (one male, one female) every month from the second month on. The question is: how many pairs will there be in one year? This was the original problem posed by Leonardo of Pisa (known as Fibonacci), which led to the famous Fibonacci code.

到了做数学题的时间啦!将一对兔子(一雄一雌)放到田里。兔子一个月大就可以进行交配,雌兔可以在交配后一个月生下一对兔子。假设这些兔子永远不会死,而且雌兔能够从第二个月开始每个月都产下一对兔子,每一对都是一雄一雌。我们的问题是:一年后会有多少对兔子?这个问题最初是由比萨的莱昂纳多(即斐波那契)提出的,由此产生了著名的斐波那契数列。

TEACHER AND PARENT GUIDE

学科知识
Academic Knowledge

生物学	家兔和野兔之间有差异；雌兔每年可以生产8次，生下多达20只小兔；为了消化，兔子有个叫盲肠的特殊消化袋，在那里细菌和酵母菌可以用酶将纤维素分解掉；盲肠大概要比胃大10倍；盲肠便，也叫夜间排泄物，富含矿物质、蛋白质和维生素，会被兔子吃掉；深海热泉周边的生物有机体的密度要比其他地方高出10万倍。
化 学	盲肠便的表层能让营养素在通过兔子的胃时不受胃酸腐蚀；兔子的尿液含氮量非常高；咀嚼食物的作用是把食物磨碎，将其与消化酶混合；肠胃胀气在所有长着肠道的动物身上都会发生；细菌可以消化卷心菜、西兰花、洋葱、豌豆、韭菜和大蒜中存在的亚硫酸盐；发酵作用是用细菌将糖转化成酸、气体或酒精的代谢过程。
物 理	合成气体发酵作用在低温低压下发生，并不需要一氧化碳、二氧化碳和氢气之间的特定比例。
工程学	炼钢炉、石油化学工厂和废物管理公司产生的富含碳的工业废气被转化成日用燃料；德莱赛眼刺激实验包括向兔子眼中滴入化妆品来检验其毒性。
经济学	在上海宝钢集团公司试验后，朗泽科技和首钢正式推出一家合资公司，极具竞争力地从炼钢炉排放的废气中生产出数百吨乙醇。
伦理学	假如我们把科学作为目标并准备好去冒险，当炼钢厂和炼油厂排放的废气全部都能被利用时，为什么还会把它看作一种肇事责任呢？
历 史	天然的产乙酸菌的出现早于地球上的蓝藻细菌或藻类；希腊人和罗马人把兔子描述成最多产的动物，象征着生命力和生育力。
地 理	产乙酸菌在深海热泉（地球表面的岩溶裂隙，通常在火山附近）排放的气体中旺盛地生长；在新西兰北岛发现了一些最有名的间歇喷泉。
数 学	斐波那契最初在1202年研究的问题是兔子在最理想的情况下能繁殖得多快。
生活方式	兔子越来越多地被养作宠物；人们日益抵制在实验室实验中使用兔子。
社会学	最好和最快的学习来自我们的父母；"龟兔赛跑"来自伊索寓言。
心理学	我们有时候对某些事物有成见，结果却发现其实是完全不同的（比如盲肠便和便便）；我们经常认为有些事情是个问题（比如炼钢厂排放的废气），却忽略了它可能创造出解决方法。
系统论	排放的废气可以转化成燃料，利用的是地球上最古老的生物化学反应之一。

教师与家长指南

情感智慧
Emotional Intelligence

兔妈妈

兔妈妈非常有耐心和爱心。她花费时间抢在她的孩子们开始担心自己的牙齿之前去解释并提供事实。兔妈妈介绍并分享发酵背后的过程。她教孩子礼貌，并提出更好的词汇选择。她小心地提供自己的智慧，并说明了消化作用是如何运作的。当兔宝宝们观察失误时，她并没有生气；她鼓励孩子们去学习，提示了共生以及学会与其他事物（如细菌）合作的重要性。她显示出对兔子世界的了解，说明了兔子们是如何教会和启发其他同类的。兔妈妈的哲学思维超出孩子们的想象。

兔宝宝

兔宝宝急于学习，并且渴望让妈妈做最好的老师。兔宝宝充满热情，但也有担心。兔宝宝关注细节，充满好奇。他善于观察，并不因为提出尴尬的问题而感到难堪。兔宝宝不能确定但不隐瞒，将他关于自己的行为和对人类的态度的疑惑讲出来。兔宝宝通过一连串问题，使自己从对生命感到不确定逐渐变为拥有自信，并准备和他人分享新获得的知识。

艺术
The Arts

在16世纪，阿尔布雷特·丢勒画了一幅美丽的兔子图。在我们拥有家用电脑和打印机之前，这幅画就已经被复制过许多次，并被用于装饰欧洲的很多墙壁。然后还有兔八哥，来自华纳兄弟影业的非常受欢迎的卡通形象。你是怎么看待艺术作品中的兔子呢？如果你来画兔子，背景会是什么呢？你准备在艺术作品中如何重新创作你喜欢的兔子形象？

TEACHER AND PARENT GUIDE

思维拓展
Systems: Making the Connections

　　温室气体排放量持续增加——即使在多年文山会海后，各国政府还是没有能力阻止温室气体排放。20世纪90年代，提出的解决方案之一是生产生物燃料。许多国家支持这一策略，但只有很少数国家在一开始就意识到生物燃料需要农作物，因此会与粮食生产竞争。过去用于生产食物的土地被留作生产燃料，导致基本食物的价格上涨，超出了贫困人口的承受能力。此外，生产生物燃料需要大量的水，这样也会与人类需求相竞争，而且大量的排水需求会提高能源成本。全世界用了十多年时间才意识到这显而易见的疯狂，并停止了使用农作物生产生物燃料。

　　从生物质能生产出乙醇是一个发酵过程。生物质能可以使合成气发酵。合成气由一氧化碳、二氧化碳和氢气混合而成，是由炼钢炉、水泥窑、市政固体废物处理设施和石油化工厂大量排放出的，这也是温室气体的主要来源。这种气体的发酵与糖类的发酵截然不同，它是地球上最古老的生物化学反应之一：一种被称为产乙酸菌的细菌族将合成气转化成乙醇和其他副产品。这种方法为生物燃料的生产带来了一个完全不同的新视角：污染空气的废物和排放物可以被转化成可再生燃料。这一过程在地球上早于蓝藻细菌和藻类的出现，现在已经得到工业化。对兔子的消化系统和胃肠道胀气的研究为发酵途径提供了深刻借鉴，这是一种不需要跟食物竞争而生产生物燃料的突破性进展，不用强制将耕地用于生产能源，而且将过量的温室气体转化成好东西。

动手能力
Capacity to Implement

　　我们想用黑烟制造燃料。要说服最大的污染者们接受这项新技术会有难度吗？问问你自己他们为什么不早点这样做。区分借口、理由和原因三者的不同。使用这三种分类，列出所有不能治理温室气体的理由。然后再决定你将如何强化"把黑烟转化成燃料"这一策略的逻辑性，并召集所有潜在的合作伙伴，让他们了解这个机遇。

教师与家长指南

故事灵感来自
This Fable Is Inspired by

肖恩·辛普森
Sean Simpson

肖恩·辛普森的学术生涯横跨生物学和生物化学领域。他从英国提赛德大学获得理学学士学位。然后他从诺丁汉大学获得植物遗传学硕士学位，最终在约克大学获得植物生物化学博士学位，圆满完成他的学术训练。毕业后，他进入了瑞士和奥地利的药物生产行业，然后在日本筑波大学研究细胞结构，最后定居于新西兰，在那里他研究了如何将硬木转化成乙醇。他在某些特殊兔子品种的消化轨迹中识别出微生物，这种微生物能够使气体中的碳发酵，将含有一氧化碳和二氧化碳的排放物转化成燃料。由于斯蒂芬·廷代尔的开创性支持，后来是维诺德·科斯拉爵士的加入，辛普森博士在新西兰合伙成立了朗泽科技公司。他现在是朗泽科技的首席科学家。

图书在版编目(CIP)数据

冈特生态童书.第二辑修订版：全36册：汉英对照 /
(比)冈特·鲍利著；(哥伦)凯瑟琳娜·巴赫绘；
何家振等译.—上海：上海远东出版社，2021
书名原文：Gunter's Fables
ISBN 978-7-5476-1759-5

Ⅰ.①冈… Ⅱ.①冈…②凯…③何… Ⅲ.①生态
环境-环境保护-儿童读物—汉、英 Ⅳ.①X171.1-49

中国版本图书馆CIP数据核字(2021)第213075号

著作权合同登记号图字09-2021-0823

策　　划　　张　蓉
责任编辑　　祁东城
封面设计　　魏　来　李　廉

冈特生态童书
兔子的燃料
[比]冈特·鲍利　著
[哥伦]凯瑟琳娜·巴赫　绘
王菁菁　译

记得要和身边的小朋友分享环保知识哦！
八喜冰淇淋祝你成为环保小使者！

Energy 62

一起摇摆！

Swinging Together!

Gunter Pauli

[比]冈特·鲍利 著
[哥伦]凯瑟琳娜·巴赫 绘
王菁菁 译

上海远东出版社

丛书编委会

主　任：贾　峰
副主任：何家振　闫世东　林　玉
委　员：李原原　祝真旭　牛玲娟　梁雅丽　任泽林
　　　　王　岢　陈　卫　郑循如　吴建民　彭　勇
　　　　王梦雨　戴　虹　翟致信　靳增江　孟　蝶

特别感谢以下热心人士对童书工作的支持：

匡志强　宋小华　解　东　厉　云　李　婧　陈　果
刘　丹　熊彩虹　罗淑怡　旷　婉　杨　荣　刘学振
何圣霖　廖清州　谭燕宁　韦小宏　李　杰　欧　亮
陈强林　王　征　张林霞　寿颖慧　罗　佳　傅　俊
胡海朋　白永喆　冯家宝

目录

一起摇摆！	4
你知道吗？	22
想一想	26
自己动手！	27
学科知识	28
情感智慧	29
艺术	29
思维拓展	30
动手能力	30
故事灵感来自	31

Contents

Swinging Together!	4
Did You Know?	22
Think about It	26
Do It Yourself!	27
Academic Knowledge	28
Emotional Intelligence	29
The Arts	29
Systems: Making the Connections	30
Capacity to Implement	30
This Fable Is Inspired by	31

一只橙子在观看肥皂的制造过程。他向一只正等着被混合到肥皂中的椰子抱怨。
"你知道吗？等榨出我的果汁后，他们就会蒸煮我的果皮。"橙子说。
"很疼吗？"椰子疑惑地喊道。

An orange is watching soap being made. He complains to a coconut, who is waiting to be added to the mix.

"Do you know that after they squeeze out my juice, the peel will be steamed?" says the orange.

"Does that hurt?" the coconut wonders out loud.

就会蒸煮果皮

The peel will be steamed

这还不是最糟糕的

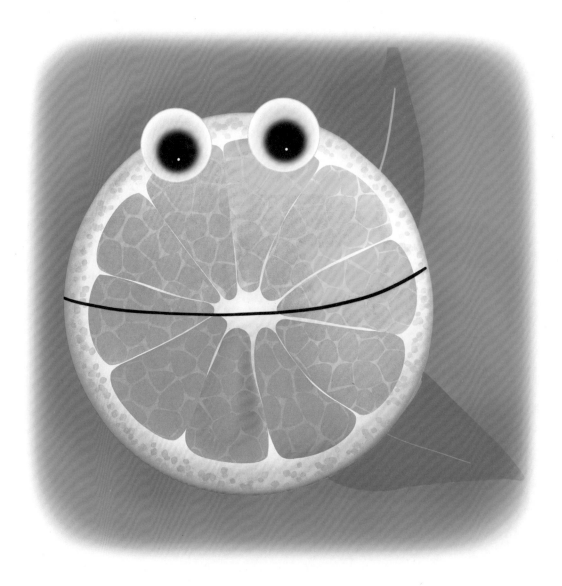

The worst is yet to come

"不。幸好我们没有神经,所以我们感受不到疼痛。但这还不是最糟糕的。"
"还有什么能比压榨你的果汁、蒸煮你的果皮更糟糕的呢?"椰子沉思后说道。

"No. Thank goodness we don't have nerves, so we don't feel pain. But the worst is yet to come."
"What could be worse than having your juice pressed out and your peels vapourised?" muses the coconut.

"噢，最好的清洁剂隐藏在我的果皮中。需要不断地转啊转，至少转一个小时才能开始提取出来。"橙子叹息道。

"你还含有清洁剂？"

"Well, the very best cleaning product is locked into my peel. And it will have to turn and turn for at least an hour before it's ready to do the job." The orange sighs.

"You also contain a cleaning product?"

你还含有清洁剂？

You also contain a cleaning product?

但不是可持续的

It was not sustainable

"是呀。你看，当人类意识到棕榈油可生物降解时，就有越来越多人想用。于是，数百万公顷热带雨林遭到了破坏。那时人们才发现，尽管这是可再生的，但不是可持续的。"

"真是糟糕的消息！"椰子惊叫道。"我原本被种植在世界各地用于榨油，但是非洲棕榈树能产出更多的油，以至我都不受欢迎了！"

"Yes. You see, when humans realised that palm oil was biodegradable, more and more people wanted it. And then millions of hectares of rainforest were destroyed. That's when people found out that although it is renewable, it was not sustainable."

"What bad news!" exclaims the coconut. "I was planted around the world for oil, but the African palm produces so much more of it that I became unpopular."

"噢，我是在巴西的橙汁生产商们意识到他们可以通过我的果汁和果皮赚钱后，才被当作清洁剂广受欢迎。而这意味着可以赚更多钱！"

"这么说，你是用废物制造出的清洁剂……"椰子评论道。

"Well, I only became popular as a cleaner after Brazilian orange-juice makers realised they could make money from my juice and my peel. And that means making more money!"

"So you are a cleaning product made from waste …" observes the coconut.

用废物制造出的清洁剂

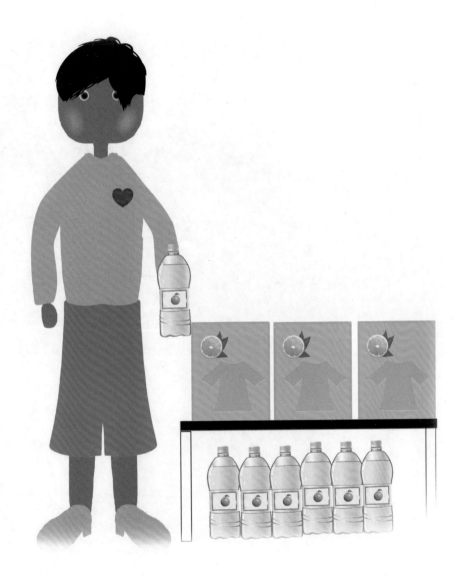

Cleaning product made from waste

我的整个果实都可以用

They use my whole fruit

"我不是废物。我的用途比人们原来认为的更多。我可以被用来制造果汁、果冻或者果酱。"

"噢，我也不只是产油这么简单。"椰子回答道。"他们用我制造100多种产品。我的整个果实都可以用。"

"I'm not waste. I just have more to me than people think. I can be used to make juice, jelly or jam."

"Well, I am also more than just oil," responds the coconut. "They use me to make more than a hundred products. They use my whole fruit."

"咱们能提供这么多东西,真是太好了。要是人们能够充分利用我们,而不是只利用某个部分去赚钱就更好了。"

"不是有很多人在找工作吗?"椰子问。

"It's wonderful that you and I can offer so many things. If only people used us for all we are, instead of trying to make money off only one part."

"Don't they have a lot of people looking for jobs?" asks the coconut.

很多人在找工作

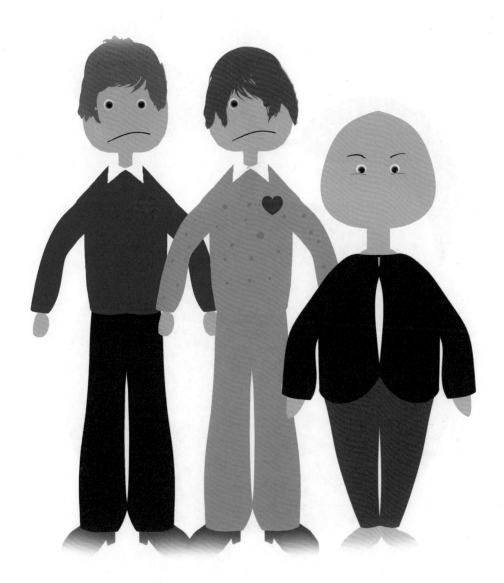

A lot of people looking for jobs

我们被转得晕头转向

We get dizzy from all that swirling around

"数百万人失业，数百万人想使用天然产品，而所有人都想节约能源。这就是我不明白为什么他们要扔掉这么多，还继续把我放在桶里转来转去的。"

"你是什么意思？"

"嗯，为了制造优质的清洁剂，他们需要将所有原料混合，而这个过程太漫长了。转来转去的，我们被转得晕头转向。"

"Millions of people are jobless, millions want natural products, and everyone wants to save energy. That is why I don't understand why they throw so much away, and keep on turning me around in vats."

"What do you mean?"

"Well, to make a good cleaning product, they need to mix all the ingredients and it takes so long. We get dizzy from all that swirling around."

"那么他们为什么不采用我听说过的一种新的强有力的摇摆?"

"你是指水流的自然摇摆,也就是漩涡吗?"

"是的。这很好玩!我们一起跳进去旋转,只要几分钟的狂野探戈,我们就都摇匀了。你应该尝试一下!"

……这仅仅是开始!……

"So why don't they use this new powerful swing I've heard about?"

"You mean the natural swing of water, called the vortex?"

"Yes. It's fun! We all get into a twist together, and within minutes of our wild tango we're all shook up. You should try it!"

...AND IT HAS ONLY JUST BEGUN!...

……这仅仅是开始！……

...AND IT HAS ONLY JUST BEGUN! ...

Did You Know?

你知道吗？

Citrus peel contains the ingredients for cleaning products, perfumes, insecticides and solvents for 3D printing, but is usually thrown away.

柑橘皮含有制造清洁剂、香水、杀虫剂和3D打印溶剂的成分，但是它通常都被扔掉了。

Palm oil is biodegradable and renewable but not sustainable since it is mainly farmed on land that was rainforest. This conversion from forest to agricultural land has destroyed the habitat of many primates.

棕榈油是可生物降解的，也是可再生的，但不是可持续的，因为它主要的种植地区过去原本是热带雨林。这种将森林转变成农田的行为已经破坏了许多灵长类动物的栖息地。

Oranges are the most cultivated fruit trees on Earth. Oranges keep fresh for up to 12 weeks after harvest and therefore were the preferred fruit on long voyages in days gone by.

橙子是地球上种植最多的水果。橙子摘下后可以保鲜长达12周，因此在过去它们是首选的用于长途旅行的水果。

Brazil produces more oranges than any other country in the world and exports 99%, most of it with ships the size of oil tankers.

巴西是世界上生产橙子最多的国家，其中99%用于出口，大部分用油轮一样大的船只运输。

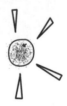

Coconuts are part of the daily diet of a billion people. It is a unique fruit that has more water than any fruit and provides oil for cooking, and making soap and cosmetics.

有10亿人将椰子作为日常饮食的一部分。它是一种奇特的水果，在水果中含水量最高，能提供用于烹饪的食用油，还能用于制作肥皂和化妆品。

The name coconut was given by the Portuguese who thought that the shell of this nut looked like "coco", a witch in their folklore.

椰子（coconut）这个名称是由一位葡萄牙人起的，他认为这个坚果（nut）的外壳看起来很像他们当地传说中一位叫做Coco的女巫。

Soap is made from a vegetable oil, lye (sodium hydroxide or caustic soda, NaOH) and water. Homemade soap needs 4 weeks to cure, just like cheese.

肥皂是由植物油、碱液（氢氧化钠，化学式NaOH）和水制成的。家里自制的肥皂需要4周时间来硬化，就像制作奶酪一样。

To perfectly distribute all ingredients, oranges must be mixed and blended in a vat with an agitator, which takes time and energy. A vortex uses the force of moving water to mix.

要让所有成分完全均匀地分布，橙子必须被放入一个大桶里用搅拌机混合调匀，这个过程消耗时间和能量。漩涡利用水流的力量来搅拌混合。

Think about It

想一想

Is it healthy to drink orange juice in the morning on an empty stomach?

早晨空腹喝橙汁是有益于健康的做法吗?

你会用一个女巫的名字,还是用原产国的名字给水果命名?

Would you name a fruit after a witch, or after its country of origin?

Why would one only press out juice from an orange when there is more money in the peel?

橙子的果皮含有更高的经济价值,为什么人们还只用橙子榨果汁呢?

你想要种哪一种树:橙子树、油棕榈树还是椰子树?哪一种能够带来更多的好处?

Which tree would you prefer to plant: an orange, an oil palm or a coconut? Which one brings more benefits?

Do It Yourself!
自己动手!

This is a quick and easy one. Where in your house do you generate a vortex at least 7 times per day? Well, usually we go to the toilet 7 times a day, and each time you flush, you generate a vortex. This is an ingenious way to cleanse and transport waste: the process exploits the force of water flowing not more than 30 centimetres from the cistern. That is all that is needed to keep the toilet clear. Where else can you observe a vortex in your home? Go and check – you will realise that vortices are to be found all around. You might even have seen one through the window when a storm is raging outside.

这次是个快速简单的任务。在家里的什么地方你每天能制造至少7次漩涡？通常我们每天要去7次厕所，而你每冲一次水，你就制造了一次漩涡。这是清洁和运输排泄物的好方法：利用水从不到30厘米高的水箱流下时产生的动能，这就是保持厕所清洁所需的能量。在你家里还有什么地方能观察到漩涡？去找找看，你会发现漩涡无处不在。你也许曾经在暴风雨肆虐的天气里隔着窗户看到过。

TEACHER AND PARENT GUIDE

学科知识
Academic Knowledge

生物学	柑橘类水果极易通过不同种类的杂交创造出新品种；在人类体内，椰子油中的脂肪酸直接从胃部进入肝脏，然后转化成酮；可生物降解、可再生和可持续这三个概念是有区别的。
化学	橙子是酸性的，其酸碱度（pH值）低至2.9，富含维生素C，不容易变质，可以预防坏血病；椰子富含中链饱和脂肪酸；橙皮里面的白色部分富含果胶。
物理	肥皂使不能溶解的颗粒变得可溶解；肥皂具有亲水性，可以吸收湿气；钠皂是一种硬皂，用于洗衣；含钾肥皂是一种软皂，会产生更多肥皂泡，用于刮胡子；肥皂在水中发生电离反应形成碱性离子，使肥皂滑滑的；漩涡是自然界中水流运动的基本形式；科里奥利效应和地心引力决定了漩涡的旋转方向。
工程学	可以用蒸馏的方法从橙皮中提取天然化合物；漩涡是一种闭合的水动力系统，对水泵、螺旋桨、生物反应器和循环水管的设计有重大影响。
经济学	随机变量和概率分布对于一些商业领域的发展非常关键，如电信行业（多少个电话）、销售管理（每天多少名顾客）、机动车（多少辆轿车经过一个十字路口）以及生产率（需要多少酵母菌使酸奶发酵）。
伦理学	当有生产肥皂和食用油的替代来源时，我们怎么还能破坏猩猩的栖息地——热带雨林来种植棕榈树呢？
历史	柑橘类水果最早种植于公元前4000年左右的东南亚，公元前2500年开始在中国栽种；橙子是由克里斯托弗·哥伦布引进南美洲的；航海探险家们沿着航线一路种植橙子树；法国哲学家、科学家勒内·笛卡尔和英国物理学家开尔文勋爵都研究过漩涡。
地理	多普勒雷达被用于寻找龙卷漩涡信号（TVS）；美国中部地区被称为龙卷风走廊。
数学	随机变量和概率分布（如泊松分布）；平均值、变异系数、方差指数、平均差、标准差、中位数和高阶矩的概念；把水流运动看作非线性运动；水利工程过去是基于传统的伯努利方程和纳维-斯托克斯方程，但现在是基于分形数学。
生活方式	个人卫生的文明始于用肥皂洗手来去除灰土和细菌；探戈舞（拉丁语意为"用针尖触碰"）兴起于阿根廷拉普拉塔河流域。
社会学	"碱性的"这个单词来自阿拉伯语，意思是灰烬。
心理学	我们倾向于遵循逻辑路线，从一个论点到下一个，这种方式被称作线性思考；漩涡迫使我们用非线性方式去思考，从而得到我们从未期待或想象过的结果；跳舞可以让人们更放松，缓解低落情绪。
系统论	有些废物其实是具有很多潜在用途的原材料；我们将废物转变成有用的产品和服务的方式需要得到大自然更多的启发，而且要更节能。

教师与家长指南

情感智慧
Emotional Intelligence

橙子对现状不满意,向椰子倾诉心事。橙子不是因为疼痛而抱怨,他是因为浪费了机会而泄气。橙子详细讲述了细节,并特别强调最困扰他的一点:混合搅拌需要很长时间。橙子具有现实主义精神,也很了解自己,知道他的潜能只是需要得到开发。橙子意识到人类的无知,而且想知道他如何才能让人类认真思考对就业和天然产品的需求。最终,橙子成功地和椰子一起找到新的解决方法。

椰子对橙子表示同情,而且尽管她认可橙子的想法,她也没有抱怨,反而想办法来安慰橙子。她了解到他们有相同的作用,即清洁,但是她并没有把橙子看作竞争对手。椰子观察到他们两者之间的差异:橙子的清洁力来自他被榨汁后的剩余物。然而,椰子很了解自己,声明她的好处不只是椰子油。椰子提出了一种更广阔的视角,即人类有就业的需求,她很贴心地让橙子解释她不清楚的问题。这种澄清激发了椰子的主动性,尽管面临艰难的处境,她仍选择以一种轻松的调子结束谈话,提议他们被混合在一起的时候一起跳舞。

艺术
The Arts

现在来发现铅笔的本领吧。画一个漩涡,直到你将漩涡的线条从起点一直画到纸边才停手。重复画这些线条,直到你画出漩涡的效果。漩涡经常是多个交织在一起,可以用一系列交织缠绕的线条来表现。交织的漩涡可以强化这种自然运动的力量。

TEACHER AND PARENT GUIDE

思维拓展
Systems: Making the Connections

当今的管理策略聚焦于具有核心竞争力的产业。这导致了产品专业化，每位生产商借此以独特工艺为基础来创建一个专门市场。任何无助于达成已严格明确的目标的事都会被看作是浪费时间和精力。这就是为什么生产橙汁的公司只对果汁感兴趣，并且会部署人力、科学和工程资源以更加低的边际成本来增加果汁产量。用橙皮的白色薄膜生产果胶、提取右旋柠檬烯和油以及生产溶解油脂的肥皂都是被忽视的机会，因为除了果汁其他东西都被看作无用的废物。这种策略会影响资源利用效率，而且意味着失去了很多工作机会。与此同时，现成的原材料被浪费了。如果有人能将以橙子加工为中心的各项活动集中起来，那么就可以创造更多价值，筹措资金提供额外的就业机会，并以更低成本提供更多产品。重新思考利用有效的可再生能源创造更多价值的策略，引出一个新问题：如果橙皮创造性地用于制作肥皂，为什么处理混合工艺要用一种过时的技术呢？液体、空气和天然气在自然界中都是通过漩涡来运输的，但是在工业中，这个过程却需要靠外力完成。创新胜过改变产品，工艺也需要改进。产品和工艺创新的结合展现了一条通向可持续发展的更快更长远的途径。

动手能力
Capacity to Implement

准备好制作肥皂了吗？找个哪怕把水溅出来一点（甚至很多）也没关系的地方。建议你戴上眼镜，以免有东西进入眼睛里。你需要一些材料：水、碱液和植物油。首先将碱液倒入水中，但是小心：千万不要反过来将水倒入碱液中！在小心地倒入碱液时，保持一段距离进行搅拌。因为这个过程会跑出些气体，一定要确保你的脸和胳膊不会沾到。当水变清后，静置一会儿。加热植物油。现在要特别小心：这个时候所有东西都非常热，所以你需要戴保护手套。将这些东西混合在一起，搅拌至少5分钟。不要觉得辛苦，也许你能制造一个漩涡出来！用一块布盖住混合物来保温，然后放置一晚让它变成肥皂。如果第二天混合物变成固体，把它翻过来，用一个月让它进行干燥和硬化。四周后，把它切下来包好，这样就不会完全干透。你可以加点草药或者芳香油，创造出自己的特色。

教师与家长指南

故事灵感来自
This Fable Is Inspired by

科特·哈尔伯格
Curt Hallberg

科特·哈尔伯格年轻时不喜欢上高中，于是决定入伍加入瑞典海军。十几年里，他周游世界，在公海航行，在船上和海滨观察水流运动。从海军退役后，他在哈尔姆斯塔德技术高中学习开发工程。他对水流的运动感兴趣。然而，线性数学无法体现他以前观察到的水流运动的逻辑和本质。他组建了一个团队，包括拉尔斯·约翰逊和莫滕·欧文森，一同研究水流的运行状况，特别是在瑞典马尔默市的实验室里研究水的净化。经过长达十年的研究、实验和技术发展，科特和他的团队创建了 WATRECO 公司，已经将两项利用漩涡原理设计的商业应用产品投放市场：一项是专为滑冰场设计的装置，能够用更少的能源生产更高质量的冰；另一项是利用漩涡的物理力将水软化，从而消除了对化学品的需求。

图书在版编目（CIP）数据

冈特生态童书.第二辑修订版：全36册：汉英对照／
（比）冈特·鲍利著；（哥伦）凯瑟琳娜·巴赫绘；
何家振等译. —上海：上海远东出版社，2021
书名原文：Gunter's Fables
ISBN 978-7-5476-1759-5

Ⅰ.①冈… Ⅱ.①冈…②凯…③何… Ⅲ.①生态
环境－环境保护－儿童读物—汉、英 Ⅳ.①X171.1-49

中国版本图书馆CIP数据核字（2021）第213075号

著作权合同登记号图字09-2021-0823

策　　划	张　蓉
责任编辑	祁东城
封面设计	魏　来 李　廉

冈特生态童书
一起摇摆！
［比］冈特·鲍利　著
［哥伦］凯瑟琳娜·巴赫　绘
王菁菁　译

记得要和身边的小朋友分享环保知识哦！
八喜冰淇淋祝你成为环保小使者！

Energy 57

来自树里的燃料

Fuel from the Tree

Gunter Pauli

[比] 冈特·鲍利 著
[哥伦] 凯瑟琳娜·巴赫 绘
王菁菁 译

上海远东出版社

丛书编委会

主　任：贾　峰
副主任：何家振　闫世东　林　玉
委　员：李原原　祝真旭　牛玲娟　梁雅丽　任泽林
　　　　王　岢　陈　卫　郑循如　吴建民　彭　勇
　　　　王梦雨　戴　虹　翟致信　靳增江　孟　蝶

特别感谢以下热心人士对童书工作的支持：

匡志强　宋小华　解　东　厉　云　李　婧　陈　果
刘　丹　熊彩虹　罗淑怡　旷　婉　杨　荣　刘学振
何圣霖　廖清州　谭燕宁　韦小宏　李　杰　欧　亮
陈强林　王　征　张林霞　寿颖慧　罗　佳　傅　俊
胡海朋　白永喆　冯家宝

目录

来自树里的燃料	4
你知道吗?	22
想一想	26
自己动手!	27
学科知识	28
情感智慧	29
艺术	29
思维拓展	30
动手能力	30
故事灵感来自	31

Contents

Fuel from the Tree	4
Did You Know?	22
Think about It	26
Do It Yourself!	27
Academic Knowledge	28
Emotional Intelligence	29
The Arts	29
Systems: Making the Connections	30
Capacity to Implement	30
This Fable Is Inspired by	31

小鹿走在回森林的路上,最近他很喜欢在那里喝新鲜的淡水。森林看起来比以前更茂盛了。人们似乎非常满足,孩子们也露出可爱的笑脸。正当小鹿在四处闲逛时,他听见松鼠们正愉快地聊天。

"真高兴能再次见到你们!"小鹿向松鼠们打招呼。

The deer is returning to the forest where he recently enjoyed drinking fresh water. The forest looks even lusher than before. People seem content and children smile. While wandering around, the deer hears squirrels chattering happily.

"What a delight to meet you all again!" the deer greets the squirrels.

"真高兴能再次见到你们！"

"What a delight to meet you all again!"

……发现了一处新的宝藏！

...has discovered a new bonanza!

"欢迎回来。你知道吗?我们这片地方发现了一处新的宝藏。"松鼠微笑着问道。

"是找到更多水源了吗?"小鹿很好奇。

"不是,宝藏是在树里!"松鼠大声宣布。

"Great that you have returned. Did you know this part of the world has discovered a new bonanza?" asks the squirrel with a smile.

"More water?" wonders the deer.

"No, it's in the trees!" exclaims the squirrel.

"难道树上长金子了？"小鹿问。

"嗯，你知道的，为了重新修复这片热带雨林，我们的人类朋友需要在每公顷土地上种植1000多棵树。"

"我记得这件事。"

"Is there gold growing on the trees?" the deer asks.

"Well, as you know, to regenerate this tropical rainforest our human friends had to plant more than a thousand trees per hectare."

"I remember that."

难道树上长金子了?

Is there gold growing on the trees?

……给本地的原生树木腾出生长空间……

… to make space for the indigenous trees to grow …

"现在他们需要移走八成已经种下的树木，给本地的原生树木腾出生长空间。"

"噢，不！人类是要把这些树卖掉赚钱吗？"

"人类太聪明了，不会这么做的。他们要烧掉大部分的树！"

"And now they have to take away up to eighty of those trees, for every hundred they had planted, to make space for the indigenous trees to grow in the same area."

"Oh no, are the people selling the trees to make money?"

"They are too smart for that - they burn most of the trees!"

"烧掉？可这是一场灾难啊！"

"不，不，这些老松树是用来焚烧发电的。然后把焚烧剩下的灰烬、尿液和便便混合在一起，就能制成生物炭！"

"生物……什么？"

"生物炭——混合着人类排泄物的炭，可以使土壤更肥沃。"

"Burning? But that is a disaster!"

"No, no, the old pine trees are burnt to generate electricity. And the ashes are mixed with pee and poo into bio-char!"

"Bio-what?"

"Bio-char-charcoal with human waste that enriches the soil."

"生物炭——混合着人类排泄物的炭，可以使土壤更肥沃。"

"Bio-char - charcoal with human waste that enriches the soil."

最大的宝藏来自正在生长的树木

but the richest comes from the growing trees

"你是说人类排泄物和树木灰烬的混合物就是所谓的新宝藏？"

"这只是宝藏之一，最大的宝藏来自正在生长的树木。"

"是要把它们砍掉吗？"

"当然不是，是给它们划个口子！"

"Are you saying that the mix of human waste and ashes from trees is the new bonanza?"

"That's one of the riches, but the richest comes from the growing trees."

"By cutting them down?"

"No, by tapping them!"

"那么,他们是在划口子等金子流出来吗?"

"从某种程度上可以这么说,因为这些树富含松节油。"

"你今天用了太多深奥难懂的词汇啦!这个松什么东西有什么用啊?"

"松节油就像燃料。我们称它为液体黄金。"

"So they are tapping gold?"

"In a sense, yes, because the trees are rich in turpentine."

"You are using too many difficult words today! What is this burp-stuff good for?"

"Tur-pen-tine is like fuel. We call it liquid gold."

松节油就像燃料

Tur-pen-tine is like fuel

松节油可以给我们的拖拉机、公共汽车和摩托车提供动力

Turpentine powers our tractors, buses and motorbikes

"可是谁想焚烧液体黄金呢?"

"不,不是这样。我们不会焚烧任何黄金,不论是液体的还是固体的。松节油可以给我们的拖拉机、公共汽车和摩托车提供动力。我们曾经认为使用生物柴油已经是前进了一步,但现在我们过滤这些金子般的树液。它既可以驱动柴油发动机,也可以驱动汽油发动机。"

"But who wants to burn liquid gold?"

"No, no, we are not burning any gold, liquid or not. Turpentine powers our tractors, buses and motorbikes. We once thought using biodiesel was the way forward, but now we filter this golden tree liquid. It runs diesel and gasoline engines alike."

"这种液体也可以使齐柏林飞艇飞起来吗?想象一下,有一天,我的家人也可以使用来自树里的燃料飞翔。我们全家都好开心,好期待这一天啊!"

……这仅仅是开始!……

"Is this liquid also good for flying a zeppelin? Imagine, one day my family could fly with fuel from trees. Wouldn't we all just love that?"

...AND IT HAS ONLY JUST BEGUN!...

……这仅仅是开始!……

... AND IT HAS ONLY JUST BEGUN! ...

Did You Know?

你知道吗?

Turpentine is a natural oil, extracted from tree resin.

松节油是一种天然油脂，是从树脂中提取出来的。

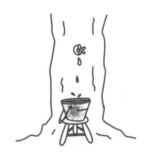

By using steam tree resin can be separated into colophon, which is used in the paper industry and turpentine, which is used in medicine.

通过蒸馏，树脂可以分离成用于造纸工业的松香和用于制药工业的松节油。

松节油可用于治疗关节和肌肉疼痛。

Turpentine oil is applied for joint and muscle pain.

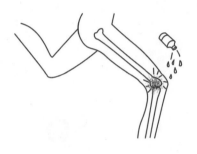

有时候，人们会通过吸入松节油蒸气，来治疗肺部疾病。

People sometimes inhale the vapour of turpentine oil to treat lung diseases.

After the Second World War, the first cars in Japan were sold with a contract to supply turpentine as a guaranteed supply of fuel.

第二次世界大战以后，日本的第一批汽车就在销售时签订了一份协议，要求提供松节油作为燃料供应保障。

Turpentine burns very well, but it needs to be highly purified or it leaves a cloud of exhaust smoke.

松节油燃烧非常充分，但必须是高度纯化的松节油，否则会排出大量废气。

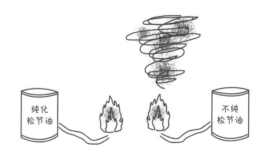

Palm oil must be chemically transformed to function as a fuel, and can only be used in diesel engines.

棕榈油必须经化学方法转化才能作为燃料使用，而且只能用于柴油机。

松节油作为一种燃料，在柴油机和汽油机中都可以使用。

Turpentine as a fuel works both in diesel and gasoline engines.

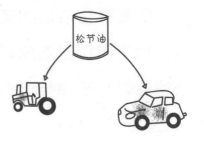

Think about It
想一想

Would you consider getting fuel from a tree – without having to cut the tree down – as a goose laying golden eggs?

你会考虑把树当成会下金蛋的鹅,不用砍伐树木而从树上直接获取燃料吗?

人们先是再造了一片森林,然后就有了免费的饮用水,现在甚至还可以免费获得燃料。这些对人们意味着什么?

What does it mean to this community that they first regenerated a forest, then got drinking water for free, and are now even getting fuel for free?

What happens to the local economy when you do not have to send money overseas to buy petroleum?

当你不需要花钱到国外去采购石油时,这对当地的经济会产生什么样的影响?

你愿意在未来的某一天乘坐使用来自树里的燃料驱动的飞行器吗?你想尝试乘坐什么样的机器飞行呢?

Would you like to one-day fly in a craft that uses fuel from a tree? What machine would you like to fly in?

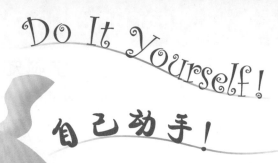

Buy some turpentine from your local pharmacy. First smell it! Does it smell like a pine tree? Now put some turpentine on your skin. What is the effect? Does it make your skin feel warm or cold? Turpentine is a disinfectant.

从你家附近的药店购买一些松节油。先闻闻它！它闻起来像松树吗？现在在你的皮肤上涂一些松节油。它有什么效果呢？它使你的皮肤感觉到温暖还是凉爽？松节油还是一种消毒剂。

TEACHER AND PARENT GUIDE

学科知识
Academic Knowledge

生物学	给树划口子；森林的树木密集度；通过砍伐森林植被为其他植物物种生长提供空间，以促进生物多样性；靠风、鸟类和哺乳动物来传播种子；排泄物作为土壤肥料所具备的营养价值；创造肥沃的表层土壤；作为食物或燃料的植物油。
化学	生产生物炭（亚马逊黑土）；使用化学制品以产生絮凝作用的净化提纯工艺；使用甲醇，通过化学反应（酯交换反应）从植物油中提炼生物柴油；不完全燃烧对空气质量的影响。
物理	齐柏林飞艇的飞行原理及其与飞机的飞行原理有何不同；使用蒸馏的方法来分离物质；过滤技术。
工程学	热电联产，即同时产生热能和电能；无需使用化学制剂的净化提纯工艺（自然沉降）；柴油发动机和汽油发动机的区别；如何将原油转化为柴油和汽油。
经济学	用本地产品替代进口产品；当资金持续在本地区流通时产生的催化剂作用；资金流通速度的重要性。
伦理学	我们应该用玉米油和棕榈油这类消耗型油品来生产燃料吗？
历史	第一批汽车在设计时是使用生物燃料和蒸气来驱动的；第一片油田是在哪儿发现的？宝藏的概念及其如何吸引当地人。
地理	世界上最大的松树林在哪里？它们中的哪些可以划口子以获取燃料？
数学	时间效率研究，絮凝作用快速但昂贵，自然沉降速度慢但便宜。
生活方式	我们已经习惯了依赖能源进口。
社会学	一个社会的适应力取决于它利用当地可获取的资源来保障其所有基本需求的能力。
心理学	具备提供饮用水、食物和燃料的自给能力可以创造自我意识和自信心；通过自我发现和不断质疑来学习，而非接受填鸭式教育。
系统论	自生系统论的概念，即一个系统一旦激活，便开始进化，不断发展，甚至创造出起初从未设想过的新成分。

教师与家长指南

情感智慧
Emotional Intelligence

小鹿

小鹿已经与松鼠们生活的地方建立起联系。他回忆起自己学到的关于森林和生产水源的知识。他对松鼠们充满了同情心，一见到他们就表达了自己的关心。小鹿对最新的发展非常好奇，他仔细地聆听，并且开诚布公地与松鼠分享自己的想法和关注点。他吃惊地听到一些新的事实，并提出问题以充分理解松鼠一开始没有解释清楚的创新点。一旦小鹿明白了这个东西是一种新形式的燃料，他就准备好尽情发挥自己的想象力（比如，驾驶齐柏林飞艇翱翔），向他的朋友们展示他的自信心。

松鼠

松鼠看到小鹿这位常客的出现非常高兴。分享这些令他激动的事实和想法让他有机会再次强化了他的自尊心。作为一名本地区的忠实公民，松鼠更像是一名代表本地区的大使，而不是一个只负责提供信息的导游。松鼠满怀信心、带有尊严地讲话，并且掌握了超出其日常词汇的专业用语。松鼠非常清楚地意识到目前取得了哪些成就，而且他急于了解、学习和分享。他具备一名教师的素质，因为他不仅能够解释新发展，而且能够让小鹿自己发现新事物，这样小鹿才会记得更牢固。松鼠能够区分教学、发现和学习三者间的区别。

艺术
The Arts

当我们用水彩绘画时，我们需要在作画后清洗干净。能够用什么来清洗呢？人们会使用天然的松节油或化学清洗剂。如果选用松节油，你可以用松节油先溶解染料，然后在棉制T恤衫上画上你的设计图案。调色不太均匀也不要紧。如果你对自己的创作不满意，可以用剩下的松节油除去大部分残余的水彩。

TEACHER AND PARENT GUIDE

思维拓展
Systems: Making the Connections

　　一旦森林开始生长，生态系统就将发生改变。森林能够降低地表温度，这样可使雨水和水分蒸发得更慢，从而更多地流向土壤。一旦森林能够完全覆盖一片区域，这个地区的温度就会下降，从而导致降雨量的增加。由于林下植被生长越来越茂密，更多的细菌、真菌和昆虫就会遍布从树冠到根系的整个森林空间。有一种生长在哥伦比亚雨林的松树通过产生树脂来保护自己免受昆虫和真菌的侵袭，这种树脂能够覆盖任何暴露的树皮裂缝。不断产生的树脂会增加火灾隐患。降低这种隐患的最好办法就是将每个工人都训练成消防员。还有一种方法就是在树上划个口子，将树脂转化成一种经济作物。故事中提到的树林和工厂就坐落在哥伦比亚的拉斯加维奥塔斯，人们使用蒸馏法就地将树脂加工成松香和松节油。一直以来，涂料和造纸工业对松香都有极大需求，而松节油一直被认为仅仅是一种具有医疗价值的副产品，直到有一天有人意识到它被用于内燃机的潜在价值。关键的成功因素是去除杂质。尽管过滤系统能够很好地完成这项任务，但是当地没有过滤器，因此采用了最简单的去除杂质系统：依靠物理定律，让直径达到10微米的细小颗粒随着时间的流逝沉降在松节油缸的底部。这种净化提纯过程需要花费的时间更少，比提取石油、经过裂化和提纯后再运到拉斯加维奥塔斯更节省时间。这种无需依赖进口化石能源的独立自主性和无需任何化学制剂即可生产自给的水源的做法为当地创造了可观的收入，为当地的可利用资源提供高附加值，并为区域经济发展提供了先决条件。

动手能力
Capacity to Implement

　　让家长或者老师陪你一起去松树林。带上一些小玻璃罐还有一些取松节油的工具。在树皮上找到裂口，收集40—50克松节油滴。请一定要非常小心，因为松节油极其易燃，绝不能接触明火。当你回到家或者回到教室时，将少量的松节油倒入耐热杯中。然后在松节油上倒入少量热水。持续加热。当你搅动松节油和水的混合物时，会发生什么现象？你看到较轻的透明部分停留在杯壁的上部，而较深的部分则停留在杯壁的下部吗？这两部分分别是什么呢？

教师与家长指南

故事灵感来自
This Fable Is Inspired by

保罗·卢加里
Paolo Lugari

保罗·卢加里青少年时期经常随父亲四处旅行。受到他父亲的工作和研究的启发，他没有开展任何学术研究，而是周游世界，其中有较长一段时间生活在印度。他带领团队设计并实现了第一个，也许至今仍是世界最大的太阳能热水系统，供哥伦比亚波哥大和麦德林的（公共）住房使用。他用实践证明了在 8 000 公顷的区域内，通过重新修复森林来为奥里诺科河流域的居民提供生计的可行性，这片森林所在地曾经覆盖着 2 000 万公顷的热带草原，区域涵盖哥伦比亚、委内瑞拉和巴西。在成功种植了上百万棵树并因此而产生水源后，这一地区因此开启了范围广阔的一系列发展，并于 2012 年达到顶峰，从松树的松节油中生产出了燃料，第一次为骑摩托车的当地人提供无碳燃料作为动力，并为农业设备提供能源。

图书在版编目（CIP）数据

冈特生态童书.第二辑修订版：全36册：汉英对照 /
（比）冈特·鲍利著；（哥伦）凯瑟琳娜·巴赫绘；
何家振等译 . —上海：上海远东出版社，2021
书名原文：Gunter's Fables
ISBN 978-7-5476-1759-5

Ⅰ.①冈… Ⅱ.①冈…②凯…③何… Ⅲ.①生态
环境–环境保护–儿童读物—汉、英 Ⅳ.①X171.1-49

中国版本图书馆CIP数据核字（2021）第213075号
著作权合同登记号图字09-2021-0823

策　　划	张　蓉
责任编辑	祁东城
封面设计	魏　来　李　廉

冈特生态童书
来自树里的燃料
［比］冈特·鲍利　著
［哥伦］凯瑟琳娜·巴赫　绘
　王菁菁　译

记得要和身边的小朋友分享环保知识哦！
八喜冰淇淋祝你成为环保小使者！

Energy 58

又高又凉快
High and Cool

Gunter Pauli

［比］冈特·鲍利　著
［哥伦］凯瑟琳娜·巴赫　绘
王菁菁　译

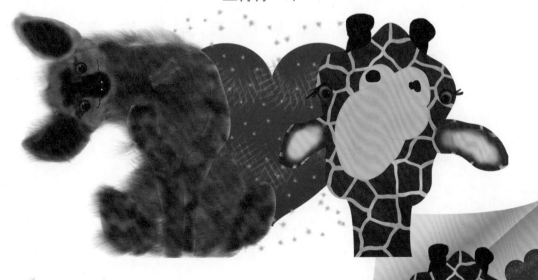

上海远东出版社

丛书编委会

主　任：贾　峰
副主任：何家振　闫世东　林　玉
委　员：李原原　祝真旭　牛玲娟　梁雅丽　任泽林
　　　　王　岢　陈　卫　郑循如　吴建民　彭　勇
　　　　王梦雨　戴　虹　翟致信　靳增江　孟　蝶

特别感谢以下热心人士对童书工作的支持：

匡志强	宋小华	解　东	厉　云	李　婧	陈　果
刘　丹	熊彩虹	罗淑怡	旷　婉	杨　荣	刘学振
何圣霖	廖清州	谭燕宁	韦小宏	李　杰	欧　亮
陈强林	王　征	张林霞	寿颖慧	罗　佳	傅　俊
胡海朋	白永喆	冯家宝			

目录

又高又凉快	4
你知道吗？	22
想一想	26
自己动手！	27
学科知识	28
情感智慧	29
艺术	29
思维拓展	30
动手能力	30
故事灵感来自	31

Contents

High and Cool	4
Did You Know?	22
Think about It	26
Do It Yourself!	27
Academic Knowledge	28
Emotional Intelligence	29
The Arts	29
Systems: Making the Connections	30
Capacity to Implement	30
This Fable Is Inspired by	31

一只长颈鹿想喝水。看着他使劲低下头才能够到水的笨拙样子，一群鬣狗歇斯底里地大笑起来。

"你们笑什么？"长颈鹿冷静地问。

A giraffe is trying to have a drink. It does not help that a couple of hyenas are laughing hysterically at his clumsy attempts to get down low enough to reach the water.

"What are you guys laughing at?" asks the giraffe, coolly.

一只长颈鹿想喝水

A giraffe is trying to have a drink

你看起来真好笑!

you look so funny!

"看你扭动着膝关节，伸出长脖子的样子，看起来真好笑。你一定经常脖子疼吧。"鬣狗首领咯咯笑着说。

"你们才是唯一让我脖子疼的家伙！我有强壮的颈部肌肉——我可以很容易地通过摇摆脖子击倒敌人。而且我的心脏是个超级强大的血泵。"

"You look so funny when you wiggle on your knees with your long neck stretched out. You must often have neck pain," giggles the leader of the hyena pack.

"You are the only pains I have in my neck! I have strong neck muscles – I can easily knock out my enemies by swinging my neck at them. And my heart is a super-strong pump."

"你可骗不了我。"鬣狗说。"世界上最强大的心脏是鲸的心脏。他号称'心脏之王',这可不是徒有虚名。"

"他当然是心脏之王,不过他生活在海洋里。"长颈鹿回答道。"由于我的头比我的心脏高了至少2米,我需要巨大的压强和非常强烈的心跳才能把足够的氧气输送到我的大脑。"

"You can't fool me," says the hyena. "The greatest pump on earth is the whale's heart. He is not known as the King of Hearts for nothing."

"Of course he is the king, but he lives in the ocean," answers the giraffe. "As my head is more than two metres above my heart, I need huge pressure and a very strong heartbeat to get enough oxygen to my brain."

……世界上最强大的心脏是鲸的心脏

... greatest pump on earth is the whale's heart

我还能靠身上的斑块来降温!

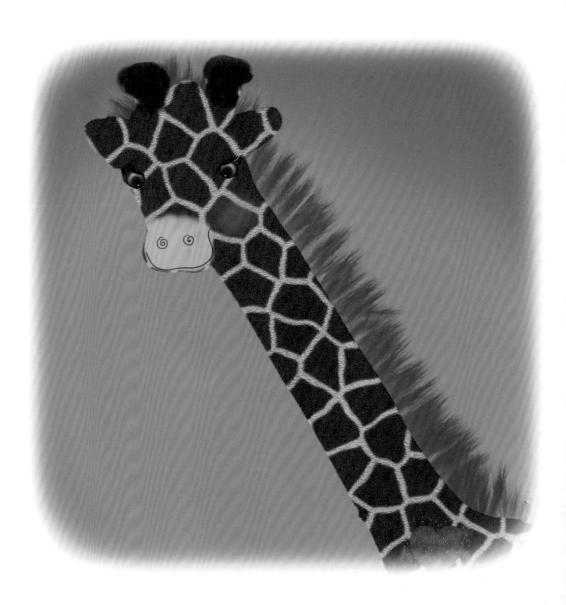

I also have my patches to keep me cool!

"我的静脉和动脉会因为这种巨大的压强而爆裂的!我们鬣狗更喜欢大一些的心脏和缓慢一些的心跳。"

"我的静脉壁比你们鬣狗的皮肤还要厚。幸运的是,我还能靠身上的斑块来降温!"

"你那些深色的斑块一定让你的身体因为太热而不舒服。"

"My veins and arteries would pop with that kind of pressure! We hyenas prefer a bigger heart and a slower beat."

"The walls of my veins are thicker than your hyena skin. Fortunately I also have my patches to keep me cool!"

"Your dark patches must make your body too warm for comfort."

"啊，我知道你们了解斑马。"长颈鹿说。"我们的朋友斑马用黑白相间的外衣来调节体温！而我皮肤上的深色斑块布满血管，这是我用来释放体热的地方。"

"这听起来像是一扇释放热气的窗户。"鬣狗说。

"的确如此，这就是我保持身体凉爽的方式。"

"Ah, I see you know about the zebra," says the giraffe. "Our friend the zebra uses his black-and-white coat to create his own air conditioning! My dark patches on my skin are full of blood vessels, and that's where I release body heat."

"That sounds like a window letting out the heat," says the hyena.

"Absolutely, and that is how I keep my body cool."

我们的朋友斑马……

Our friend the zebra …

你是如何使你的头部保持凉爽的呢?

How do you keep your head cool?

"但是你的头部总是在太阳下晒着,从金合欢树上采摘叶子吃。你是如何使你的头部保持凉爽的呢?"

"所有这些斑块,特别是脖子上的那些,可以像空调一样工作,这样即使我在太阳下,我头部的温度也总是能比我身体的温度低10度。"

"But your head is always in the sun, plucking leaves from the acacia tree. How do you keep your head cool?"

"All these patches, especially the ones on my neck, work like an air conditioner so that even though I am in the sun, the temperature of my head is always ten degrees lower than my body."

"真的吗？"鬣狗用新的眼光重新审视长颈鹿的外衣。

"是的，我的身体可以让血液通过心脏被压送到高处，还能让我的头部保持凉爽。"

"大自然不需要电池就可以做到这些，这不是太神奇了吗？"鬣狗边沉思边说。

"Really?" The hyena looks at the giraffe's coat with new eyes.

"Yes, my body is designed to pump blood upwards and keep my head cool."

"Isn't it simply amazing what nature can do without batteries?" muses the hyena.

大自然不需要电池就可以做到这些,太神奇了

Amazing what nature can do without batteries

我们是原始的"智能电网"

We are the original "smart grids"

"是的，大自然中的任何事物都不需要电池或火焰来制造能量。我们都自力更生。我们是原始的'智能电网'。"

"天哪！现在我知道当我们这么多鬣狗围着你的时候你是如何保持如此冷静的头脑了。"鬣狗轻声地笑道。

"Yes, nothing in nature needs batteries, or burning fires, to create power. We make it all ourselves. We are the original 'smart grids'."

"Goodness! Now I know how you keep such a cool head with us hyenas around you," chuckles the hyena.

"是的，记住千万别惹我。我可以用强壮的腿将你狠狠踢倒，让你甚至你的子孙都会铭记在心！"

……这仅仅是开始！……

"Yes, and remember not to mess with me: with these strong legs I could kick you so hard even your children would remember it!"

...AND IT HAS ONLY JUST BEGUN!...

……这仅仅是开始！……

... AND IT HAS ONLY JUST BEGUN! ...

Did You Know?
你知道吗？

Giraffes are the tallest mammals in the world. Even newborn calves are taller than most humans.

长颈鹿是世界上最高的哺乳动物。即使是新生的长颈鹿幼仔也比大多数人类要高。

Giraffes spend their lives standing upright. They even sleep and give birth standing up.

长颈鹿一生都保持着站立状态。它们甚至站着睡觉和生育后代。

Males fight by necking: hitting each other with their necks.

雄性长颈鹿用脖子打架，即用它们的脖子互相击打对方。

No two giraffes have the same pattern of spots. Their coats are as unique as snowflakes or fingerprints.

任何两只长颈鹿都不会拥有同样的斑块图案。它们的外衣具有像雪花或指纹一样的独特性。

The scientific name given to the giraffe, Giraffa camelopardalis, means a camel painted like a leopard.

长颈鹿的学名是 *Giraffa camelopardalis*，意思是画着豹纹的骆驼。

Spotted hyenas live in clans of up to 80 members and eat flesh, skin, bones and even animal droppings. Hyenas have their own latrines, called middens, where clan members leave their faeces.

斑鬣狗喜欢群居，群居成员多达80只，它们以肉、皮毛、骨头，甚至是动物的粪便为食。鬣狗有自己的厕所，叫作粪堆，族群成员都会在那里排下粪便。

A hyena's 'laugh' alerts other clan members to a good source of food. They have the strongest jaws of any land animal and are able to pulverise the entire skeleton of their prey.

鬣狗用"笑声"提醒其他族群成员找到了好的食物来源。它们拥有陆地动物中最强壮的下颌，能够将猎物的整个骨架咬碎。

The highest-ranking male hyena is subordinate to the lowest-ranking female. Hyenas form highly social groups and will snuggle together in a shaggy pile. When faced with a threat, a hyena may play dead.

最高级别的雄性鬣狗的地位低于最低级别的雌性鬣狗。鬣狗们构建高度社会化的群体，它们会蜷缩在一起形成毛茸茸的一堆。当遇到威胁时，鬣狗可能会装死。

Think about It
想一想

Did you ever think that an animal could laugh?

你认为动物真的会笑吗？

长颈鹿是如何通过长脖子来降温的？

How did it come about that giraffes grew tall necks to cool down?

How is it that everything in nature has power, and that no creature uses batteries?

自然界的一切事物都有自己的能量，而且任何生物都不需要电池，这是怎么回事？

如果有一群鬣狗围绕着你，你能保持冷静吗？

Would you be able to keep calm with a pack of hyenas circling around you?

Do It Yourself!
自己动手！

Let's play a clan game. If one person is considered the biggest and the strongest, how many others, working as a team, are needed to push him or her off balance? If someone is considered the smartest, how many team members are needed to find faster and better answers to the problems and challenges posed? A team that is tightly knit into a clan will always win, even against the smartest and the strongest.

让我们来玩一个集体游戏。如果有一个被认为是最高大、最强壮的人，要推倒他（她），需要一个由多少人组成的团队？如果有一个被认为是最聪明的人，需要多少队员才能比他（她）更快、更好地找到答案？即使是面对最聪明和最强壮的对手，一支紧密团结的队伍最终总会赢得胜利。

TEACHER AND PARENT GUIDE

学科知识
Academic Knowledge

生物学	鬣狗共有四大家族：斑鬣狗、棕鬣狗、条纹鬣狗和土狼；鬣狗跟狗没有亲缘关系；鬣狗是领地性动物，用富含钙质的白色粪便来标明自己的领地；斑鬣狗拥有复杂的肢体语言，包括问候仪式；鬣狗的心脏比狮子的心脏大一倍；鬣狗可以奔跑几千米，速度能达到60千米/时；长颈鹿有7节颈椎，就像人类一样；长颈鹿在雨季以阔叶和落叶植物为食，其他季节以常绿植物为食。
化 学	高钙质的饮食导致鬣狗排出白垩样的粪便；通过产生气味的细菌制造的"社会气味"帮助鬣狗们交流。
物 理	长颈鹿的脖子越长，体表释放的热量就越多；在大自然中，能量的产生都不需要电池。
工程学	"智能电网"是一种使用信息和通信技术来管理电力生产和消费的电力网络；双层玻璃窗，甚至三层玻璃窗的设计是为了在冬天将热量保存在室内，防止热量流失。
经济学	鬣狗是一种关键物种，它们无法在某个区域存活下去就说明了当地生态系统的退化，这意味着经济会受到影响，一开始是旅游业，然后就是其他重复性的生产活动，比如畜牧业、野生水果和蔬菜的采摘。
伦理学	鬣狗一直被认为与巫术有关，这种说法甚至在现在仍然被媒体与娱乐业不断强化，即使科学上已有定论，但要扭转大众的普遍观点还是非常困难。
历 史	在欧洲的拉斯科洞窟中可以找到有关鬣狗的岩画；亚里士多德和老普林尼用文字记录了鬣狗的发现；在东非的一些神话中，鬣狗是第一个将太阳带到地球，给寒冷的地球带来温暖的动物；长颈鹿曾在罗马大竞技场展出；中国人第一次看到长颈鹿是在1414年，他们以为这是神话传说里被称为"麒麟"的有蹄动物。
地 理	长颈鹿是非洲的代表性动物，也是当地旅游经济的卖点。
数 学	长颈鹿的血压很高（280/180毫米汞柱），是人类血压的两倍，其心脏跳动次数高达每分钟170次，长颈鹿颈部的长度每增加15厘米，左心室的室壁就会增厚0.5厘米。
生活方式	鬣狗的一种生活方式是偷窃——它们可以偷走狮子口中的新鲜猎物；当遇到威胁的时候，鬣狗会装死，就像负鼠一样。
社会学	鬣狗喜欢群居生活，却被大多数西方文化和非洲文化认为是愚蠢、懦弱和丑陋的。
心理学	一旦我们对某个人（或某种动物）下了定论，我们就很难改变想法，即使有科学依据也依然很难。
系统论	鬣狗可以作为检验生态系统健康状况的指标之一。

教师与家长指南

情感智慧
Emotional Intelligence

长颈鹿

当鬣狗们嘲笑长颈鹿试图喝水的样子时,他很生气。然而,长颈鹿有自知之明,解释了他的身体从颈部到心脏是如何运行的。长颈鹿在他的表述中非常明确地证实他拥有陆地动物中最强大的心脏。长颈鹿还意识到其他物种存在的作用和重要性,特别是鲸,他还认识到自己在生态系统中的地位。长颈鹿包容了鬣狗的无礼之言,并准备好解释他的独特性:他有长长的脖子,上面的斑块可以帮助他保持凉爽。当鬣狗表现出兴趣时,长颈鹿进一步解释了更多细节。长颈鹿对于自己有足够清晰的认识,将自己的能量、泵送和冷却系统形象地比喻成真正的"智能电网"。长颈鹿以对鬣狗的警告结束了对话,证实了他的自信心。

鬣狗

鬣狗对长颈鹿表现出的工程学奇迹没有表示出一点尊重,但是对于可能出现的头疼表示了关心。鬣狗非常自信,认为长颈鹿的话愚弄不了她。然而,鬣狗并没有专心听,因为她漏掉了长颈鹿清晰明确的说明。鬣狗意识到了自己的局限性,并对血泵的规格和心跳问题发表相关评论。鬣狗分享了自己的传统知识,虽然被回绝,但她很快用刚学到的新知识重新调整了自己的想法。鬣狗意识到自己掌握的原有信息与实际不符,因此表现出了对长颈鹿的尊重。

艺术
The Arts

众所周知,鬣狗会发出非常特别的声音。寻找一家声音图书馆,那里可以为你提供各种不同的大笑声和傻笑声。鬣狗还有非常明显的肢体语言。让我们试着表演鬣狗生活中的一天。学习鬣狗的身体姿态、动作和鬼脸。选择一组人专门发出声音,另一组人配合做出鬣狗的姿势和动作。让观众通过两组人发出的声音或做出的动作来判断每一次表演的含义。

TEACHER AND PARENT GUIDE

思维拓展
Systems: Making the Connections

　　生活中总是有令人惊奇的发现。鬣狗已经演变成非洲大草原上最重要的捕食者，是关键物种之一。借助其智力和毅力，鬣狗能够度过大多数艰难时刻。由于鬣狗位于食物链或者说生物金字塔的顶端，可以作为对生态系统质量的评判标准。如果鬣狗濒临灭绝，就说明在各个营养级的所有生物都受到威胁。这也就是鬣狗族群的健康状况和多样性能为生态系统的状况提供保障的原因。鬣狗的消化系统将钙质以阳离子形式返还土壤，成为一种非常适合帮助植物生长的元素。泥土结合钙质，可以产生更高的土壤孔隙度和土壤透气性，以保障排水和根部生长。长颈鹿也是一种同样重要的关键物种，它身高5.5米，是地球上最高的动物，能够轻松够到树冠，而其他食草动物却无法做到。长颈鹿和它最喜爱的食物——金合欢树开展进化竞赛。随着长颈鹿长高，金合欢长出刺和鞣酸。长颈鹿又进化出结实的嘴和长舌头来剥离树叶以免被刺伤。长颈鹿避免食用鞣酸，而且金合欢只在长颈鹿开始觅食树叶几分钟之后才会产生这种毒素，这样长颈鹿就不会将树叶全部吃光。长颈鹿和鬣狗的故事提供了独特的示范，证明了进化是如何远远超越工程学和人性化设计的。

动手能力
Capacity to Implement

　　列出一份鬣狗和长颈鹿的特征清单。注意每个物种今天的外表都是与其他物种，而不仅仅是与同一物种相互竞争、作用的结果。详细记录下使这些动物成为今天的样子的影响因素：仅仅通过观察这两种关键物种，生命的网络就变得非常清晰。现在，想一想有哪些物种需要依赖鬣狗和长颈鹿而生存。注意围绕鬣狗和长颈鹿的生命网络是如何分别慢慢融合到更大的生命网络中。每当你在思考一个项目或者创建一个新产业时，画出一张生命网络。这样会帮助你理解潜在的协同效应，让你事半功倍，帮助你更容易地进行调整。

教师与家长指南

故事灵感来自
This Fable Is Inspired by

斯蒂芬妮·梅奥
Stephanie Mayo

斯蒂芬妮·梅奥对野生动物充满热情。她出生于加拿大,在大学里学习动物保护专业,毕业后专门从事野生动物和森林保护工作。她还兼职从事写作工作。

斯蒂芬妮最早在加拿大多伦多动物园工作,把全部精力献给了动物保护事业。她希望提升全世界的动物园和水族馆对动物的保护作用。

图书在版编目（CIP）数据

冈特生态童书.第二辑修订版：全36册：汉英对照／
（比）冈特·鲍利著；（哥伦）凯瑟琳娜·巴赫绘；
何家振等译.—上海：上海远东出版社，2021
书名原文：Gunter's Fables
ISBN 978-7-5476-1759-5

Ⅰ.①冈… Ⅱ.①冈…②凯…③何… Ⅲ.①生态
环境－环境保护－儿童读物—汉、英 Ⅳ.①X171.1-49

中国版本图书馆CIP数据核字（2021）第213075号

著作权合同登记号图字09-2021-0823

策　　划　张　蓉
责任编辑　祁东城
封面设计　魏　来　李　廉

冈特生态童书
又高又凉快
［比］冈特·鲍利　著
［哥伦］凯瑟琳娜·巴赫　绘
王菁菁　译

记得要和身边的小朋友分享环保知识哦！
八喜冰淇淋祝你成为环保小使者！

Energy 59

椰子里的水
Water in the Coconut

Gunter Pauli

[比] 冈特·鲍利 著
[哥伦] 凯瑟琳娜·巴赫 绘
王菁菁 译

上海远东出版社

丛书编委会

主　任：贾　峰

副主任：何家振　闫世东　林　玉

委　员：李原原　祝真旭　牛玲娟　梁雅丽　任泽林
　　　　王　岢　陈　卫　郑循如　吴建民　彭　勇
　　　　王梦雨　戴　虹　翟致信　靳增江　孟　蝶

特别感谢以下热心人士对童书工作的支持：

匡志强　宋小华　解　东　厉　云　李　婧　陈　果
刘　丹　熊彩虹　罗淑怡　旷　婉　杨　荣　刘学振
何圣霖　廖清州　谭燕宁　韦小宏　李　杰　欧　亮
陈强林　王　征　张林霞　寿颖慧　罗　佳　傅　俊
胡海朋　白永喆　冯家宝

目录

椰子里的水	4
你知道吗?	22
想一想	26
自己动手!	27
学科知识	28
情感智慧	29
艺术	29
思维拓展	30
动手能力	30
故事灵感来自	31

Contents

Water in the Coconut	4
Did You Know?	22
Think about It	26
Do It Yourself!	27
Academic Knowledge	28
Emotional Intelligence	29
The Arts	29
Systems: Making the Connections	30
Capacity to Implement	30
This Fable Is Inspired by	31

一只金刚鹦鹉看到他的猴子朋友正在将树上的椰子一个个扔到沙滩上。每个椰子里都有满满的水。

"这些水是怎么到树上的椰子里去的呢？"金刚鹦鹉疑惑地大声喊道。"我没有看到树根有水泵啊。"

A macaw watches his monkey friend throwing coconuts from a palm tree onto the beach. Every coconut is full of water.

"How does the water get up into the coconut?" wonders the macaw out loud. "I see no pump at the bottom of the tree."

将树上的椰子一个个扔到沙滩上……

Throwing coconuts from a palm tree ...

……你认为需要水泵来运送……

... you think you need a pump to pump ...

"这些水不是用水泵运送上来的。"猴子回答道。

金刚鹦鹉感到很迷惑。"但是万有引力会把所有东西都拉向地面。所以,没有水泵是不可能运送上去的!"

"你的错误在于,"猴子说,"你认为需要水泵来运送。"

"The water moves up without a pump," answers the monkey.

The macaw is confused. "But gravity forces everything down. So without a pump, nothing can go up!"

"Your problem," says the monkey, "is that you think you need a pump to pump."

"好吧，但是如果没有水泵，那么是谁或者是什么将这些水推送到树的顶部的呢？还是有什么东西用吸管把它吸上去了？"

"看看那些大片的树叶。"猴子指着树叶说道。"它们产生了一种力，可以将汁液吸上来。"

"OK, but if there is no pump, who or what is pushing the water upwards? Or is something sucking it up with a straw?"

"Look at those big leaves." The monkey points to the palm fronds. "They create a force that sucks up the juice."

一种力可以将汁液吸上来

A force that sucks up the juice

叶子在白天释放氧气，在夜间释放二氧化碳

Oxygen during the day, carbon dioxide at night

"你的意思是这些叶子通过在风中上下移动,就能起到水泵一样的效果?"

"不,当然不是。"猴子笑道。"叶子可以吸收阳光,然后释放什么呢?"

"哦,当然啦!叶子在白天释放氧气,在夜间释放二氧化碳。"

"You mean they work like a pump by moving up and down in the wind?"

"No, that's not it," laughs the monkey. "Leaves capture sunlight, and release what?"

"Oh, of course! Leaves release oxygen during the day, and carbon dioxide during the night."

"那么树里有空气吗?"

金刚鹦鹉想了一会儿。"嗯,如果树里有空气,那也一定是小气泡。树木不像我们一样有血管。"

"And is there air in the trees?"

The macaw thinks for a moment. "Well, if there is air, it must be in tiny bubbles. Trees don't have veins like we do."

那么树里有空气吗?

Is there air in the trees?

那么水会不会被卡住？

Wouldn't the water get stuck?

"你说得完全正确！这些水被树叶释放的水蒸气拉上去，同时这些微小的气泡也会把水往上推。这些水通过树中那些非常狭窄的管道被输送到顶部。"

"可是，如果这些管子这么窄，那么水会不会被卡住？"

"You're exactly right! So the water is pulled up by the air being released by the leaves and also pushed up by the tiny air bubbles. The water travels up very narrow pipes in the tree."

"But if the pipes are so narrow, wouldn't the water get stuck?"

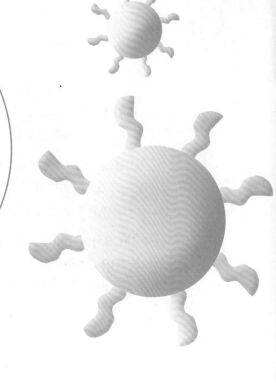

"那只是你的想象。实际上这些管道的管壁非常特别,几乎没有任何东西会粘到上面。"

"啊哈!"金刚鹦鹉兴奋地点点头。"你的意思是这些水流动时可以不受任何阻力,就像沙鱼一样?"

"是的,就像沙鱼可以在沙漠中的沙丘里游泳,黑甲虫可以让露水从背部流走一样。"

"That's what you think. But the walls of the pipes are special. Nearly nothing sticks to them."

"Aha!" The macaw nods happily. "You mean the water flows without friction just like sandfishes do?"

"Yes, just like the sandfish swims the desert dunes, and like the black beetle gets dew off its back."

水流动时可以不受任何阻力,就像沙鱼一样

Water flows without friction like sandfishes do

根部保持凉爽，叶片保持温暖

Roots are cool and the leaves are warm

"而这对树来说也一样适用！"兴奋的金刚鹦鹉说道。"这样会有助于树木的根部保持凉爽，叶片保持温暖吗？"

"确实有帮助。"猴子说。"当水温上升，水就会上升。当水温下降，水就会往下流。"

"哇！所以苹果不需要飞就能长到树顶上，而椰子也不需要水泵就可以充满椰汁！"

"And that works for the tree as well!" says the excited macaw. "Does it help that the roots are cool and the leaves are warm?"

"It does," says the monkey. " Whenever water gets warmer, it will rise. Whenever it gets colder it will drop."

"Wow! So apples don't have to fly to get up in the tree, and coconuts don't need pumps to fill up with juice!"

"世界上有太多能量能够克服万有引力。"猴子陷入了沉思。"我只是不知道太阳是否与这一切有关系。"
金刚鹦鹉眼里闪着光。"你不会以为太阳只是漂亮地挂在天上，让我们感到温暖吧？"
……这仅仅是开始！……

"There is so much power that overcomes gravity," ponders the monkey. "I just wonder if the sun has anything to do with all of this."
The macaw gets a twinkle in his eye. "You don't think the sun is only there to be beautiful and keep us warm, do you?"
... AND IT HAS ONLY JUST BEGUN!...

······ 这仅仅是开始！······

... AND IT HAS ONLY JUST BEGUN! ...

Did You Know?

你知道吗？

In days gone by, the coconut monkey of Sumatra (Macacus nemestrinus) was trained to pick coconuts and worked on plantations until the 20th century.

20世纪以前，苏门答腊岛的椰子猴（猕猴属）一直被训练用来摘椰子，并在种植园工作。

A trained monkey can climb 500 coconut trees per day, 10 times the capacity of a human being. One million coconut trees require 20,000 human workers, but only 2,000 monkeys.

一只受过训练的猴子每天可以爬上500棵椰子树，比一个人强十倍。在100万棵椰子树上摘椰子需要20 000个工人，但是换成猴子就只需要2 000只。

The macaw and the monkey compete for fruit in the Amazon (Brazil).

在巴西的亚马逊地区,金刚鹦鹉和猴子会争抢水果。

A coconut requires 5 to 7 months before producing between 0.2 and 1 litre of water. After 5 months, the inside of the seed thickens into edible meat. The coconut milk pressed from this meat should not be confused with coconut water.

一只椰子需要5—7个月才能产出0.2—1升水。5个月后,种子的内部增厚,变成可食用的果肉。从椰肉压榨出的椰奶与椰子水不是一回事。

A coconut tree may yield 100 to 200 coconuts per year, and can be harvested all year. The taste differs according to the salt content of the soil and the distance from the sea.

一棵椰树每年可以产出100到200个椰子，全年都可以采摘。根据土壤的含盐量和椰子树到海边的距离差异，椰子的味道有所不同。

*E*veryone learns how the apple falls down from the tree, based on the law of gravity, but no one bothers to explain how the apple got up in the tree, defying the law of gravity, in the first place.

大家都知道根据万有引力定律，苹果是如何从树上掉下来的，但是没有人会费心去解释苹果一开始是怎样"违反"万有引力定律长到树上去的。

The flow of water (pressure and flow rate) is all that is needed to pump water up without any outside power source, using the force of gravity to overcome gravity.

无需借助外力,只需靠水的流动(压力和流速),用万有引力的力量来克服万有引力,就能将水输送上去。

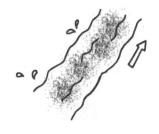

Through a combination of forces, including capillary pressure, surface tension and transpiration, water flows up trees against the law of gravity.

通过一系列作用力的整合,包括毛细管压力、表面张力和蒸腾作用,水能够克服万有引力向上流到树上。

Think about It
想一想

Have you ever wondered how water gets into the coconut without a pump?

你有没有想过,没有水泵,水是如何进入椰子的?

克服万有引力是可能的吗?

Is it possible to defy the power of gravity?

When you hear an idea you have never heard before, do you believe it?

当你听到一个你之前从来没听说过的想法,你会相信吗?

你想要如何学习和研究其他人认为不可能的想法?

How would you like to study and research ideas that other people think are impossible?

Do It Yourself!
自己动手!

Capillary action is indispensable when our eyes drain with tear fluid. If we spill something, it is capillary action that transfers the liquid to a paper towel without any mechanical action. Capillary action allows the small pores of a bath sponge to absorb a large amount of water. A fountain pen draws ink to the tip through capillary action. Look up at least a dozen more things you use that show that there are many forces and powers other than capillary action freely available around you.

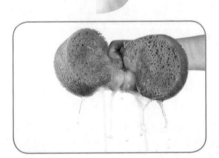

我们的眼睛会流眼泪，这与毛细作用密切相关。如果我们弄洒了什么东西，是毛细作用而不是机械作用让液体被纸巾吸收。毛细作用让浴用海绵上的小孔吸入大量水。钢笔通过毛细作用让墨水流到笔尖。查找至少12个你使用的其他东西，用它们证明除了毛细作用外，你身边还有很多其他的力和能量。

TEACHER AND PARENT GUIDE

学科知识
Academic Knowledge

生物学	椰子水是由幼嫩的种子（不足5个月）产出的，而椰奶是由生长了5—7个月的椰肉压榨出来的；从植物学上说，椰子树并不是树，而是多年生的单子叶植物；椰子树的树干实际上是茎；椰子既是果实也是种子，但不是坚果。
化学	椰子水的热量很低，富含钾，不含脂肪和胆固醇，碳水化合物和钠含量很低；椰肉是胚乳的一部分，是种子（椰子）内部的食物储存组织，含有淀粉、脂肪和蛋白质。
物理	椰子里的水是无菌的；蒸腾作用是指水分在植物内移动并通过叶和花蒸发的过程，这一过程可以为植物降温，并改变渗透压力，让水和营养物质从根部流向发芽部位；毛细作用能够让液体克服万有引力在狭窄的空间内流动；土壤水分蒸发蒸腾损失总量是指地球上的陆地、海洋表面水分蒸发和植物蒸腾作用损失的水分的总和；树根和树叶之间的温度差别越大，其渗透作用越强。
工程学	我们为电磁场设置的防护罩将来有一天会让我们发明出万有引力防护罩吗？水击作用可以60%—80%的能源效率产生高压。
经济学	椰子树的每个部分都得到了应用，因此被称作"生命之树"：它能提供饮料、纤维、食物、燃料、器皿和乐器等。
伦理学	是不是有一些新想法受到批判，并不是因为它们缺少价值，而是因为如果这些想法被证明是正确的和可行的，它们会威胁到那些对立观点的声誉？
历史	历史揭示了许多重大的科学错误：古希腊的内科医生相信肝脏使血液流通，而心脏使精神流通；托勒密提出了地心说，该观点在1 400年中一直被奉为真理，直到哥白尼提出新观点。
地理	在泰国、印度尼西亚和斯里兰卡，人们训练猴子来采摘椰子；椰子在非洲、亚洲、南美洲和太平洋地区的无霜气候中生长。
数学	使用公式来计算毛细管的力量；计算带有水锤的液压系统的效率。
生活方式	H·G·威尔斯在一个世纪前创作的小说《登月先锋》中就已经描述了万有引力屏蔽，而《丁丁历险记》的作者——比利时人埃尔热在书中让主人公穿上反万有引力服将其送到月球。
社会学	在不允许争论时，要提出创新的思想和理论几乎是不可能的。
心理学	对于我们以前从未听说过的观点，我们作出了什么反应？这些观点引起了我们的兴趣吗？还是把它们看作幻想而不予理睬？
系统论	每一个朝向某个方向的力，都有一个相反方向的反作用力。

教师与家长指南

情感智慧
Emotional Intelligence

金刚鹦鹉

金刚鹦鹉非常自信,主动加入和猴子的对话中,尽管他们是同一种食物的竞争者。金刚鹦鹉非常善于观察:他通过提问来获取新知识。猴子神秘的回答与金刚鹦鹉已知的信息相矛盾。面对这一挑战,金刚鹦鹉立刻寻找其他选择(如吸管),想证明他很努力,尽管他知道自己的答案并不正确。金刚鹦鹉准备好去冒险,去探索未知的道路(如起到水泵作用的叶子)。他放弃了自己的骄傲:接受猴子的提示,并且准备向众人坦白他对此事的无知,他在与猴子的交流中逐渐找回自信。金刚鹦鹉对真理的探索激发他去寻找现象之间的联系,这些现象乍一看与椰子没有任何关系,却让金刚鹦鹉有了更深刻的理解。当金刚鹦鹉最终了解了整件事时,他留下了一个不容易回答的开放式问题。

猴 子

猴子向金刚鹦鹉提出了一系列挑战。他的语言具有挑战性,迫使金刚鹦鹉透过表象去思考。猴子有自信,在金刚鹦鹉面前展示自己对现实的更好的理解。起初,猴子想炫耀他知道和能做到的事,但是随后他从讲述事实转为提问。从某种意义上来说,猴子甚至对金刚鹦鹉知道正确答案表示称赞,并准备回答金刚鹦鹉的问题,引导他更全面地理解水是如何流到椰子树上的。然后,猴子提出了一个更深入的问题将这场争论提升到另一个层次,让金刚鹦鹉对太阳的作用感到好奇。

艺术
The Arts

科学很难,而要简单地交流这些概念,不使用科学术语也是非常困难的。解决方案就是要有创造性。你能画一棵棕榈树,并示范水是如何向上移动的吗?在画里描述了所有的作用力后,尝试着用简单的话来解释这幅画,然后再尝试使用科学术语。不要气馁,即使是物理学专家也要反复思考才能正确地解释将水压到椰子里的各种作用力!

TEACHER AND PARENT GUIDE

思维拓展
Systems: Making the Connections

我们怎么能只教苹果是如何从树上掉落的数学知识，而不费心去问一下起初苹果是如何长到树上的呢？要应对社会（以及我们现在的生产和消费模式）向社交型和环境可持续型社会转型的挑战，我们需要学会找到所有的关联。我们经常忽视一些显而易见的事物：我们甚至不会去问在整个苹果服从万有引力定律前，组成苹果的成分是如何"违背"这一定律的。水是如何进入椰子里的？每棵椰子树可以轻松结出200个果实，这就意味着200升纯净无菌的水流到了20—30米高的树上。而且这些水仍然保持着无菌、纯净和富有营养的状态，从而可以用于静脉注射等医疗用途。

缺少求知欲以及教育系统迫使学生去死记硬背事实和理论，导致我们对这类实际问题缺少理解。如果我们的理解不完整，如果我们使用词汇不够精确，我们怎么能够期望说出准确的问题并构思出精确的解决方案呢？

这个世界是几十亿人类的家，需要能够满足所有人对水、食物、住房、能源、健康和工作的基本需求。我们现有的自然科学和工程学知识甚至无法满足世界人口的最低需求。现在我们回顾一下历史就可以发现，无论何时，当人们提出新理论，探讨新的科学数据时，我们倾向于固执己见，就像当年的人对于哥白尼证实了是太阳而不是地球是宇宙的中心时那样，要再等90年才能开始接受事实。

可持续性并不只是关于我们已经知道的有关生命之网的事实，它还与新的事实有关：生命之网中存在无数的关联，这些我们以前并不知道，也可能很难理解。这就是生命之美：不断地发现在明天看来显而易见，而今天我们却几乎无法想象的事物。

动手能力
Capacity to Implement

你需要树立起信心来讨论万有引力的力量。假设你面对这样一个拥护万有引力的团体，他们反对对这一主题的任何逻辑作改变，希望维持现状。现在你需要陈述理由，为什么地球上的生命需要依赖万有引力，以及地球上的生命如何学会克服万有引力。大胆一点，勇敢一些，不要只将你的理由建立在已经得到证实的基础上，还要用到直觉和彻底消除思想贫乏的强烈愿望。

教师与家长指南

故事灵感来自
This Fable Is Inspired by

尤金·波德克莱特诺夫
Eugene Podkletnov

波德克莱特诺夫博士出生在俄罗斯圣彼得堡,毕业于莫斯科门捷列夫化工大学,获得材料学硕士学位。他在俄罗斯科学研究院高温研究所工作了15年,直到1988年受邀赴芬兰坦佩雷大学理工学院攻读超导体制造的博士学位,毕业后他在那里的材料科学研究院一直工作到1996年。

当通过在强有力的电磁铁上方旋转来对超导的陶瓷光盘进行检测时,他注意到这些小物体就好像不受万有引力影响一样。这种力的变化幅度非常微小,仅有大约2%,但这是这种现象第一次被观察到。大多数物理学家认为波德克莱特诺夫一定是弄错了,几乎没有人愿意费心去读关于这一发现的描述。当他的文章宣称万有引力可能无效时,他受到了嘲笑,并且因为信誉受损而被解雇。波德克莱特诺夫博士持续研究他的这一令人惊奇的发现,在未来也许会成为一种新的理论。

图书在版编目(CIP)数据

冈特生态童书.第二辑修订版:全36册:汉英对照/
(比)冈特·鲍利著;(哥伦)凯瑟琳娜·巴赫绘;
何家振等译.—上海:上海远东出版社,2021
书名原文:Gunter's Fables
ISBN 978-7-5476-1759-5

Ⅰ.①冈… Ⅱ.①冈…②凯…③何… Ⅲ.①生态
环境-环境保护-儿童读物—汉、英 Ⅳ.①X171.1-49

中国版本图书馆CIP数据核字(2021)第213075号

著作权合同登记号图字09-2021-0823

策　　划	张　蓉
责任编辑	祁东城
封面设计	魏　来　李　廉

冈特生态童书
椰子里的水
[比]冈特·鲍利　著
[哥伦]凯瑟琳娜·巴赫　绘
王菁菁　译

记得要和身边的小朋友分享环保知识哦!
八喜冰淇淋祝你成为环保小使者!

石头造纸
Paper from Stone

Gunter Pauli

[比]冈特·鲍利 著
[哥伦]凯瑟琳娜·巴赫 绘
姚晨辉 译

上海远东出版社

丛书编委会

主　任：贾　峰

副主任：何家振　闫世东　林　玉

委　员：李原原　祝真旭　牛玲娟　梁雅丽　任泽林
　　　　王　岢　陈　卫　郑循如　吴建民　彭　勇
　　　　王梦雨　戴　虹　翟致信　靳增江　孟　蝶

特别感谢以下热心人士对童书工作的支持：

匡志强　宋小华　解　东　厉　云　李　婧　陈　果
刘　丹　熊彩虹　罗淑怡　旷　婉　杨　荣　刘学振
何圣霖　廖清州　谭燕宁　韦小宏　李　杰　欧　亮
陈强林　王　征　张林霞　寿颖慧　罗　佳　傅　俊
胡海朋　白永喆　冯家宝

目录

石头造纸	4
你知道吗？	22
想一想	26
自己动手！	27
学科知识	28
情感智慧	29
艺术	29
思维拓展	30
动手能力	30
故事灵感来自	31

Contents

Paper from Stone	4
Did You Know?	22
Think about It	26
Do It Yourself!	27
Academic Knowledge	28
Emotional Intelligence	29
The Arts	29
Systems: Making the Connections	30
Capacity to Implement	30
This Fable Is Inspired by	31

一只猴子捡来了很多块石头，然后他挥动着锤子，将每一块石头都砸得粉碎。在辛苦劳动了一整天后，猴子收获了很大一堆石粉。一只猫头鹰开始了他的夜间巡逻，他在上空盘旋，听到猴子在自言自语：

"它们将来会变成纸。"

A monkey takes one stone after another and crushes them. He swings his hammer and pulverises each one until, after a long day of hard labour, he has a big pile of stone dust. An owl starting out on his night flight hovers around and hears the monkey say:

"That is the paper for the day."

它们将来会变成纸

That is the paper of the day

这是纸？

This is paper?

猫头鹰怀疑自己听错了,就飞下来,问道:"你说这是纸?你一定是搞错了。这些都是石头,只不过被砸碎成了石粉。"
猴子看起来很惊讶,回答说:"哦,那么您是怎么造纸的呢?"

Not believing what he has heard, the owl flies down and asks: "You say this is paper? You must be mistaken. These are stones, ground to dust."
The monkey looks surprised and replies: "Oh, so how do you make paper?"

"噢,"猫头鹰说,"我们的纸是用纤维制成的。纤维有很多种来源,例如可以从快速生长的树木中提取。但问题是这些树木需要从土壤中获得所有的养分,还需要喝太多的水。"

"Well," says the owl, "our paper is made from fibre. There are many fibres available, from fast-growing trees for instance, but the problem is they are taking all the food out of the soil and drink too much water."

纸是用纤维制成的

Paper is made from fibres

竹子生长得很快

Bamboo grows fast

"那您有什么解决办法吗?"

"我们正在研究竹子,你知道竹子生长得很快,而且它是一种草,被砍掉之后还会再长出来。"

"So what is your solution?"

"We are working with bamboo. You know it grows quickly, as it is a grass. You cut it; it grows again."

"将竹子变成纸的过程中需要水吗?"

"当然需要,造纸需要用到水,大量的水。好消息是,竹子本身就含有水分。"

"And do you need any water to turn bamboo into paper?"

"Of course, making paper needs water, a lot of water. The good news is that bamboo will bring along its own water."

造纸需要大量的水

Paper needs a lot of water

换换老脑筋了

Think out of the box

"那太棒了。但是,我的石头纸根本不需要任何水。"

"这根本不可能!你可以骗过很多人,但我告诉你:纸张是由纤维制成的,而且需要水。"

"嗯,您应该换换老脑筋了,'聪明'的猫头鹰先生。"

"That's great. My stone paper, however, does not need any water at all."

"That can't be true! You may fool many people, but I tell you: Paper is made from fibre and needs water."

"Well, it is time to think out of the box, you supposedly 'wise' owl."

"这可不太礼貌,我年轻的朋友。我这辈子还从来没有见到过一张不是用纤维制成的纸呢。"

"我这辈子也从来没有碰到过像您这样固执的人。"

"That is not very respectful, my young friend. Never in my life have I seen a single sheet of paper made without fibre."

"And never in my life have I come across such a stubborn person."

不是用纤维制成的纸

Paper made without fibre

17

不用树木和水就能造纸

Tree-free and water-free paper

"嘿，你能对我这样的老年人稍微尊重一些吗？"

"当然，只要您能接受一些新鲜事物——这些新事物已经变成现实了！将石粉、回收的塑料瓶和一些粉笔混合在一起，我就可以制作用于书写和包装的纸，不需要砍掉一棵树或消耗一滴水。"

"不用树木和水就能造纸吗？"猫头鹰惊奇地问道。

"Hey, could you show a bit more respect for an old man like me?"

"Of course, if you could be open to new ideas – that have already been put into practice! By mixing stone dust with recycled plastic bottles and some chalk I make paper for writing and packaging, without cutting down any trees or wasting a drop of water."

"Paper that is tree-free and water-free?" wonders the owl.

"没错！这种纸还可以反复回收利用，现在种植树木的土地都可以用来种植粮食。这些以前被当作废品的小石头会有大用处。"

"更多的农场，更多能种植粮食的田地吗？我喜欢。这意味着我有更多的老鼠可以吃！"猫头鹰微笑着说。

……这仅仅是开始！……

"Exactly! This paper can also be recycled forever and ever, and all the land now used to grow trees can be used to grow food. It also creates many jobs, from bits of rock and stone, which are waste products."

"More farms, more fields growing food? I like that. It means more mice for me!" smiles the owl.

... AND IT HAS ONLY JUST BEGUN!...

……这仅仅是开始！……

... AND IT HAS ONLY JUST BEGUN! ...

Did You Know?
你知道吗？

You need 20 big trees to produce one ton of paper. To make stone paper, you do not need any trees.

生产1吨纸需要20棵大树，制造石头纸不需要任何树木。

By not cutting down 20 trees, you are allowing those 20 trees to produce enough oxygen for 40 people for a whole year.

如果这20棵树不被砍伐的话，它们产生的氧气足够40个人整整一年所需。

You need upwards of 60,000 tons of water to produce a ton of paper, but over a period of a year you only need 60 tons of water to produce stone paper. The water used as a cooling agent in the production of stone paper is recycled.

生产1吨纸需要超过6万吨的水，但生产石头纸一年只需要60吨水，在生产石头纸的过程中，这些水作为冷却剂被循环使用。

You can also recycle the water used for producing paper from trees, but then you need to pump 60,000 litres of water per ton of paper and that requires a great deal of energy.

树木造纸所需要的水也可以循环利用，但每吨纸需要用水泵抽取6万升水，这需要消耗大量的能源。

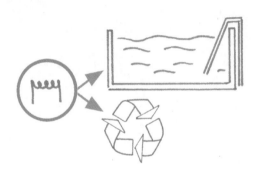

Fibres from trees can be recycled only four to five times before the fibres become too short to be used again, while stone paper can be recycled indefinitely.

从树木中提取的纤维只可以回收利用四到五次，之后纤维就变得太短，无法再使用了，但石头纸可以无限制地回收利用。

When extracting copper or gold from ore the mines need to pulverise rocks to tiny dust particles, almost to the size needed to make stone paper.

从矿石中提炼铜或黄金时，需要将岩石粉碎成微小的粉尘，这些粉尘基本上达到了制作石头纸所需要的大小。

Rock dust from mines is a health hazard. This dust can now be be used to make paper.

来自矿山的岩石粉尘会危害健康，现在这些粉尘可以用来造纸。

More money is made from the paper manufactured from mining waste than from extracting metals in the mining process.

在矿石开采过程中，相比提炼金属来说，用矿山废料造纸能赚到更多的钱。

Think about It

想一想

What would you say the first time you hear piles of rock could be turned into paper?

当你第一次听说一堆石头可以变成纸的时候,你有什么想法?

你更喜欢哪一种纸:竹子做的,还是石头做的呢?

Which do you prefer: paper made from bamboo or from rocks?

Is the use of trees to make paper a smart idea, or is it merely a tradition we inherited from our ancestors that we blindly continue to follow?

用树木造纸是一个聪明的想法吗?或者仅仅因为这是我们从祖先那里继承下来的传统才一味继续沿用?

用纤维制造的纸只能回收利用四到五次,而石头纸可以无限期地回收利用。根据这种情况,你能为林场的主人提供一些什么建议呢?

Fibre-based paper can be recycled only four to five times whereas stone paper can be recycled indefinitely. What advice would you give to owners of forests?

Do It Yourself!
自己动手!

Take some old cardboard. Cut or tear it into pieces and soak in water for an hour or longer. Stir the mixture until you have a thick soup. Now pour the paper mash onto a sheet of metal mesh, where the fibres will form a sheet on top while the water will drains away. Let the sheet dry. Now repeat the procedure but this time add some crushed calcium carbonate ($CaCO_3$). Mix up to 30% of the pulp volume with the calcium carbonate powder. Once both sheets are dry, examine them closely and describe the difference between the two sheets of paper.

找一些旧纸板，用剪刀剪或用手撕成碎片，并浸泡在水中一小时或更长时间。然后搅拌，直到这些混合物成为浓稠的纸浆。现在将纸浆倒在金属网上，其中的纤维将在网上面糊成一层，水则会流下去。等上面一层纤维干燥后，在另一片金属网上重复上述步骤，但这次在纸浆中加入30%的碳酸钙（$CaCO_3$）粉末。等这两片纤维干燥后，仔细观察，并描述这两张纸之间的差异。

TEACHER AND PARENT GUIDE

学科知识
Academic Knowledge

生物学	灵长类动物的能力是使用工具；硬木和软木的区别；草和树木的区别；桉树对土壤的影响；微细粉尘对呼吸系统健康的影响，特别是对儿童而言；植物清除空气中灰尘的作用。
化 学	纤维素、半纤维素、木质素和脂类都是树木的组成部分；使用硫酸盐的目的是消除木质素。
物 理	紫外线对木质素的影响会导致纸张褪色；吸水纸与防水纸；利用粉碎机中的旋涡可以将石头粉碎成细小的颗粒；用微米作为测量尘粒尺寸的标准量度。
工程学	从纤维素中去除木质素制造纸的技术；去除纸张中的油墨，回收利用纤维的技术；去除木质素的无化学品蒸汽爆破技术。
经济学	如何计算可以无限重复使用的石头纸的生命周期评估（LCA）；计算矿区岩石废弃物的价值，计算封闭一个矿井的成本；对纤维素纸张的需求下降后，树木还有什么其他用途？
伦理学	如果制造过程中不消耗水的石头纸能够满足市场的一半需求时，我们怎么会仅仅利用一棵树木20%—30%的纤维来制造纸张呢？仅仅少做一点坏事就足够好了吗？了解生物降解、堆肥、天然、可再生和可持续等概念之间的差异，如何正确使用这些概念，这些概念又是如何被误导使用的。
历 史	植物造纸术以及石头造纸术都是中国发明的；其他伟大的古代文明，如埃及、希腊、罗马和南美洲（玛雅）等又是用什么来书写呢？
地 理	哪些国家具有丰富的石头资源，但缺乏水，从而可以制造石头纸，以改变目前需要进口纤维素和纸张的局面？哪些国家目前还在用树木和水造纸，虽然这种纸张在石头纸普及后很可能会贬值？
数 学	计算造纸时用竹子代替桉树可以节约多少水，以及使用石头纸可以节约多少水；反弹效应：一个进程的节约（如节水）如何刺激其他需求，这样即使单位消耗下降，整体消耗反而增加。
生活方式	纸张是最常用的消费品之一，但也是最大的单一废品来源，即便它是获得社区支持的一个最成功的回收项目；大多数的纸张被进行了热回收，这意味着永久销毁。
社会学	在英国维多利亚时期，拥有一个玻璃温室种植花草和非应季水果是许多家庭社会地位的象征。
心理学	在面对事实的时候，无知和新信息如何会转换成不安全感或掌控欲。
系统论	纤维素纸张来自一种天然的可再生资源，但是它仅能回收利用几次，而石头纸来自矿物资源——当它可以被无限循环使用时，更具备可持续性。

教师与家长指南

情感智慧
Emotional Intelligence

猫头鹰

猫头鹰起初是好奇,想知道更多关于碎石的情况。他关注的是纸张的缺点,一直在研究替代方案,显示了他超越常规的开放性思维。不过,他没有理解猴子所提供的信息,并对此表示怀疑。对于石头纸,他的第一反应是抵制性的。然后,他试图利用并强调他的社会地位来为自己辩护,要求猴子尊重他的立场和年龄。而猫头鹰一旦认识到有更多田地用于种植粮食符合他的利益时,就改变了他的立场,甚至积极地迎接新的发展。

猴子

猴子具有强烈的自我意识,并具备自控能力。他非常敬业,对于分配给自己的任务十分认真。他勇敢地面对猫头鹰,毫不畏惧,最初通过提问,并很快向猫头鹰提供了新的信息。他对猫头鹰表示同情,并愿意花时间来解释。他体谅猫头鹰的情感,猫头鹰对石头纸这一新事物一无所知。猴子别出心裁,以戏弄猫头鹰的办法,让他聆听自己工作创新的奇迹。当猫头鹰感到没有得到多少尊重时,猴子让他从不同的角度看待现实,去发现他的个人利益,猫头鹰最终也这样做了。

艺术
The Arts

你是否曾经看到过20万年前的岩画?洞穴人使用什么颜色的颜料在岩石上作画?研究颜料的成分,并找出哪些颜料即使在受潮后仍能保持它们的色彩。现在,在三种不同的岩石表面绘画,看看哪种岩石的效果最好。一些岩石或许不能很好地留住油漆,一些岩石能更好地吸收油漆,它们表面的色彩也将能保存更长的时间。

TEACHER AND PARENT GUIDE

思维拓展
Systems: Making the Connections

　　对纸张的需求在不断增加，预计到2020年将上升到每年5亿吨。这也推动更多国家划拨土地种植树木，并鼓励用科学方法改善树木的基因，提高树木的生长速度，使其快于自然或人们曾经设想的生长速度。造纸需要树木，最初是从天然林中获取，现在是种植园。由于原始森林被破坏，现在已经用种植和补植的树木来取代。用树木造纸需要将木屑与水和化学物质混合，去除纤维素之外的一切杂质。剩余物"黑液"传统上会用于焚烧发电。混合物中99％是水，剩下的1％是纤维素，去除水的纯纤维素转换成纸片，最终干燥的纤维素薄膜形成片状和卷状的纸张。按照标准做法，纸张生产过程中的水会被回收。然而，这需要非常耗能的抽水设备来回收水以便工厂再利用。与树木相比，每年每公顷竹子产出的纤维是前者的好几倍。由于相关公司知识产权和投入资本的技术封锁，松树和桉树目前仍然是纸张的标准原料。和树木一样，竹子产出的吸水性纤维也可以用来造纸。竹子具有的优点是，竹林在水分循环的过程中会产生水。粉尘是采矿作业的废弃物，矿物被采完后，关闭的矿井往往会对环境造成破坏，而粉尘的利用则会消除这种破坏。粉尘是呼吸系统疾病的主要病因，尤其是对儿童而言，因此去除粉尘有利于身体健康。如果石粉能用来造纸，矿井在其作业结束后就可以焕发新生，因为用石头制成的纸张可以无限期地循环使用。

动手能力
Capacity to Implement

　　去当地销售纸张的商店购买石头纸。如果有货的话，就买几个石头纸笔记本。如果没有，询问是否可以为你订购一些。遭到拒绝后不要放弃，耐心地请求商家答应为你订购。拿出一张石头纸，在上面写字，然后将它放在水中。你的圆珠笔墨水会发生什么变化？现在想象一个回收系统：你是否会将所有的纸张运送到中心地带？或者你有另外一种方法来重新获得纸张并进行再利用？如果这一回收项目能让我们无限期地获得纸张的话，列出项目成功运行所需条件的清单。

教师与家长指南

故事灵感来自
This Fable Is Inspired by

梁石辉
William Liang (Liang Shih Huei)

梁石辉最初是一个机器制造商，1990年，他萌生了制造一种新型纸张的想法。塑料造纸技术的出现使他进一步扩展思维：用碎石头造纸。一开始，梁石辉遇到了种种质疑，但经过十年的艰苦努力后，他成功地通过混合矿物粉尘与高密度聚乙烯（HDPE）制造出了石头纸。他还证明了混合比例为80%的石头加20%的天然聚合物。又花了将近十年对设备进行调整之后，梁石辉成功实现了他的梦想：在工业规模上彻底改造纸张。2013年，第一家石头纸大型生产基地在中国沈阳正式建成投产。石头纸是一种生态纸张，不需要水和纤维素，它的成功也促使矿业公司决定在中国增加设备投资。因为石头纸可以无限期地循环使用，它通过为人们提供一种高需求的产品，给出了一种将环境成本转换为洁净生产的基本解决方案。

图书在版编目(CIP)数据

冈特生态童书.第二辑修订版:全36册:汉英对照/
(比)冈特·鲍利著;(哥伦)凯瑟琳娜·巴赫绘;
何家振等译.—上海:上海远东出版社,2021
书名原文:Gunter's Fables
ISBN 978-7-5476-1759-5

Ⅰ.①冈… Ⅱ.①冈…②凯…③何… Ⅲ.①生态
环境-环境保护-儿童读物—汉、英 Ⅳ.①X171.1-49

中国版本图书馆CIP数据核字(2021)第213075号

著作权合同登记号图字09-2021-0823

策　　划　　张　蓉
责任编辑　　程云琦
封面设计　　魏　来　李　廉

冈特生态童书
石头造纸
[比]冈特·鲍利　　著
[哥伦]凯瑟琳娜·巴赫　　绘
姚晨辉　译

记得要和身边的小朋友分享环保知识哦!
八喜冰淇淋祝你成为环保小使者!

Work 64

跟模具说再见

Metals without Moulds

Gunter Pauli

[比]冈特·鲍利 著
[哥伦]凯瑟琳娜·巴赫 绘
姚晨辉 译

上海远东出版社

丛书编委会

主　任：贾　峰

副主任：何家振　闫世东　林　玉

委　员：李原原　祝真旭　牛玲娟　梁雅丽　任泽林
　　　　王　岢　陈　卫　郑循如　吴建民　彭　勇
　　　　王梦雨　戴　虹　翟致信　靳增江　孟　蝶

特别感谢以下热心人士对童书工作的支持：

匡志强　宋小华　解　东　厉　云　李　婧　陈　果
刘　丹　熊彩虹　罗淑怡　旷　婉　杨　荣　刘学振
何圣霖　廖清州　谭燕宁　韦小宏　李　杰　欧　亮
陈强林　王　征　张林霞　寿颖慧　罗　佳　傅　俊
胡海朋　白永喆　冯家宝

目录

跟模具说再见	4
你知道吗？	22
想一想	26
自己动手！	27
学科知识	28
情感智慧	29
艺术	29
思维拓展	30
动手能力	30
故事灵感来自	31

Contents

Metals without Moulds	4
Did You Know?	22
Think about It	26
Do It Yourself!	27
Academic Knowledge	28
Emotional Intelligence	29
The Arts	29
Systems: Making the Connections	30
Capacity to Implement	30
This Fable Is Inspired by	31

一只蜂鸟在花园中飞来飞去，不停地从一朵饱含花蜜的花飞到另一朵花上。他飞过了许多美丽的雕像。一只猫正在寻觅猎物，想抓一只鸟吃。但蜂鸟飞的速度非常快，又悄无声息，猫很难抓得住他。

"我听说你为了保持身材，每天要吃掉与你的体重一样多的食物。"这只猫说道，她希望能让蜂鸟多停留一段时间，以便可以抓住他。

A hummingbird is flying around a garden, flitting from one flower that is rich in nectar to another. He flies past beautiful sculptures. A cat is out hunting, looking for a bird to eat. The hummingbird is so fast and flies so silently that it will be hard for the cat to catch him.

"I hear you eat your whole bodyweight every day to keep in shape," says the cat in an attempt to keep the hummingbird still long enough so she can catch him.

蜂鸟飞的速度非常快

The hummingbird is so fast

高能量的饮食

High-energy diet

"没错，但不是任何食物都可以。我需要高能量的饮食。"

"我相信你的心脏是所有动物中跳动最快的。"猫呜呜地说。

"That is right, but not just any food. I have a high-energy diet."

"And I believe your heart is the fastest beating heart of any animal," purrs the cat.

"噢，我的心脏每分钟只能跳动120次，但我的蓝哥哥的心跳要比我快10倍以上。我的翅膀每秒种可以扇动10次。"

"哇！10倍以上。我想知道这是如何做到的？"猫沉思道。

"Well, my heart only beats 120 times per minute, but my blue brother's heart could pump ten times more. My wings flip-flap ten times per second."

"Wow. Ten times more. I wonder how that works?" ponders the cat.

我的蓝哥哥的心……

My blue brother's heart ...

我想知道你是如何长成这样的?

I wonder how you are made?

"是这样的,我们每小时吃15顿饭。在休息的时候,我们的体温会下降,这样我们就不需要太多的能量来保持活力——特别是有你在周围寻找食物时。"

"真是太神奇了!我想知道你是如何长成这样的?"猫一边说,一边越来越靠近蜂鸟。

"Well, we have fifteen meals an hour, and when we rest our temperature drops so that we do not need too much energy to remain alive – especially with you around, looking for food."

"That is fantastic. I wonder how you are made?" says the cat, moving closer and closer to the hummingbird.

"我们就像所有的鸟一样，从蛋里孵化出来，然后一天天长大。"

"那和这些雕像的情况完全不同。它们最初是一大块石头，然后凿去不需要的部分，变得越来越小。有一些雕像是利用模具浇铸而成的，它们首先被制成蜡模，然后再被热融的铜代替。"

"We are made like all birds, starting with an egg, and then we grow."

"That is different to these sculptures. These are made from a big block of stone, which is then chipped away so it becomes smaller and smaller. Or some statues are cast in moulds, where they are first made from wax, which is then replaced by hot, molten copper."

那和这些雕像的情况完全不同

That is different to these sculptures

……仅仅使用粉尘

...from just dust

"呵呵，这些人类啊！他们根本不知道如何制作：他们采矿、浇铸、加热、锤打，他们需要这么多的水和能源。而到了最后，又会遗留很多的废物。"蜂鸟感叹道。

"哦，也不完全是这样。"猫说，"现在的人们可以不使用模具来制造蜂鸟，而仅仅使用粉尘。"

"Oh, those human beings! They don't know how to make things: they mine, they mould, they heat, they beat and then they need so much water and energy. And in the end they leave so much waste behind," laments the hummingbird.

"Oh no," says the cat. "There are now people who can make hummingbirds without moulds, from just dust."

"我从来没有听说过这种事情。"

"嗯,现在人类可以利用一些奇妙的材料,通过计算机直接做出就像你一样漂亮的造型。"

"I have never heard of that."

"Well, these days humans can make beautiful shapes, just like yours, in wonderful materials – straight from a computer."

通过计算机

Straight from a computer

这种鸟没有颜色……

It is a bird without colour ...

"这样做出的鸟是什么颜色的?"

"这种鸟没有颜色,也没有心脏,它的翅膀也不会扇动。但我的天啊!它看起来漂亮极了!让我大流口水。"猫舔着她的嘴唇说。

"And what colour would such a bird be?"

"It is a bird without colour; it has neither a heart, nor wings to flap. But my goodness, does it look gorgeous! It makes my mouth water," says the cat, licking her lips.

"当心,如果你用嘴咬它的话,可能会崩掉你的牙齿。"蜂鸟说完,迅速飞走了,躲开了这只猫。

……这仅仅是开始!……

"Watch it, if you bite into it, you might lose a tooth," says the hummingbird, quickly taking off to escape the cat.

... AND IT HAS ONLY JUST BEGUN!...

……这仅仅是开始!……

... AND IT HAS ONLY JUST BEGUN! ...

When the Spanish returned from the Americas with hummingbirds, people at first thought that a hummingbird was a cross between an insect and a bird.

当西班牙人在美洲发现蜂鸟并将它们带回时，人们最初认为蜂鸟是昆虫和鸟类的杂交品种。

Hummingbirds were originally found only in the Americas. That is why they are absent from the fairy tales and legends of Europe, Asia and Africa.

蜂鸟原本只存在于美洲，这就是欧洲、亚洲和非洲的童话和传说中没有出现蜂鸟的原因。

According to a Mayan legend the hummingbird is the sun in disguise, trying to court the moon.

根据玛雅传说，蜂鸟是太阳的化身，它还试图向月亮求爱。

The Aztecs believed that every warrior who died in a battle would reincarnate as a hummingbird.

阿兹特克人认为，在战斗中死亡的每名战士都将重生为一只蜂鸟。

3-D printing melts thin layers of metal powder using the heat from a laser beam to create and shape metal parts.

3D打印通过激光束加热熔化金属粉末薄层，创建和塑造金属部件。

3-D printing with plastics uses small beads to form layers as the material hardens, immediately after extrusion from the nozzle.

塑料3D打印使用小珠形成薄层，材料从喷嘴喷出后会立刻变硬。

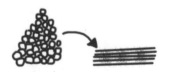

发明家进一步扩展了3D打印技术，将材料从塑料和金属延伸到了陶瓷，并将打印技术和艺术相结合。

Inventors have expanded the technology of 3-D printing from plastics and metals to ceramics, combining printing technology with the arts.

3D打印现在已经延伸到食品领域，可以制造具有独特形状的巧克力、馄饨、玉米片、冰糖甚至比萨饼。

3-D printing is now extending to food, offering creative shapes to chocolate, ravioli, corn chips, sugar candies and even pizzas.

Think about It

想一想

Instead of starting with a block of metal one now starts with metal dust. Do you mind that we lose the artisans as we embrace this new technology?

以前是用一大块金属塑造雕像，现在开始使用金属粉尘，你是否担心这种新技术出现后会造成工匠的消失？

以前处理金属废料的方式一般是丢弃（特别是当它们与水和油混合在一起时），现在人们开始回收废旧金属。3D打印将不再产生任何需回收的废物，你认为这是一种进步吗？

Metal waste used to be discarded, especially when blended with water and oil, now waste metal is recycled. Is it progress when with 3-D printing there will no longer be any waste to recycle?

The hummingbird conserves energy by lowering its body temperature. Could you do that?

蜂鸟能够通过降低体温保存能量，你能做得到吗？

增强蜂鸟心脏功能的高能量饮食需要在一小时内进食15次，食物总量相当于蜂鸟体重的三分之二。蜂鸟的新陈代谢与我们相比，谁更有效率？

The high-energy diet to strengthen the heart of the hummingbird requires 15 meals an hour and an amount of food equalling two thirds of its bodyweight. How efficient is their metabolism compared to ours?

Do It Yourself!
自己动手!

Let us calculate the efficiency of locally producing a dental prosthesis. A dentist makes an imprint, which is shipped to China to be manufactured, and then it is shipped back. After minor adjustments it is placed in a patient's mouth. Calculate the time and the cost of shipment, and factor in the discomfort of having to wait for weeks. Now imagine that you can take a 3-D image, and have a technician make the prosthesis based on the digital image, which is sent to an online 3-D metal (titanium) printer to manufacture it. Estimate how fast the prosthesis will be delivered and how much transport cost will be eliminated. What is your guess of the total cost of the one compared to the other? Now that you have built up your arguments, make a list of the reasons why insurance companies (government or private) should back local 3-D printing of prostheses instead of the global sources based on cheap labour.

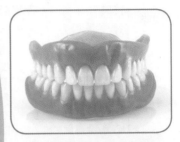

计算一下本地制造假牙的效率。牙医先做好印模,再运到中国进行生产,然后运回。经过小幅调整后,将其放置在患者的口中。计算时间和运输费用,并将需要等待几个星期而感到不适的因素考虑在内。现在想象一下,你可以拍摄一个3D图像,由技术人员根据数字图像制作假牙,其方法是将数字图像发送到在线的3D金属(钛)打印机进行制造。估算一下假牙交付的速度,以及可以节省多少运输成本。你估计两者的总成本分别是多少?现在你已经建立了自己的论点,归纳一张清单,列出为何保险公司(不论国有还是私营)应该支持在本地进行3D打印假牙,而不是利用基于廉价劳动力的全球资源。

TEACHER AND PARENT GUIDE

学科知识
Academic Knowledge

生物学	蜂鸟是物种多样化的代表之一；心脏的功能；心脏跳动和翅膀扇动所需的能量来自哪里；蜂鸟的翅膀在肩部比较灵活，但在前端不够灵活；比较鸟类和其他动物的消化过程；蜂鸟的食谱主要是糖（花蜜），但它也需要蛋白质（昆虫和花粉）来增强肌肉；蜂鸟的舌头与花的形状协同进化；蜂鸟被归类为滤食动物；蛰伏是一种为了应付食物匮乏期而降低代谢活动的状态；蛰伏和冬眠的区别。
化 学	花蜜如何提供能量，蛋白质如何增强肌肉；蜂鸟通过晒太阳将梳理羽毛的油转换成维生素D，同时也放松紧张的肌肉。
物 理	蜂鸟无需使用颜料即可呈现缤纷的色彩（虹彩），因为光以不同于空气中的折射率穿过羽毛，并且部分被反射回第二界面；生活在水中的一些动物通过翅膀的扇动在空气中形成一个涡流，从而将猎物吸入微小的口中，蜂鸟是唯一掌握这一技术的空中动物。
工程学	在生产过程中将大量和大范围热量的应用转化为微量激光热的应用，从而在提供定制产品的时候不再需要水。
经济学	在当地生产急需的组件会减少大量的运输费用，降低成本；变革的阻力来自那些利用传统商业模式赚钱的人；供应链管理。
伦理学	蜂鸟也会打架，但从来不会造成彼此的伤害，当食物较少时，它们就会减少争斗，专心寻找食物。
历 史	北美洲文明（普韦布洛人、莫哈韦沙漠人、霍皮人）、中美洲文明（阿兹特克人）和南美洲文明（玛雅人和印加人）都赋予蜂鸟独特的能力和历史。
地 理	蜂鸟起源于西半球；秘鲁南部的纳斯卡平原有一幅蜂鸟的雕刻画，这幅画非常巨大，只能从300米的高空进行观测；从阿根廷的最南端一直到阿拉斯加都有蜂鸟分布；厄瓜多尔（163种）和哥伦比亚（135种）是蜂鸟种类最多的地区。
数 学	计算自己每天的心跳次数，并与蜂鸟的心跳次数进行比较；计算有多少升血液流经吸蜜蜂鸟的心脏（其重量还不到2克），计算流经自己心脏的血液又有多少升。
生活方式	我们通过血压和心跳来衡量我们的健康状况。
社会学	蜂鸟的命名说明了这些小型鸟类是如何吸引社会大众的：阳光天使（领蜂鸟），金属尾（黑辉尾蜂鸟），宝石羽（古氏蜂鸟），精灵（紫冠仙蜂鸟），芒果（黑胸芒果蜂鸟），木之星（小林蜂鸟），璀璨之光（黑喉辉蜂鸟），翡翠（安第斯蜂鸟）和蓝宝石（蓝头红嘴蜂鸟）。
心理学	先有鸟还是先有蛋？
系统论	工业领域不能再简单地提倡减少能源和水的消耗，而是需要强制实施新的标准，例如将能耗减少10倍，并不再使用水。

教师与家长指南

情感智慧
Emotional Intelligence

猫

猫有一个明确的目标：抓住蜂鸟。她吸引蜂鸟的注意，向他提问，并给予赞美。一旦蜂鸟开始与她对话，无论蜂鸟说什么，她都会表示赞同，以便在她向蜂鸟靠近的时候吸引他的注意力。当猫说到通过计算机直接用金属制造一只鸟的时候，暴露了她的真实意图，这使得蜂鸟有理由飞到安全地带。

蜂鸟

蜂鸟意识到危险，注意保持着安全距离。蜂鸟提供信息，并进行了详细描述，以此来保持对猫的密切关注。当猫开始没话找话（"我想知道你是如何长成这样的"）并对人类制造产品的方式进行批判时，蜂鸟陷入了危险境地。当蜂鸟得知鸟类可以由计算机制造的时候，他有一瞬间放松了警惕。不过，当蜂鸟注意到猫垂涎欲滴的样子，就趁机逃脱了。

艺术
The Arts

现在是诗歌时间。去找一些著名诗人艾米莉·狄金森和 D·H·劳伦斯写的诗，阅读和欣赏他们对蜂鸟的赞美。观看一些关于蜂鸟的视频，尝试自己写一首诗。不组建团队，而是自己独立完成这项任务。想象蜂鸟的美丽和各种令人惊讶的特点，围绕这些内容创作你的诗。

TEACHER AND PARENT GUIDE

思维拓展
Systems: Making the Connections

对于一个遵循供应链管理的生产系统，其目标就是不断降低成本，为厂商创造更多的利润。医疗服务就是一个很好的例子，由于人口老龄化，假体的销售量持续增加。牙齿固定装置的成本很高，医疗保险往往不能完全覆盖。这样一来，大家都在寻找各种降低成本的方法，这导致假牙市场成为全球化经济的一部分，牙医用蜡制作一个牙齿模型，实际的生产会在中国进行，最后由牙医进行细微的调整。然而，整个过程包括两次长途运输。这种对客户的延迟服务是降低成本尝试的一部分，然而，一旦使用3D金属打印，在牙科技师设计了假体之后，牙医提供的快速3D图像就被送到制造中心。牙科技师不需要进行手动操作，而是在计算机上进行假体的数字化设计，将高分辨率的图像通过宽带互联网发送到3D制造中心。该服务可以在一夜之间完成。此外，假体的核心材料是钛，不需要水，而且机器的能耗强度仅相当于两个熨斗，从而可以在城市中心区开展这项生产，这将有助于报废假体的回收，意味着材料和能源效率显著增加了大约8倍。由于减少了运输环节，加上废旧固定装置再利用的潜在可能性，将降低这项全世界所有人都可能会用到的医疗服务的碳排放。然而，这种生产技术和材料使用的转变也需要在学历教育方面进行根本性的转变。这种新商业模式最重要的变化是：它意味着资金将投入在本地，并创造经济价值，而中间商和运输商将被淘汰出局。但是，这将产生一个问题：在牙齿问题的解决方案中，运输商和中间商所产生的价值是什么？

动手能力
Capacity to Implement

列出一个物品清单，看看你每天使用的产品哪些可以通过3D金属打印生产。然后再把会因此受到影响的行业列成表。这将催生多少新的企业？又有多少老的企业因不适应新的生产状况而陷入困境？

教师与家长指南

故事灵感来自
This Fable Is Inspired by

马里奥·弗勒林克
Mario Fleurinck

马里奥·弗勒林克声称自己是从地球母亲学院毕业的，他用这种婉转的方式指出自己并没有大学学位。他最初在家乡比利时进入工业领域。1995年，在波音公司的一个特别项目中，他在美国桑迪亚国家实验室通过现在被称为叠层制造的流程，接触到了复杂涡轮部件的实验生产。在航空航天工业领域积累了几年经验后，2003年，他担任了一家小型工具制造商的负责人。这家公司原本是模拟生产的地方领导者，现在被改造成用数字化方式提供复杂机械解决方案的开拓者。在他的领导下，这家公司成为用3D金属打印化工行业、牙科和医疗市场关键部件强有力的竞争者，最近还涉及了消费类产品（如眼镜）的定制生产。他是世界上最早完全用数字化生产义齿的人，将其对环境的影响降低了8倍。

图书在版编目（CIP）数据

冈特生态童书.第二辑修订版：全36册：汉英对照／
（比）冈特·鲍利著；（哥伦）凯瑟琳娜·巴赫绘；
何家振等译.—上海：上海远东出版社，2021
书名原文：Gunter's Fables
ISBN 978-7-5476-1759-5

Ⅰ.①冈… Ⅱ.①冈…②凯…③何… Ⅲ.①生态
环境－环境保护—儿童读物—汉、英 Ⅳ.①X171.1-49

中国版本图书馆CIP数据核字（2021）第213075号

著作权合同登记号图字09-2021-0823

策　　划	张　蓉
责任编辑	程云琦
封面设计	魏　来　李　廉

冈特生态童书
跟模具说再见
［比］冈特·鲍利　著
［哥伦］凯瑟琳娜·巴赫　绘
姚晨辉　译

记得要和身边的小朋友分享环保知识哦！
八喜冰淇淋祝你成为环保小使者！

Work 65

辣椒的魔力
The Magic of Chilli

Gunter Pauli

［比］冈特·鲍利 著
［哥伦］凯瑟琳娜·巴赫 绘
姚晨辉 译

上海远东出版社

丛书编委会

主　任：贾　峰
副主任：何家振　闫世东　林　玉
委　员：李原原　祝真旭　牛玲娟　梁雅丽　任泽林
　　　　王　岢　陈　卫　郑循如　吴建民　彭　勇
　　　　王梦雨　戴　虹　翟致信　靳增江　孟　蝶

特别感谢以下热心人士对童书工作的支持：

匡志强　宋小华　解　东　厉　云　李　婧　陈　果
刘　丹　熊彩虹　罗淑怡　旷　婉　杨　荣　刘学振
何圣霖　廖清州　谭燕宁　韦小宏　李　杰　欧　亮
陈强林　王　征　张林霞　寿颖慧　罗　佳　傅　俊
胡海朋　白永喆　冯家宝

目录

辣椒的魔力	4
你知道吗？	22
想一想	26
自己动手！	27
学科知识	28
情感智慧	29
艺术	29
思维拓展	30
动手能力	30
故事灵感来自	31

Contents

The Magic of Chilli	4
Did You Know?	22
Think about It	26
Do It Yourself!	27
Academic Knowledge	28
Emotional Intelligence	29
The Arts	29
Systems: Making the Connections	30
Capacity to Implement	30
This Fable Is Inspired by	31

一只老鼠正在寻找食物,他停在一枚长长的红色果实面前,这枚果实胖乎乎的,已经熟透了。与此同时,一只鹦鹉也盯上了这枚红艳艳的果实。

"不要吃那个辣椒!它会像喷火龙一样烧伤你的舌头。"鹦鹉警告说。

A rat is looking for some food. He stops in front of a ripe, fleshy, long, red fruit. At the same time, a parrot has his eye on the same bright titbit.

"Don't eat that chilli! It will burn your tongue like a fire-spewing dragon," warns the parrot.

不要吃那个辣椒

Don't eat that chilli

它们从来不会烧伤我的嘴

They never burn my mouth

"可我实在太饿了。"老鼠反驳道,"而且看上去它是周围最好、最美味的东西。"

"还是把它给我吧。我喜欢辣椒,因为它们从来不会烧伤我的嘴。"鹦鹉哄骗道。

"But I'm so hungry," the rat objects, "and it seems the best and the tastiest thing around."

"Let me have it. I enjoy chillies, because they never burn my mouth." the parrot asks.

"怎么可能？"老鼠问道，"当我的牙齿咬上辣椒的那一刻，我感觉像是着火了一样。"

"那你为什么还要再试一次？顺便说一句，如果我吃了辣椒，会将种子散播开来。如果你吃了，种子就会死掉的。"

"How's that possible?" asks the rat. "The moment I sink my teeth into a chilli, I'm on fire."

"So why do you want to try again? By the way, if I eat the chilli, the seeds will spread, but if you eat it, the seeds are dead."

"……我感觉像是着火了一样。"

"... I'm on fire."

种植辣椒已经有6000年历史了

Farming Chillies for 6,000 years

"是吗？"老鼠沉思了一会儿，"你是说我会结束辣椒的生命循环吗？"

"别担心，人类在美洲种植辣椒已经有6 000年历史了，在亚洲也有500年了。"鹦鹉说。

"Is that so?" muses the rat. "I could put an end to the life cycles of chillies?"

"Don't worry – people have been farming chillies for 6,000 years in the Americas and 500 years in Asia," says the parrot.

"辣椒所到之处,老鼠将退避三舍。"老鼠承认道。

"你最好相信我的话!当你吃辣椒时,它会烧伤你,但并不致命。辣椒会阻止你这样的小动物拿它当食物。甚至大象也会望而却步,辣椒粉搞得他们的鼻子很痒,大象实在忍受不了。"

"Wherever there are chillies, the rats stay away," confesses the rat.

"You'd better believe it! When you eat them, it burns but it doesn't kill. Chillies stop critters like you from devouring our food. They even stop elephants. Elephants can't stand chilli powder tickling their trunks."

……辣椒粉搞得他们的鼻子很痒……

Chilli powder tickling their trunks

让人发疯

Makes people go crazy

"我曾经听说,人们甚至用辣椒粉来安抚愤怒的人。"

"是这样。"鹦鹉说,"好像非常辣的辣椒会让人发疯。我曾经看到,一个白人吃了辣椒后很快全身通红。对于人类而言,辣椒会使人心率加快、体温上升,大汗淋漓。"

"I've heard they even use chilli powder to calm down angry people."

"Yes," says the parrot. "It seems eating extremely hot chilli makes people go crazy. I once saw a white man turn red in no time. In humans, chilli makes the heart rate increase, the body temperature rise and the person breaks out in a sweat."

"这对健康有好处吗？"老鼠惊讶地问道，开始担心自己的健康。
"你坐过过山车吗？"

"Is that healthy?" wonders the rat, concerned for his own health.
"Have you ever been on a roller coaster?"

坐过过山车吗?

Ever been on a roller coaster?

……人们喜欢辣椒……

... people love chillies ...

"哦，没有，我一看到过山车就非常害怕。"老鼠哆嗦着回答。

"这正是人们喜欢辣椒的原因。它们会带来痛苦和恐惧，却对身体有好处。"

"Oh no, I get so scared just at the sight," answers the rat with a shiver.

"That's exactly why people love chillies. They cause pain and fear, but in the end they're good for you."

"那就是说到最后所有的痛苦都是无害的？"

"辣椒含有丰富的维生素和胡萝卜素，胡萝卜素就是胡萝卜中的橙色物质。"鹦鹉回答说，"它们还含有大量的钾、镁和铁，钾是你心脏跳动所必需的物质，镁可以维持肌肉和神经的活动，铁具有促进造血的功能。更重要的是，辣椒创造了成千上万的就业机会。"

"哇，看来我必须得忍受自己像一条喷火龙的感觉了！"

……这仅仅是开始!……

"So in the end all the pain is harmless?"

"Chillies are full of vitamins and carotene, the orange stuff you get in carrots," answers the parrot. "And they're packed with potassium, which makes your heart beat, magnesium to maintain muscles and nerves, and even iron to make blood. What's more, chillies generate thousands of jobs."

"Wow, it seems I'd have to suffer to feel like a dragon!"

...AND IT HAS ONLY JUST BEGUN!...

……这仅仅是开始！……

... AND IT HAS ONLY JUST BEGUN! ...

Did You Know?
你知道吗?

Chillies are from Mexico and were used in ancient Mexican cuisine as much as 9,000 years ago. Mexican cooking has also brought us tomatoes, avocados, beans, vanilla and chocolate.

辣椒原产自墨西哥。早在9 000年前,古代墨西哥菜肴中就开始使用辣椒了。墨西哥菜还会用到西红柿、鳄梨、豆类、香草和巧克力。

If you have eaten too much chilli, the burning can be eased by drinking milk. Drinking water won't help.

如果辣椒吃多了,可以通过喝牛奶来缓解烧灼感,喝水是没有用的。

To avoid burning, apply vegetable oil to your hands prior to handling chillies. If you forget to do that, wash your hands with sugar and tamarind.

为了避免烧手，在处理辣椒之前可以在手上涂一些植物油。如果你忘了这样做，还可以用糖和酸角洗手。

Ayurveda is a plant-based treatment that originated in India more than 3,000 years ago. Chilli was introduced to India only in 1498, it was quickly adopted into traditional local medicine.

阿育吠陀3 000多年前起源于印度，是一种传统的草药疗法。虽然辣椒直到1498年才传入印度，却迅速被用于当地的传统医药中。

Plants use their natural heat as a chemical weapon to selectively deter predators who are poor at spreading seeds. Birds scatter seeds far afield in their droppings and are immune; mammals, who are less efficient in spreading seed, find the pain unbearable.

植物利用自己天然制造的热量作为化学武器，选择性地阻止那些不善于传播种子的捕食者。鸟类不受这些因素的影响，能通过粪便将种子散播到远方。而传播种子效率较低的哺乳动物，却会感到灼烧难忍。

Pepper sprays were originally introduced by the US Post to protect deliverymen from aggressive dogs. Foresters adopted pepper spray to repel grizzly bears. Only then did the police discover the effectiveness in controlling masses.

辣椒喷雾剂最早是美国邮政署为了保护邮递员不被恶狗咬伤而推出的。护林员使用辣椒喷雾剂来赶走灰熊。后来警方发现辣椒喷雾剂对控制群众秩序有效。

Parrots can talk like humans, even without vocal cords. Some parrots sleep upside down, and are known as hanging parrots. It is a sign that they are feeling confident.

即使没有声带，鹦鹉也可以像人一样说话。有些鹦鹉喜欢倒挂着睡觉，被称为悬挂鹦鹉，倒挂睡觉是它们感到自信的标志。

The first roller coaster was built in St Petersburg, Russia, in 1784 under order of Catherine the Great. It was called a Russian mountain. The Russians now call roller coasters "amerikanskiye gorki", which means "American mountains".

1784年，叶卡捷琳娜大帝下令在俄罗斯圣彼得堡建造了世界上第一架过山车，它被称为俄罗斯山。俄罗斯人现在把过山车称为"美国山"。

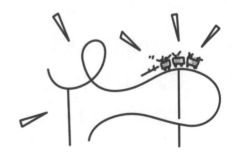

Think about It

想一想

Would you like to deter someone using pepper? What alternative do you have?

你想用辣椒来对付别人吗？除此之外还有什么其他办法？

辣椒引起的痛苦对你有好处吗？

Is the pain caused by chillies good for you?

Do you prefer your food with or without chilli peppers?

你喜欢有辣椒的食物还是没有辣椒的食物？

如果有人把辣椒放在你的食物中，他是你的朋友还是敌人？

If someone puts chilli pepper in your food, is he a friend or a foe?

Do It Yourself!
自己动手!

Are you ready to test yourself? Try tasting some hot chilli peppers – before you put any chilli on your tongue, be sure to oil your hands and have yoghurt nearby. Once it's on your tongue, concentrate on the sensation not just of the pain on your tongue, but also as you break into a sweat. How long does it take to feel normal again? And after the pain subsides, do you feel good? Decide if you want to repeat with a stronger chilli, or if you would prefer to try a softer burn.

准备好测试自己了吗？试着品尝一些辣的辣椒，在把辣椒放进嘴里之前，一定要记得在手上涂一点油，并准备好酸奶。把辣椒放进嘴里后，不仅要细心体验舌头的疼痛感，还要体验满身大汗的感觉。多长时间后感觉才能恢复正常？疼痛消失之后，你感觉还好吗？自己决定是要尝试更辣的辣椒，还是更想体验微辣的滋味。

TEACHER AND PARENT GUIDE

学科知识
Academic Knowledge

生物学	辣椒、西红柿和土豆属于同一植物科目；因为缺乏受体，鸟类和鱼不会觉得辣椒有灼烧感，哺乳动物却具有非常敏感的受体；辣椒可以用来阻止大象进入农场，防止松鼠进入鸡舍；辣椒素是从辣椒的种子中提取的，可以抑制真菌的生长；辣椒中含有丰富的维生素A和维生素C（抗坏血酸）；辣椒全年都可生长；鹦鹉可以单独活动上喙，而大多数鸟类的上喙与颅骨是一体的，不能单独活动；可以教鹦鹉说话与人进行互动交流；鹦鹉没有声带，但有鸣管，类似于人类的咽喉，鹦鹉是通过控制舌头来发声的；辣椒与蔬菜混种是一种有效的伴生种植方式。
化学	辣椒素会触发大脑产生内啡肽，这是一种天然止痛药，能提升幸福感；辣椒中的生物碱会刺激胃肠蠕动；辣椒释放的化学物质会欺骗中枢神经系统，让它以为是在炎热的环境中，因此会出汗；鹦鹉的喙是由角蛋白构成的，就像我们的指甲一样；辣椒与大蒜混合可以作为一种杀虫剂。
物理	辣椒素是能够产生热感觉的化学品，在烹饪和冷冻后仍能稳定存在；辣椒素不溶于水（疏水性），因此喝水并不能减轻灼烧感；鹦鹉喙的形状让它能够攀登和粉碎物品；在炎热和潮湿的气候中，吃辣椒会让人喝更多的水，出更多的汗，从而达到降温的效果。
工程学	辣椒可用于制造非致命性武器。
经济学	辣椒容易种植，只需要少量的水就能成长，占用的土地空间也很少，能够在任何地方生长，甚至可以在家里盆栽；辣椒有几百种不同的烹饪方式，还可以作为非致命的防御物和药品。
伦理学	鹦鹉在自然栖息地受到生存威胁，在都市丛林中却繁荣兴旺；有6种鹦鹉原产于墨西哥，但因为宠物贸易濒临灭绝。
历史	公元前7000年，辣椒就已经被用于烹饪；公元前3500年已开始人工种植辣椒。
地理	辣椒原产于中美洲和南美洲；今天，印度是世界上最大的辣椒生产国。
数学	史高维尔指标可以测量百万分之一的辣椒素。
生活方式	类似过山车这样的游戏和娱乐可以测试性格，但同伴的压力会让人们作出一起享受的决定。
社会学	墨西哥菜是最古老的美食之一，食材包括西红柿、红辣椒、鳄梨、豆类、香草和巧克力；这些食物补充了阿育吠陀医药的种类，并永远改变了欧洲的烹调风味；食用辣椒后可以少吃一点其他食物，并喝大量的水，通过这种方式在没有足够食物的时候填饱肚子。
心理学	辣椒可以让我们反思痛苦，预期当痛苦结束时会感觉好一些；吃辣椒是一种强迫性的冒险，就像坐过山车一样，在这种情况下人们可以享受疼痛和恐惧等极端情感，因为每个人都知道它实际上是无害的。
系统论	植物具有通过阻止他人而自己选择传播者的能力。

教师与家长指南

情感智慧
Emotional Intelligence

鹦鹉

鹦鹉看起来是在关注老鼠的健康,实际上却是出于自私自利的目的:他想要得到辣椒。鹦鹉想方设法去说服老鼠,不仅描述老鼠将会遭受的疼痛,还宣扬自己在传播种子方面的优势。鹦鹉想通过谈话转移老鼠对辣椒的注意力,决心将辣椒据为己有,并声称所有的哺乳动物甚至包括大象,都会避免去吃辣椒。鹦鹉太健谈了,却偏离了他原来的谈话目标。由于过于兴奋,他开始分享他对于辣椒诸多好处的见解,甚至还提倡要吃辣椒,这一点最终激励老鼠去克服即将到来的痛苦。

老鼠

老鼠犹豫不决:是承担疼痛的风险,还是应该享受美食?令老鼠感到惊讶的是,鹦鹉不会感受到这种疼痛。老鼠在学习新的知识,并没有注意到鹦鹉想将辣椒据为己有。老鼠意识到他的同胞对辣椒敬而远之,开始询问与吃辣椒相关的实际效果和风险。鹦鹉的话说明辣椒对身体是有好处的,这一观点如此吸引人,促使老鼠准备去克服自己的恐惧,并承受享用健康食物的痛苦。老鼠对即将发生的事情满怀期待,甚至使用了龙的比喻表明他将咬上一大口辣椒。

艺术
The Arts

画出一个鹦鹉的喙。所有鹦鹉的喙都长得一样吗?仔细观察,画出各个种类鹦鹉不同的喙。如果你充满信心,可以开始画一只倒挂鹦鹉的喙。如果你想尝试些不同的东西,可以头朝下躺在椅子上,画一只正常站立的鹦鹉!这将能使你以一种完全不同的视角来观察周围的事物。尽量集中注意力,这种感觉与坐过山车相同吗?或者至少有几秒钟是类似的?

TEACHER AND PARENT GUIDE

思维拓展
Systems: Making the Connections

辣椒体现了大自然设计的奇迹。辣椒是如何开发出化学鸡尾酒，来避免被不善于传播其种子的哺乳动物捕食，同时鼓励那些有利于其长期生存的动物的呢？辣椒的策略在自然界非常有效，它成功抵御的生物物种小到真菌，大到大象。但辣椒一直没能避免作为强效的治疗性植物被用于医药中。早在3 000年前，印度就出现了阿育吠陀这种基于植物的传统健康系统，达伽马1498年将辣椒引入印度后，辣椒也跻身这个医药系统之中。除了产生灼热感外，辣椒具有丰富的营养和治疗作用。除了生姜含有微量的辣椒素之外，辣椒是唯一能够合成辣椒素分子的植物。生姜也融入了阿育吠陀医药系统中，被认为是一种治疗性的香料。辣椒具有独特的阴阳两面性：浑身出汗后会让你感到凉爽，疼痛过后会由大脑产生让人感觉良好的化学物质（内啡肽）作为补偿。辣椒可以让我们了解世界上最古老的美食之一——墨西哥菜。辣椒的使用已有9 000年历史，种植的历史也有5 500年。它们体现了地球的智慧。现代遗传学和化学让人类产生了一种优越感，但小小的辣椒迫使我们学会谦逊。辣椒短暂疼痛带来的积极体验已经被西方文化（中美洲）和东方文化（最早是在印度）所认可，辣椒也是一个值得注意的生物多样性的代表，能同时提供一个更健康和更快乐的环境。鹦鹉是辣椒种子的优秀传播者，它在自然界的森林中濒临灭绝，却因为非法交易而在世界各地的城市中繁荣兴旺。每年有6万至7万只鹦鹉在墨西哥丛林中被捕获，并被运往世界各地。这一行为剥夺了400多个辣椒品种的自然传播机会。

动手能力
Capacity to Implement

自己种一些辣椒！这非常容易，但一定要有坚持的信念。请记住，一旦你撒下了种子，未来的9到10年你都将收获辣椒。这种具有阴阳两种属性的植物将是你的伙伴，向你提供热量和健康。

教师与家长指南

故事灵感来自
This Fable Is Inspired by

涅玛拉·奈尔
Nirmala Nair

涅玛拉·奈尔出生于印度喀拉拉邦。她在新德里的贾瓦哈拉尔·尼赫鲁大学学习社会学，并在海牙社会学研究院获得发展研究硕士学位。她曾在拉贾斯坦邦的赤脚学院工作，从事了30年的可持续发展研究。奈尔积累了大量关于健康（阿育吠陀）、性别和种族问题、社区发展和青年改造等方面的专业知识，并从传统文化和自然界原生系统中汲取灵感。奈尔从她的母亲那里学习了天然药物和烹饪的知识。2012年，她创办了可持续实践学校。奈尔将自己看作是生活的学生，谨慎地生活，积极地参与。奈尔曾经入选可持续领导者网络的杰出成员，她将美国的环境保护先驱德内拉·梅多斯（Donella Meadows）视为自己的行为榜样。

图书在版编目(CIP)数据

冈特生态童书.第二辑修订版:全36册:汉英对照/
(比)冈特·鲍利著;(哥伦)凯瑟琳娜·巴赫绘;
何家振等译.—上海:上海远东出版社,2021
书名原文:Gunter's Fables
ISBN 978-7-5476-1759-5

Ⅰ.①冈… Ⅱ.①冈…②凯…③何… Ⅲ.①生态
环境－环境保护－儿童读物—汉、英 Ⅳ.①X171.1-49

中国版本图书馆CIP数据核字(2021)第213075号

著作权合同登记号图字09-2021-0823

策　　划	张　蓉
责任编辑	程云琦
封面设计	魏　来　李　廉

冈特生态童书

辣椒的魔力

[比]冈特·鲍利　著
[哥伦]凯瑟琳娜·巴赫　绘
姚晨辉　译

记得要和身边的小朋友分享环保知识哦！
八喜冰淇淋祝你成为环保小使者！

Work 66

蚕丝鞋
Shoes from Silk

Gunter Pauli

[比]冈特·鲍利 著
[哥伦]凯瑟琳娜·巴赫 绘
姚晨辉 译

上海远东出版社

丛书编委会

主　任：贾　峰
副主任：何家振　闫世东　林　玉
委　员：李原原　祝真旭　牛玲娟　梁雅丽　任泽林
　　　　王　岢　陈　卫　郑循如　吴建民　彭　勇
　　　　王梦雨　戴　虹　翟致信　靳增江　孟　蝶

特别感谢以下热心人士对童书工作的支持：

匡志强　宋小华　解　东　厉　云　李　婧　陈　果
刘　丹　熊彩虹　罗淑怡　旷　婉　杨　荣　刘学振
何圣霖　廖清州　谭燕宁　韦小宏　李　杰　欧　亮
陈强林　王　征　张林霞　寿颖慧　罗　佳　傅　俊
胡海朋　白永喆　冯家宝

目录

蚕丝鞋	4
你知道吗?	22
想一想	26
自己动手!	27
学科知识	28
情感智慧	29
艺术	29
思维拓展	30
动手能力	30
故事灵感来自	31

Contents

Shoes from Silk	4
Did You Know?	22
Think about It	26
Do It Yourself!	27
Academic Knowledge	28
Emotional Intelligence	29
The Arts	29
Systems: Making the Connections	30
Capacity to Implement	30
This Fable Is Inspired by	31

一只蚕宝宝在桑树上开心地吞食着树叶。他只有一根手指大小,正准备变成蚕蛾。桑树目睹了整个过程。

"你从前只是一枚小小的卵,就像纸上的一个小圆点那么大。"桑树评论道。

A silk caterpillar is happily devouring the leaves of a mulberry tree. He is the size of a finger, and is getting ready to turn into a moth. The mulberry tree watches the whole process.

"You were once a tiny little egg, the size of a dot on some paper," comments the tree.

吞食着桑叶

Devouring the leaves of a mulberry tree

你的皮肤会经历四次变化

You change your skin four times

"桑树啊，我们蚕和你们相伴成长已经有差不多5 000年了。你的祖先见证了我们第一次孵化，长出毛茸茸的皮肤。"蚕宝宝回答说。

"你是一种很少见的毛虫，你的皮肤会经历四次变化，最后变得柔软而光滑。曾有一阵子，我以为你会变成美丽的蝴蝶。但是，你却变成了飞蛾。"

"We caterpillars have been growing with you for nearly 5,000 years, Tree. Your ancestors saw me hatch with my hairy skin from day one," the caterpillar responds.

"You are a rare caterpillar. You change your skin four times until it is soft and smooth. Once upon a time I thought your type would turn into a beautiful butterfly. But no, you became moths."

"你当时感到失望吗?"蚕宝宝问道。

"也许有一点儿吧,可现在我非常佩服你能吐丝结成这个椭圆形的白色蚕茧。"

"你知道的,一旦我们破茧而出,就不再进食,直到我们中的雌性蚕蛾产下蚕卵。然后,我们就进入了下一段生命历程。"

"Were you disappointed?" asks the caterpillar.

"Perhaps, but now I so admire how you spin this oval white cocoon."

"You know, once we're out of the cocoon, we don't eat until our ladies have laid their eggs. Then we move on to the next life."

……我们中的雌性蚕蛾……

... our ladies ...

大量的唾液用来制作蚕丝

A lot of saliva to produce it

"令人惊讶的是你们一直向人类提供美妙的蚕丝,这实在令人惊叹。"

"我们消耗了大量的唾液来制作蚕丝。"

"哦!我还以为你们像蜘蛛一样,从尾部吐丝呢。"桑树咯咯笑着说。

"Amazing how you keep supplying people with wonderful silk."

"It takes a lot of saliva to produce it."

"Oh! I though it came out of your rear, like the spider," chuckles the tree.

"我的嘴唇上有一个吐丝器。"

"你用嘴唇吐丝?我听说,你吐出的丝足有1000多米长呢。"

"I have a spinneret on my lips."

"You make this silk with your lips? I've been told you can spin a thread more than 1,000 metres long."

你用嘴唇吐丝？

You make silk with your lips?

不会滋生真菌和寄生虫

Free of fungi and parasites

"还可能更长呢。但让我感到奇怪的是,人类只关心我吐出的丝。我的茧里剩下的东西也是非常有用的呀。"

"用来堆肥吗?"

"堆肥当然也行,但还可以有更棒的用途。我的茧不会滋生真菌和寄生虫。"

"It could even be more. But it surprises me that humans are only interested in my silk. What is left over from my cocoon is very useful as well."

"To compost?"

"No, that's fine but you can do better. My cocoon is free of fungi and parasites."

"这太棒了,不过,作为一棵树,这对我来说没什么用。"

"你的确用不着,但是人类不但要穿丝绸衣服,还需要穿鞋子呀。"

"那是当然!"

"That's great, but as a tree I have no use for any of that."

"You don't, but humans who wear silk also wear shoes."

"Sure!"

这太棒了!

That's great!

时髦的鞋子会让他们的脚变形

Modern shoes deform their feet

"他们穿运动鞋的时间太长,就会导致脚趾甲被真菌感染,患上脚癣。还有,穿那些时髦的鞋子会让他们的脚变形。"

"那么你有什么建议吗?"

"And their feet spend too much time in sports shoes that give them toenail fungus and athlete's foot. Or they wear modern shoes that deform their feet."

"So what do you suggest?"

"只要用我的茧来制作鞋底，就不会滋生真菌。我觉得可以在鞋厂旁边建造丝织厂。"

"如果在制作鞋底时加一点咖啡，还能去除异味；如果我们再添些荨麻，就可以用纯天然的方式织造鞋底了呢。"

……这仅仅是开始！……

"It's time to make shoe soles out of my cocoons, and the fungus will be gone. I imagine silk factories next to shoe factories."

"And if we add some coffee, the smell will be gone, and if we add some nettles, we can tie it all up nicely in pure nature."

...AND IT HAS ONLY JUST BEGUN!...

……这仅仅是开始！……

... AND IT HAS ONLY JUST BEGUN! ...

Did You Know?
你知道吗?

Over its lifetime, a caterpillar can eat 25,000 times its body weight, while its body mass increases by 1,000 times.

蚕在一生之中吃掉的食物是自身体重的 2.5 万倍，它自身的体重会增加 1000 倍。

After it has hatched, a caterpillar's first meal is its own eggshell. Caterpillars can develop up to 4,000 muscles (humans have 639), each of which is connected to a neuron.

孵化出来之后，蚕宝宝的第一餐是自己的卵壳。蚕宝宝能发育出 4000 块肌肉（人类只有 639 块），每一块肌肉都和一个神经元相连接。

Caterpillars have 12 eyes, but cannot see well. They have 8 pairs of legs, but only 3 pairs are real legs – the other 5 pairs are pro-legs, which help them to hold on to a plant.

蚕宝宝有12只眼睛，但视力不好。它有8对足，但只有3对是真正的足，其余5对是腹足，这些腹足能帮助它们附着在植物上。

Caterpillars live almost everywhere, from sandy beaches to meadows to mountain forests, worldwide. There are even caterpillars in the Arctic.

蚕在世界各地都有分布，无论沙滩、草甸还是山林，甚至在北极也有蚕的身影。

𝒪ne cocoon is made of a single threat nearly 1,000 metres long. A beautiful long dress may need as many as 1 million cocoons.

一个蚕茧是由一根大约1000米长的丝线构成的。制作一条美丽的长裙可能需要多达100万个蚕茧。

𝒮ilkworms have been domesticated and no longer live in nature. They have lost the ability to fly.

蚕已经被驯化,不再生活在大自然中。它们已经失去了飞翔的能力。

Silk products create an antimicrobial and antifungal environment and do not absorb humidity like cotton does. Silk is ideal for sutures, wound dressing and underwear.

蚕丝产品创造了一个抗细菌和抗真菌的环境,不像棉制品那样吸收湿气。蚕丝是制造手术缝合线、绷带和内衣的理想材料。

Caterpillars are boiled to death in their cocoons so as to leave the long silk fibres intact. However, traditional Bhutanese silk is only harvested after the caterpillar has transformed into a moth.

蚕宝宝最后被煮死在自己的茧中,这样做的目的是获得完整的长蚕丝纤维。然而,传统的不丹蚕丝是在蚕宝宝变成蚕蛾之后才收获的。

Think about It
想一想

Could you imagine the patience and dedication required to spin a 1,000-metre thread?

你能想象吐出1000米的蚕丝需要多少耐心和奉献精神吗?

你会为了获得容易纺织的长纤维而杀死蚕蛾,还是会乐意使用短纤维,为蚕蛾留一条生路?

Would you kill a moth to get a long fibre that is easy to spin, or would you be happy to work with short fibres and give the moth a chance to live?

Should industry look for efficiency and profit only, or should engineers also consider quality of life?

工业应该只追求效率和利润,还是工程师也应该考虑生活品质?

利用蚕茧、荨麻和咖啡制作鞋子只是一个梦想,还是说它能够成为现实?

Does a shoe from cocoons, nettles and coffee seem like a dream or a reality?

Do It Yourself!
自己动手!

Take a piece of paper and trace around your bare feet. Ask your mom, dad and other family members if they are prepared to stand on a piece of paper so that you can trace their feet. Don't tell them why you are doing this. When you have traced 5 pairs of feet, ask if you can borrow everyone's nicest shoes – the ones they wear at special events or at work. Put these shoes on the feet you have drawn and report back: shoes that are smaller than their feet are torturing their feet.

拿一张纸，在上面画出你赤脚的脚印。如果你的妈妈、爸爸和其他家庭成员愿意，请他们也站在纸上，让你画下他们赤脚的脚印。不要告诉他们你为什么这样做。当你画完5双脚之后，询问是否可以借用大家最漂亮的鞋——他们在特殊场合或工作时穿的。将这些鞋放在你画的脚印上，将它们进行比较，然后向大家报告：比自己脚印小的鞋子，会让脚很受罪。

TEACHER AND PARENT GUIDE

学科知识
Academic Knowledge

生物学	蝴蝶和飞蛾是有区别的，蚕宝宝和蠕虫也不相同；变态或完全变态是指通过细胞生长和分化彻底改变体态特征；蚕蛾一生的四个阶段分别是胚胎、幼虫、蛹和成虫；昆虫生下来就有翅膀和功能性的生殖器官；蚕宝宝的体重在35天内会增加1万倍；蚕宝宝最多可以吃掉一棵树上50%的叶子，而不会伤害到树；桑树是一个先锋物种，蚕宝宝能够生成表层土；穿不合脚的鞋会造成爪形趾和锤状趾；脚癣是因为温暖和潮湿的脚趾在鞋中互相挤压而导致的真菌感染，脚癣具有传染性。
化学	蚕丝的主要成分是两种蛋白质，即丝胶蛋白和丝心蛋白；丝心蛋白是折叠片结构的氨基酸；蚕丝具有耐酸性，但硫酸和氯漂白剂除外；蚕茧是由非织造的天然蚕丝复合材料构成的。
物理	蚕丝受潮后会失去弹性；蚕丝是弱导电体，从而容易造成静电吸附；蚕丝具有压电性能，这表明它具有发电的能力。
工程学	去除丝胶蛋白后，蚕丝可以成为理想的非吸收性医用缝合线；3—10股天然蚕丝可以纺成一根纤维；蚕茧可拉伸、可压缩并具有透气性，是良好的非织造复合材料；鞋通过挤压脚趾和缩小脚的宽度、减少脚后跟动作、绷紧足底筋膜和将距骨球向前推动，重新塑造我们脚的形状。
经济学	3 000只蚕消耗104千克桑叶才能产出1千克蚕丝；大约90%的树叶被蚕排泄成为粪便，形成表层土，使半干旱的土地变得肥沃。
伦理学	我们能够一边赞美漂亮的丝绸衣服，一边对这背后蚕宝宝被煮死的事实感到心安理得吗？
历史	东罗马帝国（拜占庭）用黄金换取中国的丝绸；甘地反对通过杀死蚕的幼虫获取蚕丝的工艺，提倡不杀生的丝绸，用残破的茧纺纱。
地理	中国生产了全世界50%的丝绸，其次是印度，只占15%。
数学	蚕宝宝按照8字形移动吐丝，将自己包裹在蚕茧中；根据地图的比例尺，计算古代中国的长安（今天的西安）和君士坦丁堡（今天的伊斯坦布尔）之间的距离，要考虑行进的速度。
生活方式	由于合成纤维和更便宜的纤维的兴起，丝绸已经风光不再；高跟鞋和太紧的鞋子会导致神经损伤和脚的变形，可能需要手术来矫正。
社会学	在古代中国的文化中，缠足是美丽和地位的象征，因为它仅适用于那些不需要工作的富裕女性。
心理学	为了美丽，人们能够忍受多大程度的痛苦？
系统论	蚕宝宝制造了表层土，将先锋树种——桑树的叶子变成了一个重建生物多样性的健康基地，特别是在粮食生产方面。

教师与家长指南

情感智慧
Emotional Intelligence

蚕宝宝

蚕宝宝熟悉自己的历史和背景。蚕宝宝注意到，当发现破茧而出的是一只蚕蛾，而不是多彩的蝴蝶时，桑树有点失望。蚕宝宝花时间解释了自己的繁殖顺序和蚕丝生产体系。这表明蚕宝宝对自己的食品提供者抱有同情和感激之情。蚕宝宝分享了自己的担忧，尤其是不认同大家只对蚕丝感兴趣而忽视蚕茧的做法。蚕宝宝并没有提出蚕茧问题的快速解决方案，但与桑树探讨了一种想法。这一行为成功地激发了更多的想法，激励了桑树，并增进了彼此间的友谊。

桑树

桑树表现出了一种慈爱之情，就像祖父母在观察一颗卵孵化为一个蚕宝宝。桑树比较关注细节，计算了蚕宝宝蜕皮的次数。她承认自己一度有点儿失望，但又立刻表达了她的赞美。桑树乐于学习更多的知识，并分享她对新信息的惊叹。她敏锐地发现了更多的解决方案，并学会了如何把这些方案与当前的背景情况相结合，还探索了以前从未尝试过的新方法。在提出这些突破之后，桑树又在酝酿新的想法，并信心十足地去推进。

艺术
The Arts

通常我们用画笔作画。现在，我们要尝试用我们的脚来作画。准备一大张纸。最好是在室外找一片空地，在纸张上分散着滴一些颜料。然后，用你的脚将这张纸变成一幅美丽的图画。注意只能用脚，要保持双手清洁。

TEACHER AND PARENT GUIDE

思维拓展
Systems: Making the Connections

　　蚕在生态系统中发挥着关键作用。中国很早就发现了蚕与桑树的共生关系，只需经过几十年的种植，数百万棵桑树就能将半干旱的土地转变为肥沃的土地。蚕丝不仅能吸收二氧化碳，还能制造表层土。人们认识到蚕丝对健康和生活品质的独特贡献，也发现提取完整的蚕丝需要在蚕蛾破茧而出之前杀死它们。为了提高生产率，纺织行业开始煮茧，以便抽出1 000米长的蚕丝，并织成可供商用的纤维。尽管已经认识到蚕蛹的痛苦，却很少有人意识到抽丝剩下的蚕茧（通常被丢弃）是由非织造的蚕丝复合材料构成的，具有包括透氧性在内的诸多特性。蚕茧（或复合丝）的现成可用性给人们提供了合理利用这种天然材质的多重性能开发新产品的机会。现在市面上许多鞋的创意设计只是基于时尚考量，而很少考虑脚的健康问题。不幸的是，人们也愿意牺牲足部健康来满足虚荣心。我们的脚遍布神经节点，因此非常敏感。脚的变形已经成为现代化的象征，但这种情况不应该再继续下去了。现在可以用蚕丝制作鞋子，用蚕茧制作鞋底，这样做的好处是：鞋底很柔韧，容易贴合脚的确切形状和样式，而蚕茧还可以降低患上脚癣的风险，给人们带来更多的快乐和健康。

动手能力
Capacity to Implement

　　获得蚕茧可能会有点儿困难，所以我们需要一种替代方案。首先找到一些纸板，最好是瓦楞纸板，它的两层纸之间有一个收缩板，可以使空气流通。将纸板打湿，这样就可以对它进行塑形。在脚上包裹至少10层湿纸板，去草地或柔软的地面上行走，让它们贴合你的脚、脚跟和脚踝，确保你的脚趾可以活动自如。在脚底增加更多的填充物，以提供额外的保护并增加强度。最困难的是要耐心等待它干燥。你可以耐心等待几个小时？如果你耐心有限，注意让包裹的纸板不要高于脚后跟太多，且不需要盖住脚面。当"鞋"逐渐成形后，小心地抽出你的脚。想象一下用蚕茧来贴合你的脚的情形。你认为人们会愿意成为你的客户吗？

教师与家长指南

故事灵感来自
This Fable Is Inspired by

吉列姆·费雷尔
Guillem Ferrer

　　吉列姆·费雷尔是个彻头彻尾的马略卡人，在西班牙的马略卡岛出生、成长，致力于该岛的可持续发展。他长期从事鞋子的设计工作，供职于马略卡岛帕尔马的露营者公司。当露营者公司不再认同他坚持研发可持续鞋子的想法后，吉列姆另辟蹊径，开始从事关于最健康鞋子的设计实践。除了设想让鞋子贴合脚的形状和样式，从而不会造成任何不适或损伤，他还认为鞋子应该是有利于世界、人类与环境的。在不丹廷布举行的一个经济发展研讨会上，吉列姆了解到蚕丝的奇妙以及蚕茧的可用性，因此受到启发去构思一款未来之鞋。

图书在版编目(CIP)数据

冈特生态童书.第二辑修订版:全36册:汉英对照 /
(比)冈特·鲍利著;(哥伦)凯瑟琳娜·巴赫绘;
何家振等译.—上海:上海远东出版社,2021
书名原文:Gunter's Fables
ISBN 978-7-5476-1759-5

Ⅰ.①冈… Ⅱ.①冈…②凯…③何… Ⅲ.①生态
环境-环境保护-儿童读物—汉、英 Ⅳ.①X171.1-49

中国版本图书馆CIP数据核字(2021)第213075号

著作权合同登记号图字09-2021-0823

策　　划	张　蓉
责任编辑	程云琦
封面设计	魏　来　李　廉

冈特生态童书

蚕丝鞋

[比]冈特·鲍利　著
[哥伦]凯瑟琳娜·巴赫　绘
姚晨辉　译

记得要和身边的小朋友分享环保知识哦!
八喜冰淇淋祝你成为环保小使者!

Work 67

稀土不稀奇

Rare on Earth

Gunter Pauli

[比] 冈特·鲍利 著
[哥伦] 凯瑟琳娜·巴赫 绘
姚晨辉 译

上海远东出版社

丛书编委会

主　任：贾　峰
副主任：何家振　闫世东　林　玉
委　员：李原原　祝真旭　牛玲娟　梁雅丽　任泽林
　　　　王　岢　陈　卫　郑循如　吴建民　彭　勇
　　　　王梦雨　戴　虹　翟致信　靳增江　孟　蝶

特别感谢以下热心人士对童书工作的支持：

匡志强　宋小华　解　东　厉　云　李　婧　陈　果
刘　丹　熊彩虹　罗淑怡　旷　婉　杨　荣　刘学振
何圣霖　廖清州　谭燕宁　韦小宏　李　杰　欧　亮
陈强林　王　征　张林霞　寿颖慧　罗　佳　傅　俊
胡海朋　白永喆　冯家宝

目录

稀土不稀奇	4
你知道吗？	22
想一想	26
自己动手！	27
学科知识	28
情感智慧	29
艺术	29
思维拓展	30
动手能力	30
故事灵感来自	31

Contents

Rare on Earth	4
Did You Know?	22
Think about It	26
Do It Yourself!	27
Academic Knowledge	28
Emotional Intelligence	29
The Arts	29
Systems: Making the Connections	30
Capacity to Implement	30
This Fable Is Inspired by	31

一只猫从兽医那里回来,深深地叹了口气,跳进了她最喜欢的沙发。狗注意到了她闷闷不乐的样子。
"兽医伤害到你了吗?"

A cat comes back from the veterinarian and sighs deeply, plunging into her favourite sofa. The dog notices her worried look.

"Did the vet hurt you?"

兽医伤害到你了吗？

Did the vet hurt you?

激光实际上是很好的东西

Lasers are actually very good

"没有，我只是用激光做了一些治疗。"猫回答说。

"你是说那种非常强烈的光？这对你的眼睛非常危险。"狗低声说。

"嗯，激光实际上是很好的东西。这种冷激光有助于我的细胞生长，可以让我看得更清楚，甚至还能清洁我的肾脏。"

"No, I just had some treatment with a laser," responds the cat.

"You mean that very intense light? It's dangerous for your eyes," murmurs the dog.

"Well, lasers are actually very good. This cold laser helps my cells to grow. It could make me see better and even clean my kidneys."

"你的意思是不用吃任何药片？"狗好奇地问道。

"没错，只用光就可以。这是一种特殊的光，它可以通过吸收更多的能量来变得更强。"

"这听起来相当危险。"

"You mean without any pills?" wonders the dog.

"Exactly, only with light. A special light, which gets stronger by pumping in more energy."

"That sounds quite dangerous."

吸收更多的能量

Pumping in more energy

……祛除体毛

...get rid of body hair

"现如今,激光无处不在。我们用它来看电影、听音乐和切割金属,有些人还用它来祛除体毛。"

"都是用同样的激光吗?"

"These days, lasers are everywhere. We use them to watch movies, listen to music, cut metal and some people use it to get rid of body hair."

"All with the same laser?"

"不是,激光有许多不同的类型。你知道现在有可能在不使用模具的情况下,利用激光和等离子对金属进行塑形吗?"

"我不知道等离子是什么,但我想知道为什么在讲了这么多关于激光的好消息后,你还是这么难过?"狗强调说。

"No, there are many different types. Did you know it's now possible to make metal shapes with laser and plasma without moulds?"

"I have no idea what plasma is, but I want to know why, after all this good news about lasers, you're still sad," insists the dog.

利用激光和等离子对金属进行塑形

Make metal shapes with laser and plasma

你得有稀土金属

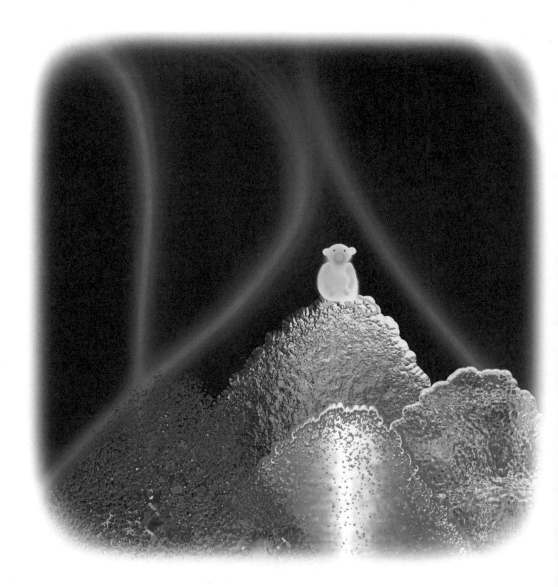

you need rare earth metals

"嗯，你得有稀土金属才能制造激光。人类认为这种金属比较稀有。"

"那就危险了。"狗评论说，"如果人类必需的某些资源比较稀有的话，他们就会为此发动战争。"

"Well, to make lasers you need rare earth metals. Humans think we don't have enough of these metals."

"That is dangerous," remarks the dog. "When there isn't enough of something people believe they need, they wage war over it."

"没错，但实际上这些稀土金属足够大家使用了。"

"抱歉打断一下，那人类为什么还称它们为'稀'土金属呢？"

"Exactly, but there actually is enough of all these rare earth metals for everyone."

"Excuse me, but then why do they call them 'rare' earth metals?"

这些稀土金属足够大家使用了

Enough of all these rare earth metals for everyone

人类并不一定总是正确的

People are not always accurate

"你知道的,人类并不一定总是正确的。他们称蚕宝宝为蠕虫;他们说蝴蝶在飞翔,但实际上它们是在做冲浪运动;还有,椰子其实并不是一种坚果。"

"好吧,但为什么他们会称这种并不稀奇的金属为稀土金属呢?他们甚至可能会为了获取它而发动战争。我是不是有什么没搞明白啊?"

"You know, people are not always accurate. They call caterpillars worms, they say that butterflies fly when they actually surf, and a coconut is not a nut."

"OK, but why do they call this metal rare when it's not rare at all. They might even go to war about it. Am I missing something?"

"这一切都和经济问题有关。"

"我听说过为了赚钱而进行计算。但把并不少见的东西说成是稀有的,听起来更像是一个骗人的故事。这完全就是错的。"

"这就是希望获得便宜产品和廉价工资的自由市场经济吗?"猫陷入了沉思。

……这仅仅是开始!……

"It all has to do with economics."

"I hear about calculations for making money, but saying the something is rare when it isn't sounds more like a trick. It's just not true."

"And is it free-market economics that wants cheap products and low wages?" ponders the cat.

... AND IT HAS ONLY JUST BEGUN!...

……这仅仅是开始!……

... AND IT HAS ONLY JUST BEGUN! ...

Did You Know?

你知道吗？

Rare earth metals (REM) are not rare. Cerium is the 25th most abundant element on Earth. Thulium, the least abundant REM, is 200 times more common than gold.

稀土金属（REM）并不稀有。铈在地球上含量最丰富的元素中排第 25 位，铥是最少见的稀土金属，但也比黄金常见 200 倍以上。

Ytterby, Sweden, is a small hamlet on an island in the Stockholm Archipelago that has 4 elements on the periodic table named after it, thanks to the early research on REM undertaken there in the 18th century.

于特比是瑞典斯德哥尔摩群岛上的一个小村庄，元素周期表中有 4 种元素因它而得名，这是由于 18 世纪有关稀土金属的早期研究是在这里进行的。

于特比

REMs are used in modern technology applications for industry and consumers, including wind turbines, laser scalpels for surgery, hybrid cars and DVD-players.

稀土金属应用于工业和消费品行业的现代技术中，包括风力涡轮机、激光手术刀、混合动力汽车和DVD播放器等。

China has 50% of the world's REM reserves and produces 95% of the world's supply since production in California, USA, was slowed, mainly for environmental reasons.

中国拥有世界上稀土金属储量的50%，满足世界上95%的市场需求。其中一个原因是美国加利福尼亚州出于环境方面的考虑，减少了稀土金属的生产。

Lasers are light bundles and are an everyday part of modern life, from medicine to entertainment, metal processing, data transmission and automotive industries.

激光是一种光束,是现代生活中不可或缺的一部分,其应用涵盖从医药到娱乐、金属加工、数据传输和汽车行业等领域。

The market economy depends on expectations. The price is not only set by supply and demand, but by people's expectations of supply and demand.

市场经济依赖于预期。价格不仅受供求关系影响,也由人们对供给和需求的期望所决定。

预期

The market of supply and demand functions if everyone has equal, full access to all information. Unfortunately, not everyone has access to information and this allows people to speculate with insider knowledge.

如果每个人都可以平等、完全地获得所有信息，供需市场就会发挥作用。不幸的是，不是每个人都能够获得所有信息，这促使人们利用内部消息进行投机。

In the market, people may believe rumours, which could lead to a crash in the market, a correction in the prices and the loss of wealth.

在市场上，人们可能会相信谣言，这可能导致市场崩溃，造成价格的动荡和财富的损失。

Think about It
想一想

Who do you blame when you were made to believe something is rare only to find out it is not?

当你相信某件东西是稀有的，结果却发现并非如此时，你会责怪谁？

当你预测某件东西在将来可能会卖出高价时，你是否准备现在买下它？

Would you be prepared to pay for something now that you could expect to sell for more in the future?

How do you view someone who has earned their wealth by spreading rumours and speculating on the market?

你如何看待那些通过散布谣言和市场投机而赢得财富的人？

在你相信某件东西有效之前，你是否需要证据？

Do you need proof before you believe that something works?

Have a look at the workings of the stock exchange. You'll see that prices rise and fall during the day. Have a look at the stock price over the last few years of a famous enterprise like Apple or Alibaba – do you see the hills and the valleys? When you see the stock rise in value, do you think it will continue to rise? Try to understand the basic principles of supply and demand, and how expectations influence the price people are prepared to pay and the correction that is made afterwards when the reality becomes apparent.

观察一下证券交易所的运作。你会注意到价格在一天中不断波动起伏。看一看某家知名企业（如苹果或者阿里巴巴）的股价在过去几年中的表现，你是否看到了高峰和低谷？当股价上涨时，你认为它还会持续上涨吗？试着去了解供给和需求的基本原则，以及预期会如何影响人们准备购买的商品的价格，当情况变得明朗之后价格又如何修正。

TEACHER AND PARENT GUIDE

学科知识
Academic Knowledge

生物学	稀土金属在生物体中的含量很少，因为它们是不溶于水的，不利于新陈代谢；有一种非常特殊的细菌，能生长在pH值2—5、温度50—60℃、含有高浓度稀土金属的水中。
化 学	稀土金属可以通过分步结晶法进行分离；元素周期表中有一个包括17种稀土金属的特殊分区。
物 理	激光是单色的；一般的光会向四面八方传播，但是激光只会向一个方向发射，接近平行，生成一个高能量的焦点；等离子体是物质在固态、液态和气态之外的第四种状态；稀土金属几乎不溶于水；稀土金属具有发冷光和磁性等特殊性能。
工程学	激光的形成需要铈、钕、铕、铽、镝和铒等稀土金属；激光可以携带电视和电话信号、切割和焊接金属、切割面料以及与等离子结合，塑造金属。
经济学	宏观经济模型研究工人、消费者和管理者的预期；博弈论；对于投机而言，亏损的风险被巨大收益的机会完全抵消；投资和投机的区别；投机者据说能提供更多的市场流动性，但具有很高的亏损成本。
伦理学	当投机变成一个自我实现的预言，一个人就能为了确保个人经济利益在市场上散播谣言吗？
历 史	关于稀土金属的原创性研究开始于18世纪的于特比，那是瑞典斯德哥尔摩群岛上的一个小村庄，它也是元素周期表中4种元素的名字来源：钇（Y）、铽（Tb）、铒（Er）和镱（Yb）。
地 理	中国内蒙古的包头市是当今稀土金属的主要产地。
数 学	建模预期和不确定性下做出的选择；预期风险和纯粹机遇之间的差异，后者取决于一个完全随机的结果。
生活方式	我们现在如此依赖于金属和矿业，并没有意识到微量的稀土金属如何改变了我们的现代生活方式。
社会学	正如约瑟夫·熊彼特所描述的那样，市场经济对旧的经济结构形成了"创造性破坏"，任何人都可以推动创新突破，从工人到零售商、从未经培训的企业家到高瞻远瞩的工程师。
心理学	在信息时代，投资者似乎无法进行冷静的推理，因为往往需要在一刹那作出决定；快速决策会使我们忽视总体目标，有时候我们的决策带来的结果与我们的期望完全相反。
系统论	现代社会对金属的需求很大，但是却忽视了环境成本和产业的战略重要性。

教师与家长指南

情感智慧
Emotional Intelligence

猫

猫若有所思，而且显得有点闷闷不乐，她好不容易才把自己的想法解释清楚，并与狗分享她的烦恼。猫展现了她的智慧，并对技术及其应用有详细的认识。当猫告诉狗她所了解的知识后，通过分析稀土金属虽然有足够的储量却被赋予误导性的名字这一事实，表现出了她的聪慧之处。猫认为这将刺激投机行为，因此坚持认为需要对数据和信息进行解读，从而可以形成自己有根据的观点。

狗

狗有点儿担心他的猫朋友，并表现出了极大的同情。他担心猫在遭受痛苦。当狗听到关于激光的事情时，他不仅关注激光本身，而且也关心对猫健康的防护。狗渴望了解更多的知识，提出了很多问题，相信猫会回答这些问题。在了解到更多的事实后，狗又回到他最初关心的问题：猫为什么闷闷不乐？狗敏锐地预见到了更大的远景，并得出结论：一个重要产品的短缺可能会造成麻烦，甚至引发战争。狗惊讶地听说稀土金属并不少见，而且不明白为什么它们的名字与事实不符。狗可能不具备专门的技术知识，却能对问题进行一个清晰的分析。

艺术
The Arts

稀土金属就像是俄罗斯套娃。有多种金属隐藏在内部，每当剥去一个外层，就会获得更多的发现。有两种类型的稀土金属"套娃"：铈组具有7种附加的稀土金属，钇组具有8种附加的稀土金属。让我们来庆祝这一伟大的俄罗斯艺术，同时学习化学和物理知识！

TEACHER AND PARENT GUIDE

思维拓展
Systems: Making the Connections

采矿应该像手术那样，只在必要的情况下进行，伤疤会随着时间的推移淡化，最终消失不见。包括贵金属及稀土金属在内的许多金属，是制造业、工业和医疗设备不可或缺的材料，然而，采矿需要大量投资，可能需要几十年才能回收成本，投资者将承担更大的风险，在这种情况下，管理层热衷于将成本限制到最低。采矿将产生大量废弃物，因此公司的盈利能力很大程度上并不是取决于金属的实际价格，而是决定于处理废弃物的成本。这导致了矿业公司都不愿意承担健康问题的责任和闭矿成本，闭矿成本是指恢复采矿点生态环境的必要支出。市场在矿业公司的成功中发挥着重要作用。首先，"淘金热"释放了大量的资本，因为金融家期望获得高投资回报率，他们率先将自己的股份投资到土地上，以垄断矿石资源。其次，经营矿业公司需要评判其财务业绩，衡量指标包括矿石储量、总收入和运营成本，最重要的是闭矿成本。投机者知道闭矿成本是一个未知因素，往往会受到法规变化的影响。同时，它也是未来的成本，会降低一个公司时下的价值。投机者一直在寻找隐性资产或成本，这种知识并没有在市场中进行共享。但它们却是决定出售或购买该公司股份的参考因素。就像各种热潮往往是被谣言所推动那样，金属的储备和闭矿成本会根据传闻而变化，但它却会创造或粉碎一个企业的未来和管理，并影响一个村庄，甚至一个国家的民生。

动手能力
Capacity to Implement

记录矿石的处理过程。计算稀土金属的生产量，并估计每年产生的废弃物。哪里可以存储这些废弃物？谁有可能会用到这些矿渣？这些废弃物是否有可能不再是一种成本，而成为一个收入来源？谁有可能会用到磨碎的矿石？组织一个智囊团并保持一种积极乐观的态度：我们的目标不是要批评或谴责，而是要找到服务于社会并使行业更具竞争力的解决方案。

教师与家长指南

故事灵感来自
This Fable Is Inspired by

普丽哈尼·萨特亚帕尔
Prishani Satyapal

普丽哈尼·萨特亚帕尔是土生土长的南非彼得马里茨堡人。她曾在南非纳塔尔大学学习社会科学和商务,并在英国巴斯大学获得硕士学位。在从事了一段时间金融和商业咨询工作后,普丽哈尼加入了世界最大的黄金矿业公司之一,根据她在加纳的经验,她开始建议企业计算真正的环境和社会成本,以及考虑在目前的企业运营中如何为将来的闭矿做好准备。2013年,她决定创建自己的公司——可持续发展实况调查公司,并担任了国际经济学商学学生联合会(AIESEC)南非董事会的主席,激励下一代管理者学习那些不会在大学里讲授的策略。

（CIP）数据

...书.第二辑修订版:全36册:汉英对照/
...特·鲍利著;(哥伦)凯瑟琳娜·巴赫绘;
...家振等译.—上海:上海远东出版社,2021
书名原文:Gunter's Fables
ISBN 978-7-5476-1759-5

Ⅰ.①冈… Ⅱ.①冈…②凯…③何… Ⅲ.①生态
环境–环境保护–儿童读物—汉、英 Ⅳ.①X171.1-49

中国版本图书馆CIP数据核字(2021)第213075号
著作权合同登记号图字09-2021-0823

策　　划	张　蓉
责任编辑	程云琦
封面设计	魏　来　李　廉

冈特生态童书
稀土不稀奇
[比]冈特·鲍利　著
[哥伦]凯瑟琳娜·巴赫　绘
姚晨辉　译

记得要和身边的小朋友分享环保知识哦!
八喜冰淇淋祝你成为环保小使者!